*Dreaming of Dry Land*

# Dreaming of Dry Land

*Environmental Transformation in Colonial Mexico City*

Vera S. Candiani

STANFORD UNIVERSITY PRESS
STANFORD, CALIFORNIA

Stanford University Press
Stanford, California



This publication is made possible in part from the Barr Ferree Foundation Fund for Publications, Princeton University.



Printed and bound by CPI Group (UK) Ltd, Croydon, CR0 4YY

Library of Congress Cataloging-in-Publication Data

Candiani, Vera S., author.
Dreaming of dry land : environmental transformation in colonial Mexico City / Vera S. Candiani.
pages cm
Includes bibliographical references and index.
ISBN 978-0-8047-8805-2 (cloth : alk. paper)
1. Drainage--Mexico--Mexico, Valley of--History--17th century. 2. Flood control--Mexico--Mexico, Valley of--History--17th century. 3. Hydraulic engineering--Mexico--Mexico, Valley of--History--17th century. 4. Public works--Mexico--Mexico, Valley of--History--17th century. 5. Human ecology --Mexico--Mexico, Valley of--History--17th century. 6. Land use--Mexico--Mexico, Valley of--History--17th century. 7. Mexico--History--Spanish colony, 1540-1810. I. Title.
TC978.M6C36 2014
625.7'34097253--dc23

2013048823

ISBN 978-0-8047-9107-6 (electronic)

Typeset by Bruce Lundquist in 10/12 Sabon

*Para Fabrizio Mar Candiani Estrada y Carlos "Beto" Estrada*
*A la memoria de Jorge W. Candiani, mi padre*
*y de Bernardo Tolcachir, mi abuelo*

# Table of Contents

# Images

*All images are available in their original color and in enlargeable digital form in the section dedicated to this book at:*
http://scholar.princeton.edu/candiani/

## Figures

## Tables



## Maps

# Preface

When I first visited the City of Mexico as a student, I was overwhelmed. The torrential rains of July, the imposing architecture, the delicious sizzling smells of street taco stands, the filth, the cobalt blue and bright green of dinnerware, the ceremonious courteousness of men, the pushing and shoving of policemen who separated metro travelers into gender-segregated cars during the rush hour, the strange affection toward an indifferent active volcano ("el Popo"), the calculated familiarity of the vendors in the *tianguis* beckoning me to their fragrant produce *("¿Qué le doy, güerita? ¿Qué va a llevar?")* It was impossible to ignore the glorious and terrifying contradictions of the city. Beauty and stench; kindness and brutality; well-dressed people building and buying condos and cars as though all this had a future.

I soon found out that both the magnificence and the decay of this city had a cause—the desiccation of its lakes. This is how the pharaonic public work known as the Desagüe de Huehuetoca, the main character of this book, came into my life. Captivated by its remote origins in the sixteenth century, I set out to explain not just how the lacustrine region of the central Mexican plateau disappeared, but also why, by whose hand, and for whose benefit.

The research that followed coincided with my son's early childhood. I split my time between reading about hydraulics, Renaissance technology and craftsmanship, and Mexican history and learning about cognition, evolution, ecology, art, urbanism, and other topics that seemed important to parenting a precocious boy born into difficult times. I had no idea how that reading in such apparently divergent topics would influence the ways I would interpret the sources. But it did: learning about how we perceive, understand, and make sense of the material world around us from birth made me curious about and sensitive to how the craftsmen and technicians I met in the archival documents understood the water in the lakes, stone, wood, and other materials they used, how they designed their tools and used energy to shape objects. Once I read *The Scientist in the Crib*,[1]

I saw parallels between how an apprentice learned these things by imitating the gestures and procedures from a master craftsman and how babies mimic sounds and movements of their parents and experiment and tinker with their surroundings. With my son modeling this behavior for me, I admired practical experience and knowledge more than ever. In turn, this helped me discover the ubiquitous presence of people with such skills even in documents written by lettered people lacking them. And, it triggered my dissent from historiographical traditions that distorted the past by forgetting that most things in it (and in the present, for that matter) were made by people who expressed themselves mainly through their actions on matter, not by the tiny minority of those who did not. I wrote the first version of this book while on this transformative parental journey but before I could fully articulate this dissent.

Soon the ideas I associated with the triumphalistic or Whig view of history as progress, with bigger economies and democracies as universal goals, and of humans as a superior species with a special destiny became untenable to me in face of the socioenvironmental crisis of the civilizational models these ideas defended. Other old and new ways of thinking about history, humans, and nature rushed in to offer me powerful alternatives. I took to heart Eric Van Young's recommendation that cultural history "colonize" economic history and the rest of the traditional fields of historical inquiry because "man the exchanger of calories with the natural environment and man the exchanger of meaning with other men are not easily separable entities."[2] I also heeded Ted Steinberg's call for a specific interrogation of nature to be shot right through the heart of traditional questions in political, economic, and social history,[3] and David Edgerton's plea for an end to the obsession with innovation, change, successes, and their actors.[4] The book became an experiment in various types of colonization, as well as an effort to define what this overused word—colonization—means.

During this process, colleagues in other fields frequently asked me to explain my observations about the Desagüe in ways that made them intelligible to students of Atlantic, European, or North American historical processes, as well as to historians of science. I obliged, at first frankly irritated, then progressively intrigued by what I was seeing. I looked at the Desagüe through the wide lens offered by Atlantic history as evinced in J. H. Elliott and Stuart B. Schwartz's latest works and through the exciting things I was discovering in the literature on science, technology, and expertise. Conversely, through the context of the Desagüe, I recognized some of the current weaknesses of both these fields.

By doing the former, I found that the Desagüe confirms the more balanced picture of the Atlantic that is emerging from recent scholarship on how practitioners and the learned operated in the Spanish realm.

The men who shaped the Desagüe were less like the dull-witted dinosaurs of the Black Legend and more like the sharp-eyed, dynamic predators of natural knowledge populating the pages of works on both the Spanish realm by Jorge Cañizares-Esguerra, Daniela Bleichmar, María Portuondo, Paula DeVos, Antonio Barrera-Osorio, and others, and on the European continent as well. In fact, by looking at the Desagüe in light of the early modern European studies of knowledge production and expertise by scholars such as Karl Appuhn, Eric Ash, Tara Nummedal, Pamela O. Long, Mario Biagioli, Chandra Mukerji, and Pamela H. Smith, a single phenomenon came into focus for me. Rulers and entrepreneurs on *both* sides of the Atlantic during the early modern period seemed keen on procuring, consuming, and deploying specific kinds of knowledge—practical, vernacular, and yet subject to theoretical explanation, as Eric Ash describes it for Elizabethan England.[5] Vested in "expert mediators," or "peritos e inteligentes"—that is, men with both practical and reasoned knowledge of their métier—the expertise coveted by early modern sovereigns and entrepreneurial elites was increasingly useful, not academic. They all, regardless of nationality, religion, or location, patronized individuals or institutions that might assist their worldly colonization of ecosystems, matter, and energy anywhere according to their own and their allies' class priorities.[6]

The similarities in the actors added up to more substance than the differences, confirming for me a rather equitable international distribution of dynamism. Clearly, it was worthwhile joining Hispanist and Hispanoamericanist scholars in their advocacy of the analytical possibilities that open up when the Iberian realm is taken as an integral—as opposed to marginal—part of early modern cultural history. I did so gladly. At the same time, I realized that I differed from these colleagues in that I sought not to show what the Americas taught early modern Europeans, but rather what studying the Americas can teach us about Atlantic early modernity as a whole.

One of the most important and in my opinion least understood phenomena of early modernity is colonization. By looking at the current weaknesses of Atlantic studies through the Desagüe, I became troubled that rich as Atlantic historiography was, oddly I could not find a definition of colonization in the many books aspiring to examine the phenomenon in this wide perspective. I was myself guilty of dragging the word up and down the pages of my manuscript, and I was ashamed of being a colonialist who could not explain with precision what colonization was. This is where I believe this book can make a difference. In linking the phenomenon of early modern colonization with that of knowledge-making in the Atlantic, writing this history of the Desagüe helped me identify and test some of the limits that still inhibit our understanding of both.

The definition of colonization I use in this book came from heeding those who encouraged me to explain the Desagüe in terms of Dutch, Italian, French, English, and other drainage projects of the epoch and realizing that like the Desagüe, few of these environment-changing public works were good for all. This is exactly what makes such projects perfect vehicles for understanding the social priorities and logic embedded in every single material and organizational structure surrounding us and how these priorities endure or mutate through time. Among these projects, drainage and desiccation works are especially revealing in that they transform the very building blocks of ecology—land and water—and therefore fundamentally and irreversibly alter how humans will relate to nature. That is why it was important to look more closely at who the agents were and what colonization looked like on the ground, as it altered the relationships among people, water, land, and biota.

In this light, colonization in the Americas appeared more and more not as the movement of some "peoples" or "nations" over others and their territories, but of alliances of some social classes and groups over others, both of which may also happen to be predominantly of one ethnicity or another. The usefulness of the "metropolitan"—in the sense of "mother country"—versus "colonial" divide that implacably splits the Atlantic realm of study waned for me. The cleavage seemed both too sharp and too capacious: it became increasingly difficult to force settlements of Spaniards in Mexico, Peru, or the Río de la Plata, for instance, in the same "colonial" pigeonhole as the coastal footholds the Portuguese set up in Brazil, or the agro-commercial enclaves of the British in North America. This is why to answer its central questions—why intertwined built and natural environments look the way they do, in general, and in the Valley of Mexico during the process of colonization specific to the region, in particular—this book will rest on an understanding of early modernity as a continuum straddling the Atlantic, not separated by it.

I am not certain what could result if scholars and teachers spent less time trying to explain how *peoples*—"Spaniards" or "Europeans"—or *nations*—"Spain" or "France" or "England"—colonized "America" (or "Africa," for that matter) and instead experimented with a different explanatory framework. Using an environmental framework of analysis, Kenneth Pomeranz and Edmund Burke III have offered the notion of a globally "shared developmentalist project," which cutting across epochs, continents, and systems allows us to see that the mode of utilization of nature that used to be associated solely with the rise of European capitalism was in fact not peculiar to it. I agree that looking closely at human-environmental relationships can reveal that any colonization process, even more so than imperialism per se, "often looks different in degree but not in kind from the kind of 'civilizing' projects that states carried out on their own populations and landscapes."[7] But I suspect that a more satisfactory explanation of colonization might be reached by asking why and how the different *classes* that participated in the formation of early capitalism—or the developmentalist project, for that matter—colonized water, land, and biosphere, and the relationships among them, as well as what each meant to humans in Europe, America, and beyond. Might this better explain who created nation-states and empires at the same time and as the result of the same dynamics, not the latter as a consequence of the former?

In considering the environmentally transformative activities and knowledge of Europeans on both sides of the Atlantic as part of the same process whose dynamics and actors were at least as heavily influenced by their class characteristics as they were by philosophical outlooks, I reiterate an invitation I extended to other scholars in an earlier article about the Desagüe.[8] I hoped that Europeanists and other scholars of the global north, including those working on science, technology, and expertise, would cross-fertilize with historiography from and of the global south. This cross-fertilization could take many forms—it might trace circulation as some Atlantic scholars have done; it could recover analytical tools such as the theory of value, uneven and combined development, and mode of production to better understand the phenomena they study in light of the "colonial" world; it might test the understanding of colonization proposed here. There are many ways to cut the vast cloth of early modernity, and there will be much less wasted material if the tailors study how their patterns fit together.

Like other authors of history books, besides asking questions, I do argue and claim things. But my main goal is to explain the origins of an important aspect of our present civilizational and ecological crisis and to demonstrate how such explanations can be found. Asking the kinds of questions I pose here can help us see the social and historical logic embedded in our built and "natural" environments, offering us clues on how change that logic. It is probably too late for the basin of Mexico, but it is still possible to change the social logic of the material things that steer life elsewhere.

# Acknowledgments

Many people shaped the thinking from which this book arose. At Berkeley, Margaret Chowning and Tulio Halperín Donghi taught me how to think historically. Their mentorship was crucial to me, and Margaret's insights, friendship, and support give me shelter to this day. What I learned in seminars with Jan de Vries and Carla Hesse and from William B. Taylor's critiques also became central to how I work now. Thanks to them all, my unattainable models for doing history became Fernand Braudel, Stanley J. Stein, J. H. Elliott, Carlos Sempat Assadourian, Enrique Tandeter, Warren Dean, and Halperín himself.

Other people inside and outside of academia also challenged me to develop my ideas. Before our parting of ways, for many years Argentine activist Carlos Petroni showed me how to think dialectically and to dissect class motivations and understand their role in past and present. In 2004, John M. Staudenmaier, SJ, and the first writing and publication workshop of the Society for the History of Technology, on the one hand, and Bernard Bailyn and participants of the Atlantic History Seminar, held in the CRASSH Center of the University of Cambridge, on the other, critiqued early drafts of Chapters 4 and 5 of this book.

While appreciative of the intellectual atmosphere created by all of my colleagues at Princeton, I have some special debts. Tsering Wangyal Shawa, head of Map and Geospatial Information Center at the Peter B. Lewis Library, gave me a crash course in the basics of GIS that made the maps in this book possible. The late Mike Mahoney turned me toward a deeper engagement with what technicians actually did and knew. Jeremy Adelman has been a simultaneously frank, caring, and generous mentor—exactly what I needed. He read this manuscript at different stages, helped me think about its broader implications, and has just been there for me always. Besides giving me their warmth and support, Stanley J. Stein and Dirk Hartog pushed me to clarify what I wanted to say, particularly about indigenes and the peasantry. Tony Grafton's, Michael

Gordin's, and Keith Wailoo's critique of the manuscript opened my eyes to its explanatory weaknesses, its Whiggish excesses, and its relevance to the history of science, technology, and expertise. My chair William Chester Jordan has been extraordinarily supportive not just of me but of all junior faculty, while regaling me with his knowledge and humor about things rural to boot. I thank Rubén Gallo and colleagues at the Program of Latin American Studies for their feedback, support, and funding. The talented group of Latin American history graduate students challenged me as a teacher and rewarded me with their humanity. I hope my dear friends and colleagues Mariana Candido, Ekaterina Pravilova, Bhavani Raman, and Wendy Warren know what they have meant to me without me confessing it here. Not least, the funding of the University Committee on Research and the Barr Ferree Publication Fund made all this possible.

I also owe much to friends and colleagues in Mexico. The late Marcos Mazari Menzer, physicist and soil mechanics expert, introduced me to what lay beneath the grand architecture of the city. Francisco Platas arranged a tour of the entire Porfirian Desagüe, accompanying me as engineer Julián Zarco-Herrera, documentarian Isaías Soriano, and field workers of the Sistema de Drenaje Profundo showed me every nook, cranny, and twist of the works. These men answered my questions, told their own frightening (and to me heroic and eerily reminiscent of Adrian Boot's report in Chapter 2) tunnel inspections amidst toxic effluents and gases, made funny hydraulic jokes, filled me in on drainage politics, lent me their raingear, and then rumbled me back up the road to eat delicious *barbacoa de horno* at a favorite truck stop. I thank them all deeply for one of the most thrilling days of my life (*mil gracias por una de las jornadas más apasionantes de mi vida*), and Miguel Carmona Suárez, Director of Drainage, Treatment, and Reuse in the Sistema de Aguas de la Ciudad de México, for allowing all this. Alejandro Tortolero Villaseñor's critiques saved me from embarrassments. Besides giving me her hospitality, Priscilla Connolly rendered conscious for me the key issue of rentier land use. Israel Sandré Osorio, Director of the Archivo Histórico del Agua, helped me understand the Cuautitlan River basin and modern water politics. I am also grateful to Teresa Rojas Rabiela, Jacinta Palerm Viqueira, and Luis Aboites Aguilar for their expertise on water management, to Michel R. Oudijk for help with images, and to the staff of the Archivo Histórico del Distrito Federal, the Fondo Reservado of the Biblioteca Nacional, and the Archivo General de la Nación for their professionalism.

In Spain, Ignacio González Tascón of the Fundación Juanelo Turriano in Madrid treated me with great kindness and answered my questions about early modern Iberian hydraulic traditions. It was a blow to learn of his untimely death. Horacio Capel Sáez of the University of Barcelona helped me understand the military engineers. The staff and officers at the

Archivo Central Militar, the Servicio Geográfico del Ejército in Madrid, and the Archivo General Militar in Segovia gave me their interest and resources, while the personnel at the Real Academia de la Historia, the Biblioteca Nacional, and the Archivo General de Indias efficiently handled every request. Asmaa Bouhrass secured AGI publication permits. The Fundación Carolina enabled this with a research grant.

Eric Van Young has been a detailed as well as elegant critic of this manuscript and the work leading up to it. In persistently inviting me to present my work in different venues, the inexhaustible Jordana Dym stimulated me to think in new ways about images and maps and their value as sources and about teaching. She too was an insightful critic of the manuscript for this book. Cynthia Radding, María Elena Díaz, John Soluri, and Latin American environmental historians of SOLCHA have been sharp and supportive interlocutors, often over really enjoyable meals. Eric Engles, my superlative editor, is responsible for this book being clearer than it would have been otherwise. Editors Norris Pope, Stacy Wagner, and Emily Smith at Stanford University Press patiently guided me. This book comes out as my analyst Barbara Cohen and I end our sessions, but years of healthful insights with her suffuse these pages. Everyone I have read or talked to about history, politics, evolution, and ecology along the way has influenced how I think, so those I do not mention by name please know that I thank you nonetheless.

Above all, this book exists because of my son. With his birth, Fabrizio changed me in every way. Through his sparkling eyes and mind, I vicariously relived the pleasure of discovery in the nooks and crannies of everyday life. As I parented Fabrizio, the thrilling explorations of my own childhood—with all the searching for causes and effects in nature, objects, and actions—with my father, Jorge W. Candiani, a restlessly creative Argentine chemical engineer, came back full force. The difference was that my son took me farther, prompting me more urgently to ask questions about the origins of our surroundings and about how to make the legacy of our own and our ancestors' choices, conceits, and failures less perilous and more hopeful for Fabrizio and youngsters like him.

# Abbreviations

## Archives and Libraries

ACEG: Archivo Cartográfico y de Estudios Geográficos del Centro Geográfico del Ejército, Madrid

AGEM: Archivo General del Estado de México, Toluca

AGI: Archivo General de Indias, Sevilla

AGMS: Archivo General Militar, Segovia

AGN: Archivo General de la Nación, México

AHA: Archivo Histórico del Agua, México

AHDF: Archivo Histórico del Distrito Federal, México

AHM: Archivo General Militar, Madrid

BLAC: Benson Latin American Collection, University of Texas at Austin

BNE: Biblioteca Nacional de España, Madrid

BNM: Biblioteca Nacional de México

RAH: Real Academia de la Historia, Madrid

## Publishers and Institutions

CIESAS: Centro de Investigaciones y Estudios Superiores en Antropología Social

CSIC: Consejo Superior de Investigaciones Científicas

DDF: Departamento del Distrito Federal

ENAH: Escuela Nacional de Antropología e Historia

FCE: Fondo de Cultura Económica

INAH: Instituto Nacional de Antropología e Historia

SEP: Secretaría de Educación Pública

UNAM: Universidad Nacional Autónoma de México

# Note on the Colonial Regime

To colonize means far more than erecting a colonial regime. But it would be difficult to explain how the Desagüe helped anyone colonize anything without a glance at the colonial regime and how it affected the Desagüe over time. Readers familiar with this regime can safely pass over this introductory piece. Others will find it profitable to review how Spanish rule worked before reading about the Desagüe.

Part of the reason the Desagüe was such a challenging project is that it was planned and run by a complex administrative apparatus: major technological decisions about it were subject to the consultative method of rule prevalent in the Spanish empire under the Habsburgs. This meant that corporations such as city and church councils, the merchants' guild, the university, and other bodies representing the social and bureaucratic elites all had a voice in it, as they had in other weighty matters in the viceroyalties. University-trained bureaucrats—the *letrados* of crown and church who ran the day-to-day for the monarchs—were expected to plan and implement a permanently improvable new reality and to design a regime that reflected the monarch's priority among Iberia's multiple entities and social classes. The regime they created was born, on the one hand, out of the relationships among kingdoms and kings and cities that solidified as the Iberian Peninsula came into Christian control. On the other hand, it was decisively shaped by the legacy of the partnership formed by the Castilian crown, private enterprise, and the church for the conquest and colonization of the Americas, which unleashed hordes of people aspiring to create unfettered seigneurial privileges, a pristine new Christianity, opportunities for entrepreneurial and mercantile profit, or simply a better living. All of the partners in the regime struggled to secure their interests in the context of the crown's own patrimonial dynastic priorities. To a large extent, this mode of functioning survived into the Bourbon period.

Institutionally and philosophically, Spanish rule in America was first conceived "as a replica of its mother country."[1] In consonance with the principles of natural law that underpinned the political and juridical regime of the Habsburgs in Spain, the crown recognized certain aspects of indigenous law and custom and created an indigenous realm—the *república de indios*—endowed with a corporate content embodied in the indigenous townships (*pueblos de indios*) and distinct from the sphere where Hispanic life was to unfold—the *república de españoles*.[2] Perceptively, the Spanish grafted their system onto the structures of rule in indigenous society, which provided the warp for how the two distinct societies would interweave.

In New Spain during the formative decades of colonial rule and the birth of the Desagüe, indigenes were the basis for everything. Colonists vied with the crown and the pueblos for the labor of a dwindling indigenous population, which influenced the evolution of a variety of institutions, from the state system of labor allocation called *repartimiento* to the hacienda and the technological choices made in the Desagüe. From the mid-seventeenth century on, as indigenous populations slowly recovered, settlers and pueblos would increasingly fight over land and water rather than labor, and this too was reflected in the Desagüe. In other words, it was impossible to segregate the two *repúblicas*. In practice, they interacted on every level all the time.

To a large extent, the institutions emerging from this regime did enable the monarchy to arbitrate the relationships among social groups and classes in the fractious peninsula and in its vast empire, thus dampening social conflicts that might otherwise tear at the stability of dynastic rule. The letrados codified this rule and the interactions among social groups on behalf of the crown using four basic principles to govern the rights to nature and its resources that each republic in the Indies would have. First, all territory—land, water, subsoil, and biomass—belonged originally to the crown. Second, all water and pastures were commons, unless the monarch said otherwise; third, in promoting agriculture and mining, the crown generally privileged communal interests over private ones; and fourth, Indians were legal minors, and their welfare was a priority.[3]

In the república de indios, these principles were expressed in the persistence of pueblos as both juridical and territorial entities, with the right to be governed by their own authorities thanks to the fact that they retained land, water, and woodlands—and, in the case of the valley, marshlands. From soon after the conquest, these resources were communally owned and theoretically inalienable and indivisible, which meant that individuals or families had usufruct but not ownership rights.[4] By 1567, Spanish haciendas had been prohibited within 500 varas (415 meters) of Indian township boundaries, and livestock ranchos within 1,100 varas.[5] Known as the *fundo legal*, this endowment was ratified, refined, and increased by *cédulas* in 1687, 1695, and 1713, then encoded into the 1786

*Ordenanzas de Intendentes* that formed the backbone of the Bourbon administrative reforms.[6]

Land and water grants in favor of the república de españoles in theory followed these principles too, but other factors militated in favor of Spanish landholding, especially indigenous depopulation from epidemics. By the 1620s—the indigenous demographic nadir—between 51 and 83 percent of the then five thousand square kilometers of "dry" land available in the Valley of Mexico was in Hispanic hands, depending on the region.[7] Nevertheless, legal protections from the crown helped Indian pueblos retain access and dominion over land and water resources, creating a mosaic-like landscape.[8] Without this survival of indigenous townships, neither the drafted rotational labor (repartimiento) ubiquitous in Desagüe history nor the indirect appropriation of labor (explained later in this book) would have been possible.

Clearly, the persistence of an indigenous peasantry may not have made all Hispanics happy, but it was crucial for the crown at all times. Indeed, the resource-endowed república de indios underpinned the entire tributary system, the essential mechanism for the extraction of surpluses, and was intended to prevent the indigenous population from becoming something other than a perpetual agricultural class under the protection of the monarch, as befitted a population that was regarded as legally minor. Although constantly attacked from without and within, the *fundo legal* with its accretions—along with the peasantry it was supposed to sustain—survived into the modern era. This is key to understanding both the Desagüe and the region's uneven and combined development.

Because Indians often enough received preferred water and land access, because their agriculture in the Desagüe area was irrigated or wetland, and because their commons (*ejidos*) were often in marshlands or areas subject to seasonal flooding, indigenous agricultural and hydraulic technology designed to harness the community's ecosystemic resources remained alive through the upheavals of conquest, population loss, and the growth of Hispanic landholding. This vitality made it useful to the Desagüe's technicians. Conversely, some of the technological practices, processes, and tools among indigenous pueblos directly under the jurisdiction of the Desagüe changed as their people interacted with the public work, much as the pueblos themselves changed in their interactions with the broader market economy.

The presence of the Desagüe in a rural milieu characterized by dry land as well as seasonal wetland ecosystems had deep implications for the relationships among the different groups of people interacting within and with land, water, and biomass in the region. For several of the indigenous townships, the wetland ecosystems were essential not just for agriculture and pasture but also as a source of supplementary foods, medicines, and materials for domestic manufactures. These ecosystems, however, changed seasonally: what had been land became a marshland with the rains, and then cultivable land again. The Desagüe would profoundly affect this cycle: with the objects and regulations they deployed on the ground, its officials and technicians would redefine what was to be classified as land and what was to be treated as water, and segregate the two in the process, as will be seen.

By the time the Desagüe was built, the república de españoles was also well structured, with the somewhat overlapping jurisdictions and powers of the viceroyalty and the *audiencias* replacing *encomenderos*. Viceroys and audiencias reported to the crown, not to each other. Both enjoyed legislative and executive powers, but judicial functions were exclusively the purview of the *oidores*, *fiscales*, and other members of the audiencias, who also had legislative and executive functions. *Corregidores* organized local rule, alongside the strong Iberian tradition of town rule by city elites, whose *cabildos* legislated and ran the affairs of the municipalities.[9] Besides the church—with its regular and secular branches—there were other corporations who had important roles in colonial rule. Key allies of the crown, wholesale merchants regulated themselves through their *consulados* or guilds in commercial hubs. Seville, Cadiz, Veracruz, and the City of Mexico all eventually had consulados, as did several other cities in Spain and America, and all were deeply involved in the regime. This multilayered and overlapping method of rule might appear inefficient, but actually it facilitated crown control over its social allies and bureaucrats, who reported on and counterbalanced each other.

This is exactly how the Desagüe was run, with all these actors intervening at one time or another. Friars became its superintendants in the seventeenth century, followed by oidores in the eighteenth. The consulado of the City of Mexico regularly if not always willingly made financial contributions and administered some of the works. All important orders had to be approved by the viceroys, whose introduction to their new posts included a tour of the Desagüe shortly after arrival in the capital. *Regidores* of the cabildo and the higher bodies of the secular church were always consulted on important decisions, the former often participating in inspections, in drafting recommendations, and in approving or obstructing viceregal or audiencia initiatives regarding the public work that protected all their collective wealth and incomes. These overlaps did, however, create conflicts of jurisdiction and financial responsibility among the viceroy, the oidores, and the regidores of the cabildo of the City of Mexico, which sometimes hindered the speed of decision making in the Desagüe.[10] In broad strokes, the administrative pattern in the Desagüe and the manner in which its superintendants ran it mirrored the changes in this basic outline wrought by the end of Habsburg rule in

1700 and the ascendance of the Bourbon dynasty to the Spanish crown. From the mid-eighteenth century on, the Desagüe's oidor superintendants increased their involvement in hydraulic matters in all corners of the basin, authorizing, modifying, or rejecting projects according to their impact on the drainage. Partly the result of the absolutist enlightened reforms of the Bourbon dynasty, the oidor superintendants had broader and deeper jurisdiction over Desagüe matters also because the increasing complexity of flooding demanded it.

*Dreaming of Dry Land*

# Introduction

In the early morning of November 28, 1607, many of Mexico's most prominent men, including Viceroy Luis de Velasco the Younger, several members of the *real audiencia*, and the king's envoy, *visitador general* Diego de Landeras y Velasco, set out from the imperial City of Mexico to perform an important ceremony in the northeast quadrant of the Valley of Mexico, the large basin surrounding the city. Accompanied by a scribe to record the events of the day, the party arrived a few hours later in the Indian hamlet of Nochistongo.[1] They attended mass in a hut, and when they emerged the viceroy was given a hoe. We might imagine him pulling soft gloves over his delicate hands, securing his spectacles on the bridge of his nose, and grasping the handle of the tool. Taking a deep breath of the crisp, country air, he heaved the tool over his head and sunk its blade into the crusty ground. More than one blow would have been necessary. From some distance, fifteen hundred Indian workers watched. What a sight it must have been for them—a nobleman breaking a sweat. When Velasco was done, the little pile of dirt he had dug was blessed as cheers filled the air. Behind the severe expression on his face, the viceroy must have felt pleased—or at least relieved—to be breaking ground for this new project.

What kind of project in the countryside could require such ceremonious beginnings? The handful of dirt the viceroy had dug was in fact the first of some sixteen million cubic meters of earth that would be excavated by the end of the colonial era for the Real Desagüe de Huehuetoca, a massive hydraulic project intended to drain the lakes that surrounded the City of Mexico. At the time this project was begun, the lakes still ringed the capital much as they had under the Aztecs, even though it had been nearly nine decades since Cortés had arrived for the first time in Tenochtitlan. It was not the lakes themselves that vexed the Spaniards as much as their tendency to periodically swell with runoff from the surrounding mountains and flood the increasingly important city. Because it seemed essential to the survival of the city itself, the Desagüe was an undertaking far weightier than the construction of any single building.

By draining the lakes, urban elites believed they could free up the lakebeds to accommodate all the seasonal runoff and thus protect their palaces and streets from flooding. But draining the water away was no small task, since the basin in which the City of Mexico lay had no outlet. It would require perforating the hills that ran along the basin's northwestern margin. Doing so would be, in essence, to reverse the geological clock to before the Quaternary Period, when the basin of Mexico was still a valley, draining southward. As an engineering achievement and given the knowledge of hydrology, hydromechanics, and geology of the epoch, the Desagüe is massively impressive. That elites in New Spain would even contemplate a project that they knew was herculean speaks to the value they accorded their capital city. The City of Mexico was fast becoming the most important city in the Spanish "empire of towns."[2] Quite simply, it had to be protected from floods at all costs.

Over the eleven months that followed the launching of the drainage project, urban technicians and sixty thousand Indian laborers bored through the northwest mountains of the basin using ingenious applications of mining technology, aided with winches and draft animals, to create the thirteen-kilometer-long Desagüe. The most notable feature of this device was a tunnel seven kilometers long, running as deep as fifty-six meters under the surface, capable of conveying water captured in the northwestern lakes out of the basin altogether. It was one of the most ambitious early modern public works projects undertaken by Europeans anywhere.

There was great jubilation in the City of Mexico when the tunnel and its feeder canal performed well during the first rainy season that put them to the test, in 1608. But this initial success was far from the last word on the Desagüe. To continue to perform their functions over time, the tunnel and canal would require considerable and difficult modifications. The tunnel began to fail partially soon after its first year, and after a catastrophic flood in 1629 colonial officials became convinced that it would have to be converted into a vast open trench if it were to offer any protection from flooding at all. Although technicians were at first confident about completing the trench conversion swiftly, the project stretched on for more than 150 years, for reasons that will be explained. In addition to being an ongoing construction project for much of the colonial period, the Desagüe grew in size and complexity. While they slowly excavated the open trench, the thousands of people who worked in the drainage project over time built countless dams, levees, river diversions, silting pools, sluicegates, and other devices both within the Desagüe district and far beyond, as technicians and officials gained a better understanding of the interconnected hydrology of the basin and the multiple causes of flooding. Over time, these structures were increasingly coordinated to function in tandem with the Desagüe proper, making the Desagüe project the vastest and most complex dessication effort in the Americas during the period, commanding considerable resources for construction, administration, finance, operation, and maintenance.[3]

Looking beyond the colonial period, it becomes evident that the drainage project was never really completed. President Porfirio Díaz continued with attempts to perfect the system in the late 1800s when he ordered the construction of the Sistema de Desagüe del Valle de México, which collected both rainwater and sewage from as far as the eastern side of the city and drained it out through a canal and tunnel that ran east of the colonial Desagüe. The old Desagüe de Huehuetoca was not abandoned, as some of its structures were hitched to the newer one, which underwent further modifications and additions in the 1940s, in the 1970s, and then more recently.

Despite their huge financial and even greater human costs, both Desagües periodically failed to protect the City of Mexico from floods, much like the modern system. During the colonial period, officials sometimes wondered whether maintaining the Desagüe was worth the effort and resources. Painfully for its defenders, the only true standard by which the drainage can actually be considered "finished" is not the absence of flooding in the city, but the fact that save for a few polluted remnants in the south and north of the basin, nothing remains of the ancient lakes today.

Notwithstanding its mixed success in achieving its stated objective, the Desagüe had enormous repercussions for the indigenous and Hispanic people living in the basin, for colonial elites, and for the Spanish empire as a whole. By transforming the physical, hydrological, and biological environment of the basin, the Desagüe irreversibly changed the conditions of life for everyone in it, rendering it more amenable to Spanish patterns of production. By keeping alive into the late colonial era the method of coerced indigenous labor, it brought into play the protections conferred by the crown upon indigenous villages and thereby had a hand in sustaining the peasantry as a class. By allowing elites in the city to remain focused on rentier priorities rather than productive ones while other sectors of this class were expanding capitalist social relations and modes of production into a variety of locales and activities, it militated against "founding capitalism," to borrow John Tutino's phrase, in all of New Spain.[4] In short, the Desagüe played a central role in the process of colonization and helps explain the unique way in which that process unfolded in the basin, while also illuminating how colonization shaped early modernity in the Atlantic as a whole.

Though "merely" a collection of excavations and built structures, the Desagüe mediated the superimposition in space of two distinct forms of social—the one transplanted by Europeans vested in primarily private relations of property and production and the one surviving from the pre-Hispanic period vested in primarily communal ones. Each form of social organization had its own way of valuing land and water and their respective biota. Outwardly divided by ethnicity, Hispanics and indigenes were more fundamentally distinguished along the axis of social class. Each class and class sector had very different interests, and their conflicts played out in the Desagüe-modified terrain of the basin, regulated by a dynastic state apparatus with its own priorities. The complex and contingent process of colonization that occurred in the basin was shaped largely by this class-based conflict. Because the Desagüe was so important in mediating the environmental and technological dimensions of class conflict in the basin, it is possible to learn a great deal about the dynamics of colonization by uncovering the social relations and class priorities embedded in even the most unglamorous of structures in the drainage project.

To extract these insights from the Desagüe entails questioning a number of assumptions that have characterized writing about the drainage since the early 1800s. Consciously or not, most scholars have looked at the drainage project from the perspective of the urban elites of the City of Mexico, who were firm in their belief that the phenomenon of flooding endemic to the basin of Mexico was a problem for all and that it had to be solved. Taking for granted the ubiquity of this perspective, these scholars unwittingly generated a grand narrative under which flooding was a universal "problem" and that its solution—the Desagüe—was a universal social good. This has precluded seeing the Desagüe as primarily an expression of the interests of the elites in the capital, as a "public works" project that had differential benefits and costs for different social classes and groups.

The urban-centric orientation arose in part from an inherent bias in the sources used by scholars. Over its long history, attitudes toward the Desagüe were as diverse as the people who made it, but not all attitudes were reflected in the formal letrado sources favored by historians. The fact that this record is eloquent about the attitudes of urban elites and bureaucrats, but rather quiet about those of Indians and rural Hispanics was not considered.

Unquestioned, this record became the primary source for the great Prussian savant Alexander von Humboldt, who kept it in mind as he visited the Desagüe in 1803 and 1804 and wrote his 1808 *Essai Politique*

*Sur le Royaume de la Nouvelle-Espagne*. Born at the cusp of the enlightenment and birth of liberalism, Humboldt could not avoid infusing his otherwise reliable treatment of flooding in the lacustrine basin and the Desagüe with his generation's desires and regrets, which in turn inspired later generations of scholars, statesmen, and travelers writing the story of the Desagüe.[5] The tumultuous decades that followed the collapse of Spanish rule tempered the optimism of many authors who wrote about Mexico, but perhaps none was more insightful than geographer Manuel Orozco y Berra, who sensed the pitfalls of hydraulic projects undertaken for the benefit of a single city.[6]

But in the last three decades of that century the stable oligarchic republican regime born out of the liberal and conservative convergence resuscitated Humboldt's hopefulness among the elites. Under Porfirio Díaz, they found it possible and desirable to rescue the drainage project from the doldrums. Anxious officials and academics scoured the artifacts that colonial "creole ingenuity" had legated to them for evidence of their country's capacity for liberal modernity. In their hands, the Desagüe was buoyed to an unambivalently prideful place among the great myths about Mexico's past. Antonio García Cubas, Mexico's preeminent nation-building cartographer of the epoch, featured the Desagüe as one such symbol alongside vignettes of the Castle of Chapultepec, the Belem paper mill, and other emblems of Mexico's bounty and modernizing achievements in his 1885 *Atlas Pintoresco e Histórico de los Estados Unidos Mexicanos*.[7] Subsequently, scholars have rarely if ever questioned the assumptions built upon Humboldt. They especially avoided asking whether flooding was a problem for everyone, after all, or just for those who fantasized about their imagined nation's glorious ascent to the heights of capitalist progress.

One must abandon the urban-centric perspective and the assumptions stemming from it to recognize that, in fact, little about the Desagüe was inevitable. That is, no objective factors *forced* it into existence or to develop the way it did. The phenomenon of flooding endemic to the basin of Mexico was a "problem" only for some sectors of the population. Even if one accepts that it was inevitable that overflowing lakes would be defined as a problem, many possible options were available for addressing it, each with different costs and benefits for different social groups.

It is undeniable that the water brimming over lakes damaged public and private urban property and made life in the city difficult. But it is also true that the floods were a precondition for a large part of the basin's agricultural productivity, especially for those indigenous and even some nonindigenous cultivators who had built their practices around the seasonal rise and retreat of water over significant swaths of land (see Figures 7.5 and 7.6). To control flooding was to thoroughly compromise the ecological productivity that had supported large human populations in the basin for millennia. So to take for granted that authorities *had* to end flooding in the basin of Mexico obscures the specific interests and motivations that lay behind the Desagüe, and those that were excluded from it.

Moreover, flooding itself is a dynamic and complex phenomenon. Floods in the urban hub of Tenochtitlan-Mexico were not, like smallpox, unwittingly unleashed by the actions of the Spanish. As is well known, all its causes—the loss of ground cover, deforestation, the disturbance of the soil and silting of lakebeds—were already well under way before the Spanish arrived. But because flooding is sent to an analytical corner, so to speak, by urban-centric, Whiggish scholarship, it tends to sit there, immobilized, for the whole period, becoming a "fact" that remains unaltered in its origins, causes, and implications.

The Desagüe, as a set of physical structures and a landscape modified by those structures, both reflected the interests and values of the elites in the city and acted as a kind of stage upon which other social classes contested the expression of those interests and values. This much can be seen by simply removing the blinders imposed by the traditional Desagüe narrative starting with Humboldt's own useful assessment of the Desagüe. This assessment had two components—one "developmental," fraught with liberal desires and lamentations; the other purely hydraulic. In the pages that follow, these two components will correspond roughly to the terms "conceptualization" and "design." Conceptualization designates both how social and bureaucratic elites in the city valued water, land, biota, and energy and how this translated into a strategy of preserving wealth and rents in the city from what seemed to them to be their worst enemy—floods. Design refers to the materialization of this conceptualization in concrete structures, from the much celebrated tunnel to the most overlooked dirt dam. Humboldt noted with pain the detrimental effects of the Desagüe on indigenous survival through its history, and while lamenting the fact that the conceptualization had not included elements he thought essential to progress, he fundamentally agreed that the city had to be protected and that it was up to the elite to decide how. It would be unreasonable to expect a person of his epoch to question the right of the select few in the city to dictate the uses of nature for all or hope for anything but that these people show themselves capable of making such uses consonant with the capitalist transformation of the society they ruled for the prosperity and happiness of all. But in our epoch, when it has become evident that where realized these hopes have not quite yielded such blissful results, as it is essential that we do.

We can start by using historical analysis to dissect and make conspicuous the conceptualizations and designs of the material and social structures from the past that still inconspicuously make us march along the same, exhausted developmental paradigms. The next step here, then, is to uncover the capital's interests, and how it attempted to project them onto the hinterlands through the materiality of the drainage. How is it possible to discover the physical workings of the class conflicts and their results—to "deconstruct" the conceptualization, design, and execution of the drainage?

Obviously, physical factors such as the geomorphology of the basin and the technology of the era can explain much of the material form and functioning of the drainage. But the logic by which this technological and organizational marriage took shape is accessible only by examining the everyday actions and difficulties of technicians in shaping the earth and directing water, on the one hand, and the social context in which they operated, on the other. What links actions, material realities, and social context together are people's technological *choices*. The resulting material structures are concrete manifestations of these choices. That is why each structure within the Desagüe can reveal the priorities of the social actors who made the decisions about how those structures would be constructed, where they would be sited, and for what ends and in what manner they would function.

The concept of desiccation as a solution for flooding in the urban core reflected how its elite makers valued water, land, biota, and energy (human or otherwise). The designs the subordinate technicians created indicate their general acceptance of this conceptualization—they were not rebels, after all. But technicians had choices about how to shape the Desagüe that were neither predetermined nor as constrained as they might seem. To be sure, there were limits to the choices, but they were not purely financial, cultural, physical, or even technological, as most authors represent them. Of course these factors matter, but the true boundaries of the possible were set by mutable yet fairly stable *social factors*: the composition and priorities of the ruling urban elite and the persistence of an indigenous peasantry with a degree of control over the land, the water, and the ecosystems that the elites sought to transform.

The concrete choices that the Desagüe's architects and technicians made were choices among distinct technologies. These technologies were quite diverse. In addition to the indigenous ones, there were at least three distinct imported ones—Muslim, Iberian, and Roman. Further diversity within all four was added by the specializations in various realms of activity—agriculture, military and civil construction, mining, and so on. Each technology involved the use of particular materials (physical and biological) and particular methods and sequences of actions given meaning and utility by the cultural context in which it was embedded.

Researching such technological choices requires the proper conceptual and methodological tools.

While useful, the social constructivism advocated by Thomas P. Hughes and others is used here only in a very general way because it does not offer precise enough tools for analyzing discarded options or social effects.[8] Originally developed by anthropologists in the 1930s to study how cultures incorporate features of another's techniques, and since redeployed for other uses, the concepts *acculturation* and *transculturation* will not do here either, for two reasons. First, in the Desagüe not two cultures but many cultures confront one another, and since these concepts were developed as tools to examine cultures in colonization, and not classes in colonization, they would render the class content of the Desagüe-facilitated colonization of the basin's hinterland inaccessible, whereas the purpose here is to find it.[9] Second, while it is common for writers to refer to technology and other cultural products in colonial Latin America as "hybrid" or "mestizo," in the case of the Desagüe the tendency has been to categorize the technologies underpinning the various designs as either European or creole. All four of these descriptors have been used in a basically genetic way, in the sense that as tools of analysis they treat artifacts as though they were living organisms formed by random combinations of their parents' DNA. In addition, *mestizaje* is so heavy with associations with the foundational myths of Mexico, so synonymous with Mexicanness, that it often acts as an analytical blinder.[10]

The conceptual tool most useful to the approach of this book is the notion of technological choice used by French anthropologist Pierre Lemonnier and others. Technological choice, Lemonnier explains, "emphasizes the sorting out of possibilities on which the development of a technical system is *de facto* based, although usually in an unconscious and unintentional way."[11] This concept is useful because the record does not always provide a conscious, intentional expression of these choices in the voices of the human actors in the Desagüe. It is sometimes necessary to try to recreate the conceptual universe of technicians and officials and infer the logic of their selections from it. The conceptual universe of such men was shaped by interests and values of the classes or groups they affiliated with and by ideas of a more philosophical nature expressed in part through the written word. The most formalized of Desagüe writings were the *memoriales* or *relaciones*, which men in charge of the project or aspiring to be given commissions in it wrote for higher officials, seeking recognition and rewards or exoneration from blame. By their nature, with their self-conscious authorial voices, these texts can obscure or overemphasize certain characters and actions. The fact that most of those who worked in the Desagüe expressed themselves mainly through their actions on matter, not on paper, threatens the usefulness of these sources. For this reason, borrowing from the gaze that Peirce Lewis proposed decades ago, as updated by Don Mitchell, here the "ordinary vernacular landscapes" that the men in charge and their subordinates created through their choices will also be considered as an eloquent if "unwitting" expression of the conceptual and historical universe within which they operated.[12]

It would take a full-scale archaeological excavation to find the complete material record of the Desagüe. The alternative is to read the documentary record with an eye to the actions of Desagüe subordinates ranging from its coerced rotational labor force to its on-site wardens. As we will see, Indians were the backbone of the Desagüe, while the *guardas mayores*, *guardas menores*, and the non-letrado *sobreestantes* (the resident general wardens, the wardens under them, and the foremen) were its spinal cord. Through the comings and goings of friars, architects, military engineers, creole savants, oidores, and others whose voices prevailed in the works at one time or another, it was these less-visible actors who actually ran the Desagüe's daily functioning as a drainage device and executed much of the repair and modification work on the ground. These people are difficult to follow because while the record does name them, it only sporadically provides details of their actions, and more rarely still of their thoughts. This does not mean, however, that the Desagüe lacked a "literature" generated by its own actors.

Alongside the formal memoriales and relaciones, there is an abundance of dispatches among guardas, between guardas and superintendants, recorded verbal reports from sobreestantes, financial accounts, and so on. In contrast to the deliberate, often rhetorically ornate language of formal Desagüe writing, these mundane communications are terse, more unselfconscious and revealing. Even illiterate sobreestantes passed on their knowledge about the behavior of water, the terrain, people, and plants in their domain orally during interaction with guardas, technicians, and officials, who then folded it into the written record. This record was known as "the books of the Desagüe," and included the collected *autos* (writs), dispatches, and other important instructions and orders that arrived on the Desagüe site and sometimes duplicates of outgoing reports as well. Besides city and audiencia archives, these books were kept under the responsibility of guardas mayores in the Casa del Desagüe, the headquarters in Huehuetoca. Guardas mayores passed on these books to their successors "by inventory," so that their contents could be both confirmed there and consulted more expediently when needed.[13]

Guardas and sobreestantes provided the thread of technological continuity in practice and were also the bearers of much of the working memory of the Desagüe, the orality of which we can glimpse through their dispatches and reports. Collectively, the origins and trajectories of guardas and sobreestantes show both how fluidly specialized knowledge of the material world moved among nonspecialists and across social ranks, and how easy this is to overlook: much technological and hydrological knowledge was acquired and transmitted orally, experientially, on site, and over generations. While scholars of early modern Europe recognize the importance of such socially spread expertise for knowledge production, preservation and deployment,[14] in Desagüe histories the focus had been on the select few technicians who authored memoriales and other authorial reports, and on the bureaucrats, savants, and academicians who belonged to the universe of the city, the letrado culture of which is in many ways difficult to consider more relevant than that of the overlooked on-site subordinates. It seems terribly unjust that such men should have gone unnoticed or even disdained by the "enormous condescension of posterity," as E. P. Thompson might have called it had he turned his attention to this side of the ocean. Worse, it has blinded historians to the significance of such transoceanic similarities for a fuller understanding of early modernity.

The fact is that the Desagüe involved both Old and New World forms of "tacit knowledge." This conceptual tool allows scholars to see knowledge that is embedded and transmitted in the gestures and sequences of work and in objects themselves.[15] This kind of knowledge was the specialty of the guardas and sobreestantes and their workers. Without these men, the proposals of the most capable of letrados, technicians, or superintendants were nothing but words.

The Hispanic technicians, wardens, foremen, oidores, and superintendants who intervened in the Desagüe over the course of the seventeenth century tapped a wide range of tacit knowledge, both in their own traditions and in the indigenous ones, and recombined it in the drainage structures. Like Mukerji's Languedoc laundresses or Appuhn's *terraferma* peasants, indigenes living in the vicinity of the Desagüe works provided their expertise both voluntarily and under coercion. This expertise included the mobilization and coordination of labor; the seasonal organization of tasks and processes; familiarity with soils, minerals, and plants suitable for hydraulic engineering; and knowledge of the layout of the terrain and the behavior of water over it. Of these types of knowledge, the social dimensions that were liable to Hispanic appropriation are by now familiar. For example, we know that the repartimiento system of labor recruitment that began in 1555 for public works and eventually took groups of men from villages to deploy them in the Hispanic economy more generally were appropriations of the pre-Hispanic *coatequitl* (the summons of labor crews from subject communities).[16] But how the technological appropriations happened is a bit more mysterious.

In general, the choices technicians made among the whole universe of methods and materials available to them in order to design and shape the drainage were largely a result of how their own backgrounds intertwined with the tensions among elites in the imperial capital of the City of Mexico and between the elites and the peasantry over the same geographical space. Similarly, indigenes, as we will see, also used technology to interact with the drainage, a project that was placed among them and demanded their participation. But, since overall their ways of valuing water, land, biota, and energy were different from Hispanics, it follows that they designed their technologies differently too, even while appropriating from and interacting with what Hispanics were doing all around them.

Clearly, the Desagüe-assisted colonization of the Valley of Mexico by people from Spain was unique, distinctly different from the colonization of other parts of North and South America that took place at the same time. Other Europeans used different strategies to submit territories and their indigenes in Old and New Worlds to their wills; these indigenes had different political systems and responded in diverse ways to colonizing groups; arguably, in no other locales of the New World were massive public-works projects implicated so deeply in colonization by Hispanics as they were in the Valley of Mexico. The methodological lens used in this book, with its focus on the concrete structures of the Desagüe and on-the-ground choices by local actors, would seem, if anything, to emphasize the one-place-in-time approach of the work. Yet the Desagüe has much to say about the process of colonization in general—and therefore about such diverse topics as the relationship among colonization, empire building and nation-state formation, the origins of capitalism, the development of knowledge about nature, and the creation and deployment of expertise.

The first step in expanding the analytical frame within which the Desagüe has relevance is to understand the Iberian realm as an integral, rather than marginal, part of early modern history, from the political to the economic and the cultural. It really makes no sense to consider the Desagüe's story as illustrative only of "Spanish colonial" behavior, especially since under the Habsburgs the Americas operated and were regarded more as kingdoms of the patrimonial dynasty. And the Desagüe itself must be seen as an eminently European project, not a Spanish outlier. Its promoters and builders drew their inspiration and knowledge from far and wide and synthesized it into the project. Spanish diplomats scoured the courts of Paris, Milan, and Brussels for hydraulic experts who might be lured to Mexico, and ingenious tinkerers of the most varied callings pored over very European Vitruvian architecture treatises and mining and military engineering manuals for ideas and designs for the drainage of the lakes. The Desagüe was one of a kind to be sure, the material child of a spectacularly intrepid late Renaissance imagination. It was impossibly complex. It was almost too vast. But as an attempt to bend the landscape to the human will, it was really a typical early modern "big-think" European engineering project.[17]

Next, one must recognize that all the fundamental social processes that accompanied colonization in the Valley of Mexico had their analogues elsewhere—both in the New World and on the other side of the Atlantic. Wherever communal modes of production and social reproduction held sway and began to be challenged by the imposition of private-property relations and the rule of mercantile capital, the processes and dynamics seen working over time in the Desagüe district are to be found as well, entirely distinct in their specifics but recognizable in their overall outlines. At the most general level, these parallels exist because all through the Atlantic realm during the early modern period, the people engaged in early forms of capitalist production strove to gain hegemony over both human labor and the environment, indistinctly of whether they were in Europe or the Americas.

As demonstrated by the story presented in this book, the locally contingent ways in which this occurred are seen with exceptional clarity when one examines its operation in marshy landscapes, like the Valley of Mexico, where the literally fluid nature of the relationship between water and land was an obstacle to the penetration of private property and capitalist social forms and modes of production and actually quite favorable to autonomous and communal forms. In such locales, which include the fens of Cambridgeshire and Lincolnshire in England, the wetlands of Poitevin and the Camargue in France, the lowlands of the Netherlands, and other seasonally flooded regions in Europe and in French Acadia, early capitalists and agents of the state undertook coordinated efforts to drain, dike, and otherwise dessicate the landscape to make it amenable to commercial grain or livestock production or other uses more in line with their interests.

These drainage and desiccation projects were outwardly very different from the Desagüe, but they all fostered colonization much as the Desagüe did. It matters, of course, that the Desagüe was a Spanish project and the estuarine reclamation in Nova Scotia during the seventeenth century a French one, just as it matters that both were "colonial" while the Great Level and the works of the Association pour le dessèchement des marais et lacs de France were "metropolitan." But it is also true that these differences among these projects are not determinant. To see the parallels—and to gain valuable insights from understanding the differences—requires a conceptualization of colonization that focuses on its class content and does not get mired in the vague notion of colonization as the movement of some "peoples" or "nations" over others and their territories.

In the Atlantic realm, colonization was an intervention that varying alliances of early modern European social classes and sectors carried out on environments and humans in both the Old World and the New simultaneously, with or without legitimation from the heads of polities. Under this conception, "Spain," "France," and "England" did not colonize "America" so much as the different classes that formed part of early capitalism colonized water, land, and biota (as well as the relationship among them and what each meant to humans) in both Europe and America, creating nation-states and empires at the same time and as a result of the same dynamics, not the latter as a consequence of the former. Thus understood, colonization in the Valley of Mexico and parts of what became known as Spanish America was characterized by the superimposition in the same space of two distinct forms of social organization and production—and hence valuation of water, land, and biota—and the simultaneous intervention of a patrimonial, dynastic state apparatus with its own alliances and priorities. The different outcomes of this process throughout Atlantic early modernity and modernity were created mainly by the relationship of forces among the social groups involved—and by Whiggish hindsight.

The chapters that follow trace the everyday comings, goings, and tinkerings of all whose technical expertise shaped the Desagüe, beginning with indigenes. The role played by urban and imperial elites and the crown will be examined in order to explain the constraints within which technicians, laborers, and villagers chose one technique, material, or process over another, and the overall social priorities and logic embedded in the drainage project in this particular instance of colonization. But it is the Desagüe's wardens, foremen, Indian laborers and village hydraulicians, friars, guild architects, military engineers, and savants who are the protagonists of this book.

Chapter 1 discusses what indigenous hydraulic expertise looked like in the district of the Desagüe both before and after Spanish settlement and at the beginning of the drainage project. Chapter 2 makes use of this historical baseline in explaining specifically how the Desagüe departed from indigenous hydraulics.

Chapters 3 through 7 then explore the logic of choices, the continuities and changes in the Desagüe over the seventeenth and eighteenth centuries in the areas of design, practices, purpose, and social arrangements, and the implications of the project for land, water, the natural environment, and people, as well as for technology and science.

Chapter 8 explores how the understanding of the Desagüe built up over the preceding chapters can shed light on what colonization is and who performs it. It does this by assessing the aforementioned drainage projects in Europe through the Mexican drainage. It interrogates these projects through the class definition of colonization presented above in pursuit of the links among all such environmentally transformative colonizations, early capitalism and state- and empire-building.

The span of time encompassed by these chapters covers three periods in the Desagüe. The first period is characterized by the creation and consolidation of specific designs and arrangements in the drainage during much of the seventeenth century. This period, chronicled in Chapters 2, 3, and 4, fizzles out near the end of the century, when concerns over the incomplete state of the project and its implications for the safety of the city in an unprecedented rainy season ushered in greater scrutiny of how the Desagüe looked, functioned, and was built.

After the initial period of consolidation, there were two waves of potential challenge to the established structure and way of doing things. The first wave, which marks the beginning of the Desagüe's second period, came in the form of the increasing influence and presence of royal military engineers starting in the final quarter of the seventeenth century. This wave, which continued well into the eighteenth century, had limited or delayed impact, except for a few notable areas of practice. Chapter 5 is devoted mainly to describing this period in the Desagüe's history.

The second wave, which defines the beginning of the third period of the Desagüe, began before the ebb of the first wave of challenge, since it started with the wide-ranging reformism of the Bourbons, which had its greatest impact from the 1760s onward. Cresting in the 1790s, this wave also had uneven impact. Its story is told in Chapters 6 and 7. In the case of both the first and second waves of challenge, the failures of potentially potent forces to fundamentally change the conceptualization and design of structures and practices put in place by the friars and master architects are telling, as are the few exceptions. Chapter 8 traces the implications of these failures for colonization.

The images in this book are not illustrations. Rather, they are integral to the analysis, and therefore readers are encouraged to examine them in their original color and in enlargeable digital form in the section dedicated to this book at *http://scholar.princeton.edu/candiani/*. Included there are also additional images that were not in the book for reasons of space.

CHAPTER ONE

# Living in a Fluid Landscape

For all the people who have lived in the basin of Mexico since the first wandering bands arrived perhaps eleven thousand years ago, the enclosed nature of the basin has been both a boon and curse. Until modern humans intervened, the encircling *sierras* prevented the precipitation that fell in their watersheds from draining out toward the ocean, so this moisture collected in the lowest parts of the basin and formed wetlands and shallow lakes fortified with sediments rich in organic matter. These wetlands, like those elsewhere, were extraordinarily productive ecosystems.[1] Humans could appropriate from them large quantities of biomass to use for food, fiber, fuel, construction materials, and medicines without adversely impacting the systems' functioning. The basin could therefore support relatively large numbers of people who, over the course of millennia, parlayed food surpluses into the requisite elements of a settled and, in some locations, very urban way of life.

The basin's Indians, however, did not live in harmony with a benevolent environment. The basin's lack of an egress created the potential for buildup of salts in the soil and water and set the stage for regular seasonal flooding and occasional catastrophic floods during years with unusually high rainfall. The high water table, combined with the impermeability of the clay layers on the flat valley floor and other physical factors, meant that rainfall did not filter easily into the ground but instead pooled and spread. At the same time, the poor quality of the groundwater precluded the extensive use of wells to supply potable water.[2]

Indigenous cultures dealt with these threats and opportunities through hydraulic engineering. People savvy about the hydrological regime, from commoners in small hamlets to state technicians in the powerful city-states along the shores, designed water management systems at every scale to achieve a combination of goals: maximize food production, prevent floods in urban centers, segregate brackish waters from fresh ones, import fresh water, and ensure canoe navigability. It is crucial to note that all these works were multifunctional and built with the understanding that water and land were in constant flux, not just in terms of location but also in terms of qualities.

As Ross Hassig and others have shown, the conquest did not end indigenous hydraulic engineering traditions.[3] But it did affect them. With the decapitation of the indigenous states, large-scale projects—such as large dams, river diversions, and aqueducts—that demanded the efforts and resources of multiple communities could no longer be built or adequately maintained. But the vast hydraulic infrastructure erected by the indigenous lords remained in place, doing its job by inertia amid a lack of concerted efforts to either destroy or rebuild it on the part of the conquerors. The result was a desultory continuity in large-scale hydraulics around the urbanized core of the basin. This allowed some knowledge of water behavior and construction techniques associated with them to survive until the various entities of Spanish rule began rehabilitating and building large-scale structures in the 1550s. It was at the communal level, meanwhile, that indigenous hydraulic traditions seem to have best survived, as outlying communities made more deliberate efforts to maintain their smaller dams, ditches, and diversions.

But even in these smaller communities adjustments were necessary. The demographic decline during the postconquest decades affected all indigenous technological capabilities. As Cook and Borah suggested, communities adjusted by concentrating their efforts on lands, species, and structures that assured the maximum output for their labor.[4] Terraces and other devices that had rendered marginal lands productive were neglected, while irrigation ditches and the equipment used for planting, storing, gathering, and hunting retained their vitality. Thus when the Desagüe arrived, it showcased both the dissolution of indigenous state engineering and the survival of the hydraulics of commoners at the village level. This becomes clear once we disaggregate the lumpy category of "indigenous technology" into its finer components.

## SUBJECT AND VILLAGE HYDRAULICS

Before the arrival of the Spanish, the interconnected lakes of the basin, which are estimated to have covered about 1,500 of the basin's 8,058 square kilometers during rainy seasons at the end of first millennium, provided the watery building blocks of indigenous society.[5] In the center was Lake Texcoco, lowest in elevation. North were Lake Zumpango at about six meters above Lake Texcoco and Lake Xaltocan at about half that; south were Lakes Chalco and Xochimilco, both hovering around three meters above the central lake (see Map 1 and Table 1). Runoff from the

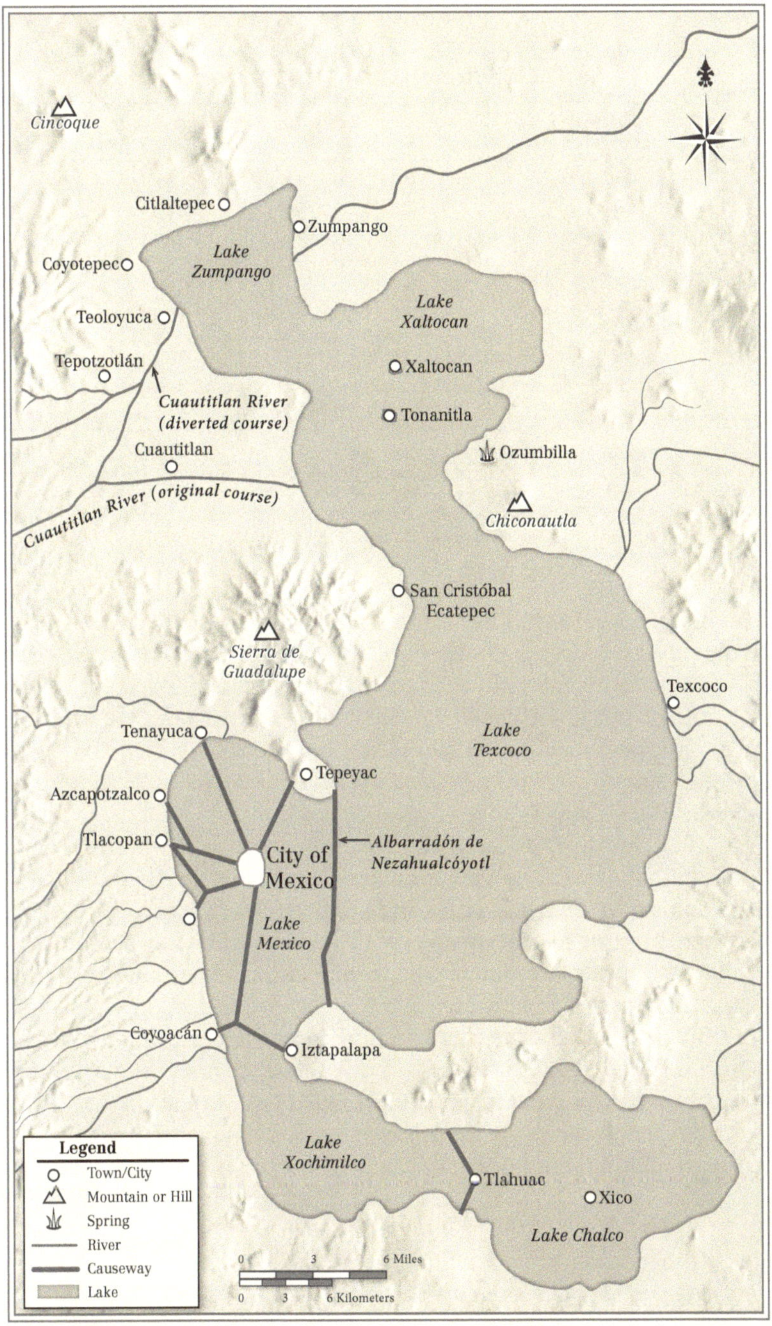


*Map 1.* The basin of Mexico and its major hydraulic structures on the eve of the Spanish conquest.

SOURCES: Gibson, *Aztecs*, map 1: "Valley of Mexico elevations." Angel Palerm, *Obras hidráulicas prehispánicas en el sistema lacustre del valle de México* (Mexico City: SEP-INAH, 1973), map 4.
CREDIT: Gerry Krieg.

*Table 1.* The lacustrine system of the basin of Mexico.

| Lake | Relative elevation | Approx. median surface, 1519 | Low-water surface II, 1861 | High-water surface, 1861 | Depth, 1861 |
|---|---|---|---|---|---|
| Texcoco | 0.000 m | | 18,328 ha | 27,217 ha | 0.50 m |
| Zumpango | 6.062 m | | 1,720 ha | 2,170 ha | 0.80 m |
| Xaltocan | 3.47 m | | 5,407 ha | 5,407 ha | 0.40 m |
| San Cristóbal | 3.597 m | | 1,103 ha | 1,103 ha | 0.60 m |
| Xochimilco | 3.139 m | | 4,705 ha | 6,336 ha | 2.40-3 m |
| Chalco | 3.082 m | | 10,448 ha | 11,417 ha | 2.40 m |
| Mexico City[c] | 1.907 m | | | | |
| Totals | | 150,000 ha | 41,711 ha | 53,650 ha | |

SOURCES: Luis Espinoza, *Memoria histórica, técnica y administrativa de las obras del Desagüe del Valle de México, 1449–1900*, 2 vols. (Mexico City: Tipografía de la Oficina Impresora de Estampillas, Palacio Nacional, 1902), vol. I, bk. I: *Descripción oro-hidrográfica y geológica del Valle de México*, 19–20; and also Manuel Orozco y Berra, *Memoria para la carta hidrográfica del valle de México* (Mexico City: Imprenta de A. Boix, 1864), 121.

surrounding mountains flowed into the lakes and remained there until it evaporated or was absorbed by the soil. By the time the Chichimec forerunners of Moctezuma arrived in the basin, the shores of these lakes were already dotted with urban centers and agricultural settlements and colonies. Several city-states vied for control of the region's land, water, and people, as can be seen in Figure 1.1, which depicts the multiple lordships that surround the lake system.[6]

For indigenous producers, water was inextricable from the land beneath it in that its seasonal recession left behind soils enriched for cultivation.[7] In this environment, water and land were not clearly distinguished from each other. Fields became food-filled lakes in the rainy season; water was not just an irrigator but also a medium critical for fertility-sustaining ecological processes; cities were solid but not dry, and so on. This applies to the entire basin. In the old alluvial plain of the Cuautitlan River and the northern lakes of Zumpango and Xaltocan, water was even more crucial because that area was more arid. Seasonal inundation was the key factor that reduced crop risk, as it sustained soil moisture, fertility, and texture, rendering agriculture less dependent on rainfall.[8] As elsewhere in the basin, indigenes in this region used water and land "ecosystemically"; that is, they hunted the animals and gathered the plants that the fluid landscape supported for the purpose of satisfying the basic necessities of life, and therefore had primarily what Marx called "use value" for land and water even after the conquest. This was not because they were somehow more reverent and less exploitative toward nature, but because they tapped into a broader spectrum of what water in combination with land encompassed. Their use of this environment ranged from simple appropriation of resources from relatively "natural" systems through hunting and gathering to intensive manipulation of the environment for food production through artificial horticultural beds raised above the surface of lakes (*chinampas*), river diversions, and canal building.

At the less manipulative end of their use of nature, indigenes supplemented their wetland agricultural systems with fishing, hunting, and gathering in order to diminish their high environmental risk (the possibility of frequent and huge variations in bio-physical conditions). These complementary activities were particularly sensitive to change in the northwest quadrant, where aridity made environmental risk higher than elsewhere in the basin.[9] Commoners used a wide variety of forms of organization and tools such as nets, hooks, fishing spears, canoe flotillas, and family groups on land to harvest fish, fowl, reptiles, insects, larva, and algae.[10] As elsewhere in Mesoamerica, in the northwest quadrant of the basin each plant and animal often had a number of uses, forming the basis of an economy that carried over into the colonial era and beyond.[11] On the eve of the conquest, the Cuautitlan region may have supported a population of up to ninety-one thousand individuals, thanks in part to the ability of these people to extract 20 to 35 percent of their foods from wetland ecosystems.[12]

*Figure 1.1.* The first Chichimecs arrive in an already densely settled and exploited pre-Hispanic lacustrine landscape.

SOURCE: Codex Xolotl, Bibliothèque nationale de France, Fonds mexicain, 1–10. Reprinted with permission.

At the most manipulative end of the continuum were agricultural systems dependent on large-scale structures. Chinampas belong to this category of works designed to maximize yield by intensifying nutrient availability within a limited space. Although there were at least five different ways of building chinampas, depending on the characteristics of the location, they all had the same basic requirements: shallow water with scant variation in level but sufficient circulation to ensure stable temperature and chemistry, plentiful vegetable organic matter and silts that did not exceed certain parameters of granularity and clay content, and the nearby presence of plants whose roots and stems could be used to help cohere the materials together and prevent the loss of soil due to erosion in case of flooding or severe rainfall. They varied in dimensions but were usually rectangular, built by filling in an armature of stakes (in freshwater lakes) or stones (in brackish water) with successive layers of silt and plant debris. They were all sensitive to lake levels, but freshwater chinampas particularly so since they were irrigated through capillarity. They were fertilized by organic materials ladled out of the surrounding waters.[13]

The Xaltocanmecas began building their chinampas in the thirteenth century. Lake Xaltocan was brackish, so the first step was to leach the salts out of the terrain where the chinampa would be erected. This process began in the dry season, when the indigenes built a walled enclosure on the lakebed high enough to ensure the chinampa would remain above the level of the water during the rainy season. This enclosure was equipped with two sluices at the bottom. Then they began the process of leaching by filling the enclosure with freshwater brought in from the springs of Ozumbilla, at the foot of the elevation of Chiconautla, approximately seven kilometers away. Once the terrain was determined to have been cleansed of its salts, the process of layering materials could begin. The Xaltocan chinampa area, centered almost two kilometers southeast of the island-town of Xaltocan, would grow to some two hundred hectares by the time of the arrival of the Spaniards.[14] The better-known Aztec-period chinampas elsewhere in the basin were apparently based on this early model in Lake Xaltocan, which Tenochtitlan expanded further as Xaltocan fell under its control in 1428.[15]

In addition to building chinampas, people in the northwest quadrant also reshaped their environment by diverting the most important river in the basin, the Cuautitlan River, and articulating a vast irrigation network to it. All of these activities required great mastery of the regional topography and hydrology. The Cuautitlan River diversion dam would become crucial to the Hispanic drainage project. Indeed, without this indigenous structure and the knowledge that went with it, there could be no Desagüe.

The Cuautitlan River ran year-round, although it was lower in the dry season. According to the *Anales de Cuautitlan*, the dam that diverted the river was originally built by Colhuas in the town of Cuautitlan long before the arrival of the Spaniards. These people had settled on the banks of the Cuautitlan River, a suitable location for much of the year. But when the river swelled in the rainy season, it tended to overflow its banks and flood the town, fanning out to the east and southeast over its alluvial plain before draining into Lake Texcoco.[16] The Colhua began addressing this problem by changing the course of the river, splitting its waters by creating a second channel for its flow. This channel was designed to handle the river during its rainy season floodstage and divert it northward. Ethnohistorians tell us that the Colhua began this rerouting in 1433 by damming it back and flanking the new channel with a structure strong enough to withstand the pressure of the river in its floodstage as it tried to regain its former drainage. This structure was a diversion dam made of rammed earth secured in place by vertical stakes driven into the ground,[17] and then it was consolidated with various live rhizomatous grasses. When this project finally ended, possibly in 1487 (8 *acatl*) the new channel drove the river through Citlaltepec into a place known as Aitictli, or "the meeting point," later known as Lake Zumpango.[18] It is probable that this structure began downstream from the locality of Atlamica but upstream from the Indian town of Teoloyuca. It may have been up to twenty-five kilometers long.[19]

As an earthen dam, this structure required specific familiarity with local soils and vegetation and knowledge of how the seasonal fluctuations affected humidity and plant growth. Cheap to build, all earthworks are vulnerable to destruction by the elements and the actions of humans and other animals. This one was no exception and required significant and constant investments of labor. The seasonal alternation of wet and dry conditions debilitated it, and the Cuautitlan River was liable to break through during peak rain events. Although we lack documentation about its problems and needs before the conquest, it is safe to assume that given the continuity in its basic form, the pre-Hispanic diversion dam was a fickle thing. Gophers (*tusas*) and snakes weakened its cohesion by riddling it with holes and tunnels.[20] It demanded constant maintenance during both the dry season, when it was inspected, and at the beginning of the wet season, when it was repaired with new grasses.

Like people in other settlements across the area moistened by the Cuautitlan River, Teoloyucans accumulated great knowledge about the interplay among land, water, and biome in their region. Much of this knowledge originated in the local populations' roles in the pre-Hispanic transformation of the northwest quadrant into an agriculturally productive region. Faced with a burgeoning urban population at the imperial center, the Mexica lords of Tenochtitlan sent colonists to the northwest quadrant over the course of the fifteenth century to create hydraulic infrastructure to render the brackish lakes and the alluvial plain of the Cuautitlan River fruitful agricultural areas. The indigenous northward diversion of the Cuautitlan River happened at this time, leaving the old bed to become the trunk for the development of a network of four canals, each feeding a myriad ditches on the alluvial plain that conveyed water to eight thousand planted hectares, making it "one of the largest irrigation systems in the Basin."[21]

This network seems to have depended on a second diversion structure that the Teoloyucan and other settlements participated in building and maintaining. This was a temporary structure, probably made of rock, branches, and earth placed across the new course of the river. It was used only during the dry season (October–May) to force the river's reduced flow into the network of irrigation canals, which fanned out from the east margin of the river. The impermanent nature of the dam, hence of its materials and construction techniques, was due precisely to the temporary need to acquire all the water in the river during dry spells. Although we lack a specific description of the dimensions and construction of this dam, it is clear that it was designed to be swept away by the bulging river when the rains arrived, allowing the diversion dam itself to bear the greater load of the water and drive the river northward.[22] Local populations thus mastered at least two different types of flow-control structures along the Cuautitlan River: one meant to last and guide the river in the rainy season, the other to be destroyed and guide the river in the dry season. This mastery included the "facts" of their region—the materials, the topography, water flows, characteristics of the soil and capacities of the lakes in the district—as well as the organizational technology that interwove practices with seasons according to both hydrology and communal needs over the course of the year and coordinated the efforts of several communities.

## THE VAST WORKS OF CITY-STATES

Of course the northwest quadrant populations did not live in a power vacuum; as a result, what they did to the local environment had both causes and effects in the rest of the basin, where water and land were hotly contested. After arriving in the basin in the fourteenth century, the Mexica ("Aztecs") rose from being mercenaries for their social betters to become feared partners of the kings of Tlacopan and Texcoco in the Triple Alliance of 1428, which established a complex, tributary empire. Each of the three great kingdoms remained autonomous within the Alliance and was sustained mainly by the labor and resource tribute from their own dependent kingdoms, provinces, and colonies that populated the lakeshores and the piedmont inside the basin, while subject regions beyond provided mainly tribute in kind. Buoyed by its conquest of such regions and its military power, by the end of the century Tenochtitlan had gained hegemony. Thereafter, its partners and other important kingdoms in the basin usually "agreed" to send artisans, laborers, and materials from their own tributary cities and provinces every time a Mexica king wanted a new temple or hydraulic work.[23]

Apace with their rise, the Mexica's urban core—the twin city of Tenochtitlan-Tlatelolco and its associated monuments such as the Templo Mayor—grew in layers like an onion, outward from the original marshy site. Using data obtained with test bores and electric cone penetration tests, geophysicist Marcos Mazari and his collaborators established that the Mexica were able to build their monumental city by creating an artificial platform twelve meters thick atop the elastic clays of Lake Texcoco. They built this platform by carrying fill-in rocks and earth to the northwest section of their isle. The platform provided a stable foundation for buildings by precompressing the clay under the water with its weight. The platform rose up to five meters above surrounding waters, but most of its thickness lay submerged beneath the groundwater level. What was exposed could be built upon without fear that it would immediately sink into the clay under its own weight, although some subsidence was to be expected. Mazari estimates that this platform supported a city that encompassed some 10,000 square meters in the early fourteenth century and more than 110,000 square meters by 1519 (following the construction boom that occurred after the conquest, the city would expand to about 380,000 square meters in area).[24]

Using vast amounts of manpower and materials supplied by subject and allied kingdoms, as well as considerable imported knowledge, Tenochtitlan engineered the environment of the basin at a scale far beyond that of the localized works of lesser polities. The Mexica developed vast hydraulic works and agriculturally colonized the lakes and shores through chinampas and other forms of land creation and usage. This growth was not spontaneous: the state was the promoter of urban design and the agricultural colonization of land and water.[25] A series of radial dam-causeways from Tenochtitlan-Tlatelolco fanned west to Tlacopan and Azcapotzalco, north to Tepeyac, and south to Iztapalapa, and two other causeways, from Mexicalzingo to Coyoacán and Iztapalapa and from Tlaltengo to Tulyehualco passing through Tlahuac, managed Lakes Chalco and Xochimilco (see Map 1). Sets of secondary causeways increasingly crisscrossed this space, while aqueducts brought in additional fresh water from more distant locations, each device articulated to a system that aimed to regulate the flow of water among the interconnected bodies of water.[26]

Farther away, the Cuautitlan River was critical to the Mexica rulers too. Every rainy season before the 1487 diversion, there was the risk that the river would flood Tlatelolco and then Tenochtitlan.[27] This is exactly what it did in 1449, prompting the Texcocan king Nezahualcoyotl to initiate the construction of a twenty-two–kilometer dike to keep at bay the floodwaters from the north. The dike served many functions. It was a tool of diplomacy, cementing Nezahualcoyotl's alliance with Tenochtitlan's Moctezuma Ilhuicamina by providing him with the necessary engineering knowledge, which Tenochtitlan also used to build the aqueduct of Chapultepec to bring freshwater into the city;[28] it connected Iztapalapa to Atzacoalco, favoring foot traffic and the circulation of goods; and most importantly, it eventually created the freshwater Lake Mexico around the Mexica capital, preventing the invasion of saline waters in the twin cities and keeping brackish water from Lake Texcoco from reaching the west and south, where *chinampaneca* communities benefited as well, even as they remained subject and tributary.

The unforeseen drawback of the dike of Nezahualcoyotl and the radial causeway-dams was that they trapped water coming in from the southern springs and from aqueducts such as Chapultepec's.[29] Barred from spreading over the large expanses of Lake Texcoco, this water flowed into Lake Mexico instead, keeping its level higher than that of Lake Texcoco. During especially rainy years, the lake filled more quickly with runoff, and its waters had nowhere to go but up—into the city of Tenochtitlan, and later the City of Mexico. When this happened, water seeped up the walls of monumental and modest construction alike, brimmed over the sides of the none-too-clean urban canals, and swamped the city streets. To solve this problem, at the end of the century the Mexica king Ahuizotl built a second barrier. Had the conquest not interrupted the historical path of the Mexica and their neighbors, they might well have eventually modified the dikes of Nezahualcoyotl and Ahuizotl and other large-scale hydraulic engineering structures so that they functioned more exclusively as urban flood defense devices that expressed the priorities of their lords, much as later works did for the Spaniards. As it was, the threat of flooding—and an awareness of the small difference between the levels of the lakes and the ground at the heart of the city—was etched deep in the consciousness of the city's inhabitants and rulers from very early on.

But this deep-seated awareness of flooding had different valences among indigenes in the basin, depending on what their relationship to it was. Although a threat to many dimensions of lordly urban existence, for the villages located all along the shores of the lakes that practiced various forms of wetland agriculture, the seasonal inundation of land was equivalent to life itself. The hydraulic and soil management practices they had developed over centuries depended on land becoming water and then land again, leaving both enriched with nutrients that sustained domesticated and wild plants and animals used for food and manufacture. At the same time, the fact that all lordly towns in the basin ultimately depended on the food produced in these villages and settlements acted as a limit on their capacity to develop too much of an enmity toward floods. This tension foreshadowed that of capital and watery hinterland under Spanish rule, but for Tenochca and other lords it was muted by the fact that for the moment their works only moved water around, as opposed to flushing it away.

Besides accommodating multiple purposes, pre-Hispanic major hydraulic engineering works were also designed to contend with both wet and dry season events. This biseasonal feature was a function of the multipurpose character of pre-Hispanic capital cities, within which consumption, agriculture, processing, administration, and exchange all took place. The famous so-called *plano de maguey*, the sole surviving preconquest map of Tenochtitlan, shows a city fully integrated with its lacustrine environment and that environment sustained a variety of economic activities, from chinampa horticulture to communications.[30] Water scarcity was as much an obstacle to life and the proper balance among all activities in the cities and their hinterlands as its overabundance. In sum, at the moment in their historical paths at which they were intercepted by European society, indigenous town and country both depended on seasonal flooding and understood land and water to be, well, *fluid*. All this would change, not with the Spanish conquest per se, but with the beginning of the Spanish colonization of town and country, of the relationship between them, and of the basin's ecosystems.

## SPANISH NOVELTIES

It should be clear by now that the growth of the Triple Alliance and its conquest of other indigenes resulted in significant hydrological and hence ecological disturbance. There were enough people living in the basin in the 1400s—conservative estimates put the number at around eight hundred thousand—that the population could not be accommodated without extraordinary food production efforts and syphoning of resources from far and wide. These ranged from the relatively benign intensive agricultural technologies of chinampas to the ecologically riskier cultivation of marginal land on the slopes, which even with terracing led to soil erosion. The requirements of Mexica hegemony also placed dangerous impositions on tributaries, which encouraged overhunting and overfishing in their territories. That populations were subjected to food stress is suggested by evidence of the increasing consumption of insects and microfauna.[31]

The problem seems not have been with the nature of Aztec and village knowledge of soils, species, and hydrology, which was sophisticated, but with the intensity of its application. Archaeological field evidence from the period has led scholars to be skeptical that "such agrosystems were sustainable in the long run."[32] Whether this "long run" had been reached in 1519, the lacustrine environment was already under pressure well before then. The species and agro-pastoral practices the Spanish settlers introduced in the basin and beyond certainly worsened these dynamics in many regards. But the postconquest landscape does not seem to have been quite the ecologically devastated, scorched caliche some scholars describe.[33] For instance, sinister for the reproduction of indigenes and their technologically reshaped environments, epidemics relieved some of the pre-Hispanic environmental pressures.[34] Still, the arrival of Europeans did mark an entirely new and unexpected epoch for Mesoamericans.

Although the most durable and dramatic changes would unfold in the first decades of Spanish settlement, the war between the Aztecs and the Spaniards and their indigenous allies was waged on and through human bodies, land, and water. As is well known, Cortés's flight from Tenochtitlan on June 30, 1520, was almost thwarted by a watery Mexica response—causeways that bridged to the mainland were broken; scores of archers on canoes showered the fleeing Spaniards with projectiles; the lakes themselves did the rest, drowning scores of men and horses. Conversely, Cortés's decision to cut the Chapultepec water supply to the city and launch a waterborne offensive facilitated the Spanish victory in 1521. Such prolonged warfare left deep scars in infrastructural and productive systems converging on Tenochtitlan.

The conquerors chose the defeated Tenochtitlan as the site for their new capital. Even without foresight of the terrible consequences of this choice, it was not an easy decision to make; Cortés went against the opinion of most in his band, who were not keen on living on the small island for any extended period of time.[35] The choice of site appears to have had many motivations. Tenochtitlan was not only the center of the vanquished empire, a politically convenient location for the new lords; it was also the location of the best-appointed accommodations and the largest demographic concentration of the area, which made sense for economic reasons. Additionally, the insular character of the new capital was not necessarily a liability for the Spanish conquerors. A water barrier probably helped in the defense of the conquered city. It has also been suggested that for Cortés it may have been "a tool for spatial discipline"—not against the vanquished but against his own mutiny-prone cohort.[36] After all, his own expedition was a product of sedition.

Although it had all these advantages, locating the Spanish city on the site of Tenochtitlan was also a temporary obstacle for the transplant of one important aspect of the European economy, as it had no fields suited to the extensive cultivation of grains and livestock breeding.[37] The difficulty of transferring the pastoral economy to a lake riddled with chinampas and other hydraulic structures was as much a factor in most conquistadores' decisions to build their palaces on more receptive ground along the perimeters of the lakes as were the strategic reasons cited by some scholars.[38] More importantly, this difficulty could not help but reinforce the specialization of the emerging Hispanic city along lines different from its predecessor on the same site, and it directly shaped the city's relationship with the hinterland, which was soon to acquire the additional responsibility of feeding urban denizens on the basis of extensively grown European grains and livestock.

In contrast to what we saw about the relationship of Tenochtitlan to the lakes and the hinterland, the Hispanic City of Mexico emerged as a more specialized administrative and consumption center, in many ways not unlike Madrid itself but with the added function of providing a beach-head for further settlement. All around this city, production during much of the sixteenth century was still carried out in large part by indigenes, who also organized a large part of intrabasin exchange. The type of hydraulic engineering that the Spanish city required was in line with this more specialized role of the urban hub, which needed to import water for consumption, to remove it after consumption, and to prevent an overabundance of it from causing floods, which could destroy a key place in imperial governance, its built wealth, and its sources of rents.

As a result, Hispanic hydraulic engineering tended to be focused on controlling wet-season events, rather than on controlling but ensuring the benefits of wet and dry season alternation. As the indigenous states and their ability to coerce their subjects into building or maintaining large multifunctional works disappeared, this contrast became even starker, as indigenous communities reverted to building and maintaining only what was strictly necessary for their own social reproduction, typically irrigation systems, lake enclosures for fishing and gathering, and chinampas. Thus, the hydraulics of the city became more flood-averse, while that of villages and settlements remained oriented toward maximizing the benefits of alternating flooding and drying.

All impoundments and diversions of rivers or streams that were initiated by city or viceregal authorities during the colonial period were mainly for urban interests, be they flood protection or water supply. The Cuautitlan River diversion is a good example. Begun as a local flood-protection and irrigation project in the first half of the fifteenth century, co-opted by Tenochtitlan and then by both irrigators and a Spanish miller locally, it would finally be subsumed into the colonial capital's Desagüe. Thereafter, the urban authorities of the Desagüe waged a permanent campaign against local productive usages for the engineered river. Unlike many other smaller and local manipulations of rivers, streams, and lakes around the basin where rural productive needs for water tended to ignore urban flood protection, here the priority of the authorities in the city prevailed.

Flooding in the basin was nothing new, of course. By the fateful year of 1555, however, four new factors had combined to aggravate existing conditions and make the alternation between dry and wet seasons more extreme. First, there was the incomplete familiarity of the first generation of Spanish settlers with the indigenous hydraulic works and technology that had both magnified the problem and sought to control it. As Angel Palerm showed, the indigenous system of causeways, dams, and aqueducts had been designed for three purposes: to increase acreage under intensive cultivation, to ensure fresh water for the chinampas (as well as for human consumption), and finally, to protect the city of Tenochtitlan. The rich Islamic legacy of knowledge about individual structures like dams, irrigation ditches, waterwheels, reservoirs, weirs, and other devices facilitated Spanish understanding of analogous indigenous structures, but it did not help the colonists apprehend the complex, interlocking hydraulic *system* of the basin, nor how it had been run under the equally complex political arrangements that had prevailed before their arrival. Thus it took some time for the colonists to understand the role of each structure in the broader context of the regional hydrology and its effect on flooding at the lowest elevations of the basin.

Second, the new modes of production introduced by the Hispanic population had altered the human relationship to land and water. The first generation of Spaniards had little reason to maintain either the property relations or the intensive agro-technological aspects of indigenous agriculture beyond their ability to nurture a tributary appropriation of surpluses or urban appetites. For those who shifted into direct productive activity, often investing in their enterprises wealth accumulated thanks to their encomiendas, land was for pasturing animals or growing nonirrigated, plow-based cereals for the emerging urban markets. What both of these uses required so far as the lacustrine area was concerned was dry land, not islandlike chinampas and the complicated networks of causeway-dams, aqueducts, and the other elements of the pre-Hispanic system. It appears, therefore, that at first Spainiards did not regard maintaining these structures to be a productive usage of labor, time, and resources—although urban flooding would soon change their minds.

Third, various Spanish practices inadvertently aggravated the potential for flooding. In the emerging productive system characteristic of the hacienda, which relied on livestock rearing and rainfed (*de temporal*) extensive grain cultivation, settlers saw pre-Hispanic irrigation canals and their water both as a source of irrigation and energy for mills.[39] Returned to rivers and lakes instead of being absorbed by the soil, this latter water exceeded the capacity of flood-control mechanisms that had assumed irrigation uses. Further, the introduction of the plow and draught animals, as well as the grazing of sheep and goats, combined with the not new but now-accelerated felling of trees for new construction timber and for fuel to dramatically increase soil erosion. As a result of both the initial neglect of indigenous hydraulic structures and of these novelties in production patterns, increasingly large amounts of sediment were washed into the lowest lakes each rainy season. This silting had begun well before the Spaniards, of course, but the earliest clear indication that colonists fully understood its link to soil loss and flooding is found in Enrico Martínez's writings from the beginning of the seventeenth century.[40]

All of this magnified a fourth factor: the swift and precipitous decline of indigenous populations left hydraulic structures uncared for and the skills related to their construction, use, and maintenance at risk of extinction. Terraces and other soil-control devices deteriorated as villages contracted or were consolidated into new ones in the *congregación* process of the late sixteenth century and first decade of the seventeenth. The result was that each rainy season the topsoil carefully accumulated over generations was swept into the lakes through neglected barriers, forcing the water up.

Tenochtitlan's lords had created a capital that grew on the basis not just of wars and alliances among powerful rivals, but also on the irrigated and terraced production of food in the wetland environment by its subjects. As a result, it was keenly dependent on the interplay between local and large-scale hydraulic engineering and could not afford to neglect either. Upon the end of Tenochtitlan's rule and the advent of a first stage of European settlement where the "plunder economy" prevailed, the capacity and desire to continue this coordination in small- and large-scale hydraulic management was suspended. Of course, none of the changes in hydraulics function and practice that occurred before 1555 were due to an active and intentional shift in Spanish water policy relative to that of Tenochtitlan's lords; they were mostly the unintended consequences of the Spaniards' very different approach to agriculture and to their evolving understanding of how this unfamiliar hydrological reality—an endorreic basin—functioned.

Large-scale hydraulic engineering and management were reborn in a somewhat different form as the City of Mexico itself became an imperial capital in ways greater than Tenochtitlan had ever been and the hydraulic infrastructure that served the capital was deemed worth maintaining. When the Spanish capital's flood protection works began and the Desagüe was built, a new stage in the reimposition of hydraulic coercion of the Indians began, characterized by the virtual absence of overlapping interests and uses that had prevailed in pre-Hispanic hydraulic engineering up to 1519. So while the changes after the conquest aggravated the preexisting flood risks at the bottom of the basin, it is not clear that the floods themselves were objectively worse under the Spaniards, only that floods were more of a problem *for* them, especially in the city, as one friar colonist admitted.[41]

Despite these changes, the Spanish conquest did not erase the underlying principles of indigenous ways of valuing, using, understanding, and living in the midst of land and water. However, these changes did mean that the conquest had imported new, more fixed views of land and water that were consistent with the European mode of production at the time, with its model of extensive grain production and livestock rearing. The conquest would also impose city-oriented forms and functions that worked in tandem with this model.

## IN THE GARDEN OF FORKING PATHS

The superimposition in the same territory of different understandings of and usages for land and water created a set of complex dynamics. As worldviews, the indigenous and Hispanic perspectives could coexist. As concrete actions on the material realm, they generally could not. Among indigenes, agriculture remained compatible with gathering and hunting in any given township's holdings because the actors remained the same, and they could sequence all actions upon or in water, land, and biomass to minimize disruption to the whole economy. For Hispanic settlers, however, water and land were entities distinct from each other and from any ecosystem. Europeans valued nature differently not because of some inherent cultural defect, as is often implied, but because of their systems of production: for them, water and land were primarily inputs in the production of "exchange values," or things they could exchange for other things.[42]

This basic difference had huge implications. It would, over time, lead to increasing divergence between the hydraulic engineering of Indians and that of Spaniards in terms of purposes, motivations, methods, materials, and scale. The more Europeans gained ground, so to speak, the less continuity that particular portion of the basin (be it composed of land, water, wetland, or lake) evinced in terms of engineering. The relative overlap in hydraulic engineering that existed in the decades following the conquest both at the "state" and "subject" levels thus declined. When and why did they begin to diverge, and what did these forking paths in the way water and land were manipulated look like?

In the urbanized core of the basin, hydraulic practices began to diverge in 1555. In 1554, Francisco Cervantes de Salazar, humanist author and twice rector of the University of Mexico, published a tract in Latin in which he sang the praises of the lakes surrounding the capital as entirely beneficent for urban life.[43] At the time, few Spaniards would have differed from this view, as there were plenty of reasons for the Spanish to appreciate the city's lacustrine setting. The lakes' canoe traffic allowed the city to be easily provisioned with foods; their extension, an easy solution to waste disposal; their complete surrounding of the city, a moat-like protection. And most of the time, the lakes presented little threat: the extant pre-Hispanic causeways and dikes contained normal seasonal runoff. It is true that the Nezahualcoyotl dike had suffered damage from war, neglect, and pilfering of materials for construction.[44] The one built by Ahuizotl had probably suffered the same fate, but scholars believe this was the work that the Spaniards reconstructed in 1556 as an arch from the Mexico-Tepeyac to the Mexico-San Antón causeways (also a pre-Hispanic legacy) and renamed it *albarradón* of San Lázaro.[45] But despite their disrepair, these structures allowed the city to function as well as could be expected in any city undergoing a change of masters and a noisy building spree.

Barely a year after Salazar published his paean to the lakes, however, exceptionally high rainfall altered the situation dramatically. Torrential rains brought lake levels up to flood the Indian barrios like Santiago Tlatelolco (formerly the twin city of Tenochtitlan) and lap around the now-Spanish core. Murky floodwaters damaged public and private urban property, impeded the adequate provisioning and functioning of the city, and threatened the lives of those trapped in it. Secular and ecclesiastic colonists fortunate enough to have obtained prized lots within the *traza* (the founding blocks of the city), where they were amassing a significant patrimony, began to fear for their property as they realized that a flood of only slightly greater proportions could very well ruin it. The flood therefore destroyed the consensus around the view that the lakes were generally beneficent.

The response of the colonists shows that while for the moment the continuity with past hydraulic practice was not shattered, a new path had opened up. On the one hand, Viceroy Luis de Velasco the Elder convoked both the main bodies of Hispanic political power and the Indian *principales* of the former powers of the Triple Alliance to discuss solutions, indicating that the Spanish had considerable respect, if not awe, for indigenous hydraulic expertise and the water management principles of the former rulers. On the other hand, some of these principles were no longer useful for the new lords of the land.

One of the manifestations of continuity was the result of Velasco's consultations: a decision to reconstruct the now derelict hydraulic devices the indigenous lords of the land had left behind. In addition, the Cuautitlan River was to be rediverted away from Lake Texcoco and into Lake Zumpango, where its water would be stored. This river had been initially diverted by the Aztecs, but when drought severely lowered the levels of Lake Texcoco during the rule of Viceroy Antonio de Mendoza (1535–50), he had ordered the river returned to its old bed so that it could replenish that lake.[46]

The city councilmen strenuously resisted paying for any of this even though it was all for their own protection, arguing that since "the Indians of the city and its environs have always built the public works of this city for which they pay no tribute to His Majesty or to other encomenderos," they should continue to do so at their own expense.[47] The cabildo refused not just rations but even the spades, sacks, and other implements that Indians would need for any new projects, alleging that in pre-Hispanic times Indians used no tools and that their chiefs, the *tlatoque*, provided none.[48] Velasco's insistence that it contribute funds and at the very least rations of meat, corn, and chili (echoing the position of the crown after the New Laws of 1542 regarding the treatment of Indians) was no more altruistic: forcing the full cost of the works on Indian subjects of the king ultimately meant the monarchy paid, one way or another. In the end, the cabildo lost, and by mid-December 1555, work was well under way on these projects, which were not intended to evacuate water out of the basin, but to lessen flood threats by altering where water was contained.

As refurbished by the Spaniards, however, these structures lost their multipurpose function—they were designed only to prevent the destruction of the city by water, much like the later Desagüe structures, and this had implications for the lake environment and its manipulation. Both these rebuilt causeway-dams and a new dam at San Cristóbal Ecatepec to stop water from the north (particularly the rediverted Cuautitlan River) from reaching Lake Texcoco by storing it in the northern lakes of Zumpango, Xaltocan, and San Cristóbal, were still fundamentally containment structures. They certainly diminished the possibility for coexistence between the Spanish city and the indigenous hinterland and "hinterlakes" but did not erase it altogether.

On the other hand, however, authorities for the first time entertained the possibility of elimination, rather than containment, of the water in the lakes for the sake of the city. This was a qualitative shift in strategy. A clear indicator of the shift was a proposal, brought forth by Francisco Gudiel in November 1555, to divert the Cuautitlan River permanently out of the basin altogether.[49] Gudiel, a *vecino* of the city since 1531 and evidently a student of the topography and hydrology of the basin, would later be a collaborator with architect Claudio Arciniega and Fray Francisco de Tembleque on the surveys conducted to assess the viability of a water supply line for the city from the springs of Churubusco. Gudiel suggested a permanent Cuautitlan River diversion linked to a canal of two leagues from San Cristóbal Ecatepec to the Tepexi River to take the water outside the basin.[50] This canal would only force the worst flood-creating offender—the Cuautitlan River, which may have contributed up to half of the total surface discharge in the basin[51]—out through Huehuetoca, thus directing "excess" water out of the basin without desiccating the lakes. Significantly, Gudiel thought that actually draining the lakes was a bad idea.[52] Although the cabildo sponsored a formal site survey and later found the proposal viable, nothing was implemented due to opposition from a variety of quarters, including the friars who tended to the Indian communities that would be affected by the considerable excavation entailed in the project.[53]

Although it remained on paper alone, historians have generally interpreted Gudiel's proposal as the first and only one to see generative value in regional waters, one that in this book would have to be considered as an effort to continue rather than break with the accommodation with the lacustrine setting that indigenes had constructed.[54] Not only did Gudiel feel that it was necessary to maintain the lakes and watery characteristics of the city, he proposed that the "drainage" integrate both navigation and irrigation. The Cuautitlan River would be the only one diverted, and the canal from the city-side edge of Lake San Cristóbal toward the Tepexi River would serve as a navigable canal to and from Huehuetoca for the transport of stone, lime, wood, maize, and other supplies for the construction site. Gudiel's proposing economically generative uses for his canal would seem to offer the technological manifestation of an economic drive beyond a merely extractive and rentier one on the part of the first generation of colonists and descendants of conquistadores and encomenderos. This would imply a substantially different conceptualization of the relationship of the city to water and land.

Unfortunately, there is no evidence that Gudiel intended the navigational function of the canal to last longer than the completion of the project; moreover, the water he wished to keep was not meant for irrigation at all but to permit the growth of pastures.[55] Besides creating pastureland for meaty animals, some vecinos felt Gudiel's plan had the potential to increase rents for the Real Hacienda by promoting the production, export, and taxation of hides and by charging Indians for irrigation water derived from a future Desagüe.[56] The evidence for attributing the drive to drain to economically generative intentions, be they under the Hispanic model or as a prolongation of the indigenous one, is thus negligible.

In any case, the fact that Gudiel's proposal to divert water out of the basin received serious consideration indicates a distinct change in thinking. Protecting the city from flooding had become so high a priority that officials were willing to consider a project that entailed enormous costs, not only to Spanish coffers but also to indigenous subsistence. An indicator of the new willingness to sacrifice the hinterlakes and the productive organization and activities they sustained for the sake of the city came in Viceroy Velasco the Elder's response to Indian complaints about plans for the new San Cristóbal dam. Consulted about the prospective dam, the guardian of the Franciscan monastery in Cuautitlan wrote to the viceroy to oppose it, using a *pintura* made by the Indians from Xaltocan to show the damages that they would incur when the dam raised the water level of Lake Xaltocan and destroyed their crops. The Xaltocanmecas complained that they should have been warned earlier. To no avail: the viceroy responded that "less harm will come about from the loss of a few cultivated plots than from a flood in [the City of] Mexico."[57] In fact, the "few plots" were in all likelihood the chinampas on the shores of Lake Xaltocan, which involved major investments of communal work over long periods of time and whose loss was thus not just of a season's planting but of accumulated labor and future sustenance. For Velasco, however, protecting the city had become the overriding concern.

The next flood, in 1580, was met with similar water-evacuation proposals and similar objections. Between 1576 and 1580, an outbreak of what was either typhus or measles killed up to two million indigenes, causing a dearth of labor and the loss of harvests in 1578, which in turn resulted in hunger and worsening mortality.[58] The royal cosmographer Francisco Domínguez de Ocampo, who had arrived in New Spain as part of Francisco Hernández's scientific expedition, buried the idea to build anything resembling Gudiel's canal by insisting that its labor demands on the Indians "would create the conditions for the few that remain in this kingdom to be extinct because the work would be great and the spirits and strength of the Indians lacking." To go ahead with the project "would be to kill them."[59]

If the events between 1555 and 1580 marked the beginning of a shift away from indigenous hydraulic principles, they also show that the colonists began to gain a deeper understanding of their own about the hydrological regime of the basin and how crucial controlling the hinterland's rivers, streams, and lakes could be to their city's fate. This process gained from and overlapped with events that were beginning to unfold in the Cuautitlan River subbasin, which fed important information to both local Spanish land grantees and authorities in the city. Since the most serious proposals for a drainage system of some sort were all looking at potential routes in the northwest quadrant, the increasing amount of information about that area coming to city authorities as a result of grants requested by Spaniards in the region was particularly useful. These petitions were often accompanied by maps drawn by local magistrates that pointed out the location and size of lands requested as well as their relationship to other productive and proprietary holdings, be they communal or private, including rivers, streams, and man-made water storage or conveyance structures, sometimes also indicating the nature of the vegetation and the soil—all important considerations for granting lands in the size and usage requested (see Figure 1.2).

In the northwest quadrant, the grant history went all the way back to the postconquest years, when Hernán Cortés had granted to Alonso de Avila a single rich encomienda containing the indigenous *cabeceras* (head towns of indigenous lordships) of Cuautitlan, Zumpango, Xaltocan, and Huehuetoca and some of their *sujetos* (subordinate townships). As part of the larger reforms to end encomendero power, this encomienda escheated to the crown in 1566, and in 1576 political jurisdiction was turned over to a corregidor in Cuautitlan (in charge of tribute collection, labor requisition, and supervision of the Indian governor).[60] Right after the escheatment, the land itself started to be granted to Spaniards in parcels apparently no larger than seven *caballerías*, with the aim that they put it in immediate production, irrigated with water from the Cuautitlan River.[61] Many of these land

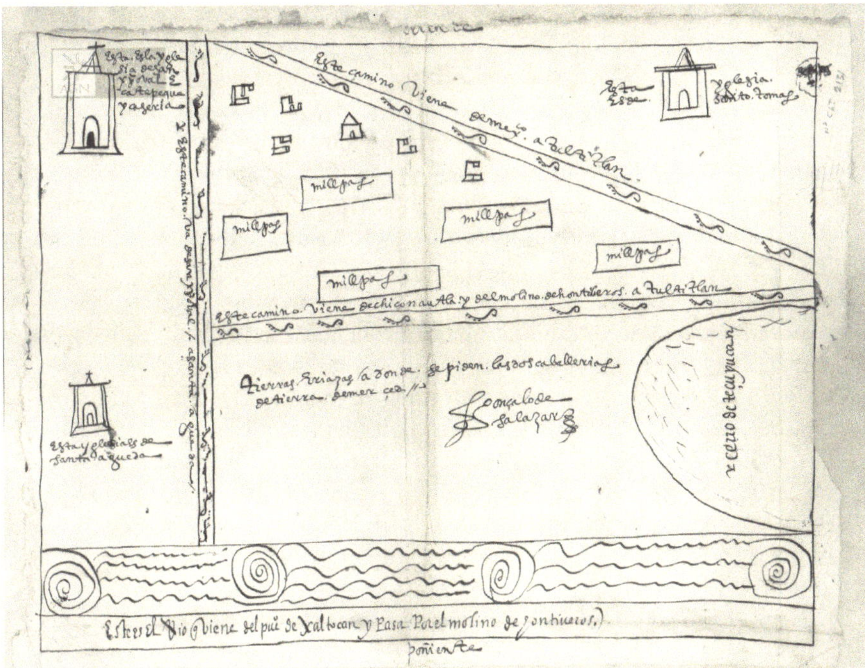

*Figure 1.2.* Magistrate of Cuautitlan maps information for a 1590 land-grant request: located west of the maize plots (*milpas*) belonging to the township of San Cristóbal Ecatepec, the plot is on "infertile lands" (*tierras eriazas*) but near water and three roads (in indigenous symbology).

SOURCE: AGN, Tierras, vol. 2777, exp. 7, f. 27. Reprinted by permission.

grants, and consequently the water to irrigate them, actually fell within the legal *términos* (boundaries) of indigenous townships and hamlets.[62] Naturally, all this generated tensions between rural indigenes and Hispanics that the incipient royal legislation protecting ancestral indigenous land and water rights was not often able to dissipate.

The changing landscape of land- and water-tenure and usage in the Cuautitlan's watershed had two consequences that are relevant here. First, as Hispanics grafted onto the irrigation system structured by both the diversion dam and the temporary dam, the incompatibilities of conceptualizations of land and water as fluid or fixed began to surface, undermining the hegemony of commoner indigenous hydraulic practices in the countryside. Second, the whole process generated struggles in and out of the courts, and this in turn created an ever-growing knowledge "map" that recorded what local indigenes and settlers knew about and how they manipulated the terrains and hydrologies they inhabited. Over time, the more conflictive the transformation of water and land proprietorship, usage, and definition grew as a result of grants, the greater the familiarity of the authorities in the City of Mexico with their hinterland.

The clearest extant documentation of one such conflict is from 1587, when the township of Cuautitlan and its subjects Teoloyuca and Coyotepec, as well as the townships of Zumpango and Citlaltepec, sought redress before the audiencia's prosecutor in charge of indigenous protection (*fiscal protector de indios*) Eugenio de Salazar, against Antonio Pérez, a Spanish grain mill operator who had usurped the water of the Cuautitlan River to power his millstones.[63] The viceroy himself accompanied the oidores of the audiencia on a site inspection of the temporary dam and canal system fanning out from the eastern margin of the river. The Indians and Pérez all testified that indigenous townships joined forces each dry season to rebuild the temporary dam and gave details about its construction and materials.[64] Based upon this testimony and documents presented by the multiple parties, a royal arbitration (*real ejecutoria*) set up a system for regulating the usage of Cuautitlan River water and specifying the rights and duties that went with that for all landholders in its basin, subsuming both the original indigenous system the settlers had grafted onto and the usages the latter had applied to it. The river's flow was apportioned among the caballerías of land it was to irrigate.

These new arrangements took material and organizational form in a new mortar-and-stone water tower (*pila*) installed in San Juan de Atlamica, on the east bank of the river. Surviving until modern times (see Figure 1.3), at the time this pila had four outlets, which fed canals that ramified to take the water to each township and grantee. Embedded in these structures is a demonstration that, unlike Mexica state hydraulics, village hydraulics would remain alive even within the overall separation from pre-Hispanic engineering traditions that began after 1555. In fact, both the temporary and the diversion dams and the old channel of the Cuautitlan River actually endured as the underlying structures of the irrigation district through the nineteenth century and right up to modern times. Because descriptions of the system documented between 1587 and 1762 closely resemble its early twentieth-century shape and functioning, it is feasible to reconstruct the system by "upstreaming" from modern sources, lending an unusual degree of reliability to modern photographs and plans of the pila and other structures shown in Figures 1.3 through 1.7.

Because the water of the river varied from month to month, rather than distributing it in set amounts, authorities divided the total water into $135^{1}/2$ parts, which corresponded to the total number of caballerías the river was meant to irrigate. They would then allocate the landholders as many parts as units of irrigable lands they held.[65] During the dry season, the temporary dam drove the river's water into two inlets. The smaller inlet, probably similar to the one shown in Figures 1.4 and 1.6, drew 15 parts of water from the $135\ ^{1}/2$ for the exclusive use of the Teoloyucans, and it may have been on the east side of the river.[66] The fact that these 15 parts fed this township's own irrigation network—itself governed by

*Figure 1.3.* Colonial-era Pila Real at San Juan Atlamica on east margin of the Cuautitlan River, as seen in 1926, with its five outlets (in Roman numerals) and intake (marked with arrow). Note mounds of silt cleaned out of it by beneficiaries.

SOURCE: AHA-AN-18-212-6/30. Reprinted by permission.

*Figure* 1.4. East margin of the Cuatitlan River with water flowing into the two sluices of El Chiflón and into Teoloyuca's inlet (*canoa*), 1926.

SOURCE: AHA-AN-18-212-5/30. Reprinted by permission.

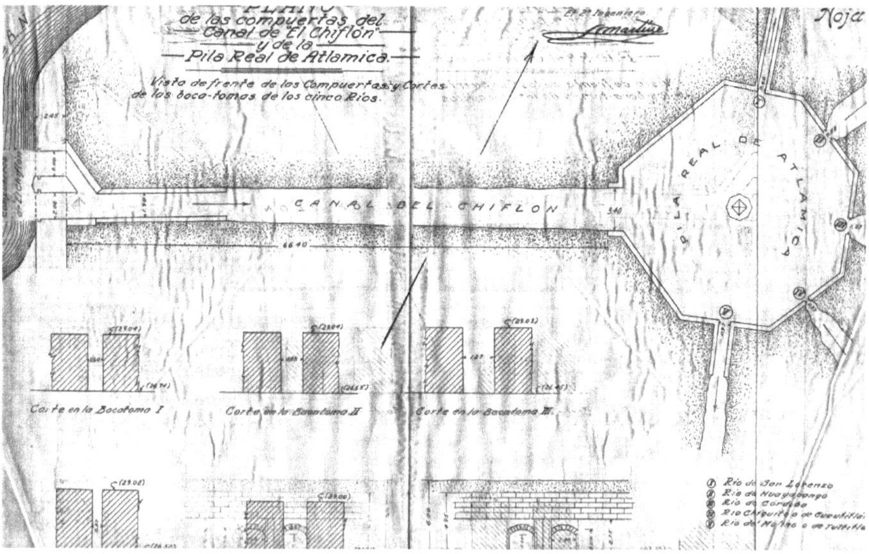


*Figure* 1.5. Plan of the gates of the Canal of "El Chiflón" and of the Pila Real of Atlamica, 1927. The canal draws from the Cuautitlan River to the left (west). The pila feeds five canals or "rivers" that ramify further.

SOURCE: AHA-AS-caja 559, exp. 8249, leg. 1, f. 201. Reprinted by permission.

*Figure 1.6.* Teoloyuca's portion of water flowing from the outlet of the *canoa*, with the township mobilized for repairs. Note the solid stakes anchoring the borders.

SOURCE: AHA-AN-18-212-7/30. Reprinted by permission.

*Figure 1.7.* Colonial-era pila in Teoloyuca, with arrows indicating inflow and outflows of water, and doubling as trough and pool.

SOURCE: AHA-AN-18-212-13/30. Reprinted by permission.

pilas as shown in Figures 1.7 and 1.9—and that its barrios, or dependencies, straddled the river, would make Teoloyuca the single most irritable neighbor of the Desagüe along the Cuautitlan River.[67]

The inlet with the greater capacity led the remaining 120 1/2 parts of water to a feeder canal (*cañón*) on the east bank of the river that in the twentieth century came to be known as El Chiflón and measured approximately 66 meters long and 3.5 meters wide. The water entered this feeder canal through a sluice and flowed to the Pila Real much in the same way as shown in the plan in Figure 1.5.[68] From there, the outlets—five, eventually—with different widths ramified into canals that distributed the water to the various beneficiaries, as seen in Figure 1.8. The old river channel apparently became the fifth canal, or *zanja del Molino*, still known as such in the twentieth century.[69] Altogether the outlets connecting to canals that made up the installations at Atlamica served a multitude of users, which by the 1580s included some ten haciendas, seven ranchos, ten pueblos de indios, three Indian barrios, and the flour mill, some of which appear in Figure 1.9, which only shows two of the *zanjas madre*, or main canals drawing from the pila outlets.[70]

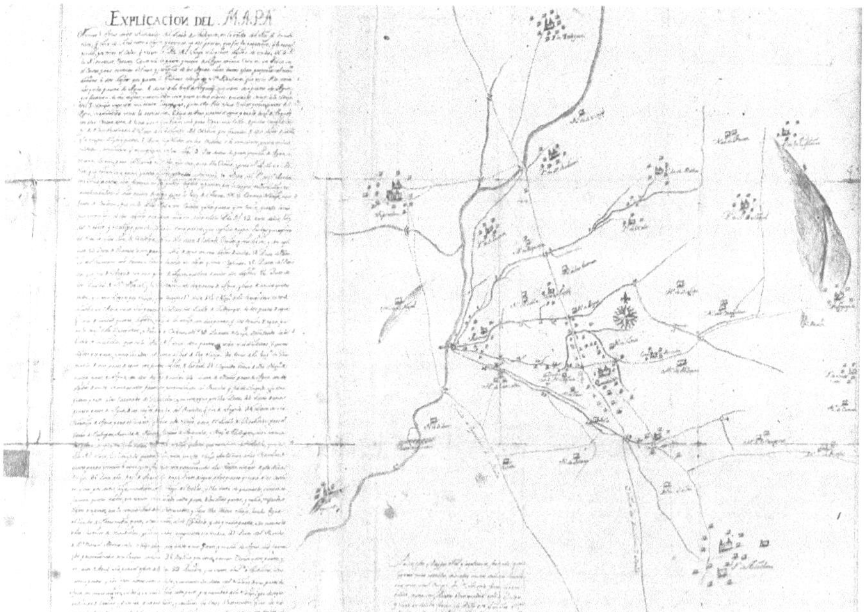


*Figure 1.8.* The Cuautitlan River's water distribution as seen by the City of Mexico, Repartimiento de Aguas del Río Cuauitlan, 1763, with detailed listing of water apportionment among Pila Real beneficiaries.

SOURCE: AGN, Tierras, vol. 2028, exp. 5, f. 118. Reprinted by permission.

All beneficiaries of the water tower were responsible for its upkeep, which meant desilting the core structure as well as the outlets and canals. They also had to clean and widen the riverbed every year, and after 1608—when the drainage began and the Cuautitlan River was diverted permanently toward the Desagüe—it became the duty for every Pila Real rightholder to maintain the sections of the diversion dam that his landholdings reached and in proportion to the number of waterparts he enjoyed.[71] Teoloyuca and other townships in the province of Cuautitlan were supposed to clean the pila at the beginning of the dry season.[72]

The main change that would affect these arrangements and the character of the clashes over the water of the Cuautitlan River was the start of the Desagüe in 1608. When the public work entered the fray, it added yet other usages for the water and land that conflicted with all prior actors together, as irrigators, while also exacerbating the incompatibility among them, as mainly producers of commodities or as mainly producers and gatherers of the essentials of subsistence. The struggles would not be just about the way all of these actors, Hispanic as well as indigenous, used water and land as inputs in production, but also about how the indigenes

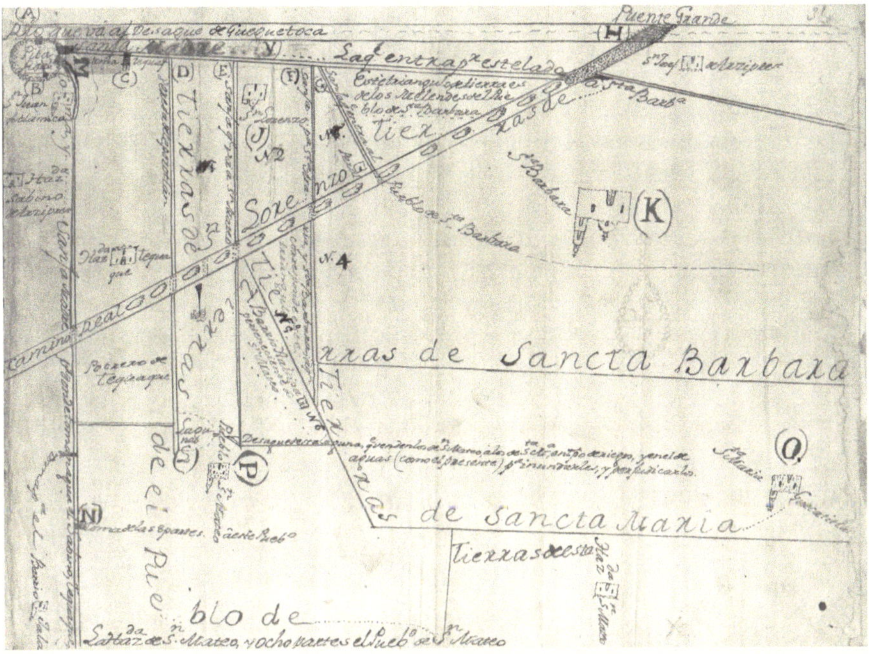


*Figure 1.9.* View of the Pila Real and the distribution of the Cuautitlan's water as seen by local users of two of pila's canals in 1756.

SOURCE: AGN, Tierras, vol. 1503, exp. 5, f. 32. Reprinted by permission.

specifically would be able to use the same exact water to keep wetlands alive as ecosystems.

In any case, these conflicts and fact-finding visits made officials aware of indigenous hydraulic skills, traditions of collective mobilization of indigenous labor, and the various official dispositions.[73] Collected locally in documents and filtered up to the City of Mexico, these three strands of information allowed Desagüe technicians to appropriate indigenous hydraulic and organizational capabilities into the drainage project. The Cuautitlan River diversion dam would become pivotal in the public work as an entire package, with all the material and organizational technological practices that went with it.

The chinampas also became important, but for completely opposite reasons: they could not be appropriated into the Hispanic mode of extensive grain production and livestock rearing, and later they could not be appropriated into the Desagüe. Indeed, they were an obstacle to both. Already affected by demographic decline, but before the Desagüe was built, the Xaltocanmecas resettled from the isle to the shore of Lake Xaltocan, presumably to better sustain their chinampas.[74] During that time, in 1535, the Cuautitlan River had been allowed to return to its old channel toward Lake Texcoco, since Spanish residents in the city had no evidence of its potential dangers. Authorities decided to divert it toward the northern lakes in 1556, which prompted a strong reaction from the Xaltocanmecas. If the Indians were sensitive to the *level* of the lake it was because the chinampas were still alive: an excess of water would rot them and expose them to dangerous resalinization, as much as a deficit would dry them.

In 1599, eleven pueblos and barrios that were subject to Xaltocan as cabeceras all used chinampas within their brackish lake. They built and protected these structures collaboratively.[75] This suggests another important element of continuity in practices and skills: they were inextricable from the intercommunal form of social work. Since similar struggles would later erupt in the 1740s over the Ozumbilla springs and the level of water in their lake, it is probable that the Xaltocanmecas either continued with their chinampas all along or tried to restart them in the eighteenth century.[76] Whichever is the case, these events point to the survival of indigenous hydraulic skills and knowledge of the terrain and its contours—and possibly even their further refinement—within the same region in which the Desagüe would be built.

Even as they incorporated Hispanic tools and practices, indigenes of the northwest quadrant continued to extract protein-rich foods and raw materials from lakes and wetlands in ways the Hispanic population did not.[77] But it is also true that the settlers incorporated some of the life-forms used by Indians into their own techniques. For example, just like chinampa horticulturists, Hispanic farmers applied *lama* (water grasses and decomposed plant matter taken from the water)[78] to their fields to replenish their fertility.[79] Like Indian agriculturists in the region, they also spread nutrients in the form of silt and aquatic vegetation over their fields with their own version of seasonal inundation called *entarquinado*, whereby *ladrones* or small breaches in the diversion dam allowed the river to temporarily flood fields.[80] But indigenous and Hispanic practices were actually opposite in their underlying principles even if they were formally similar.

The continuity of hydraulic practices in the Valley of Mexico after 1519 was possible because the urban Spaniards simply rode on the coattails of extant state-built indigenous structures so far as water supply and control in the city were concerned and because much of the city's nonurban hinterland in the basin was still in control of Indians at the ground level, particularly the lakes themselves. There was no need for the Spaniards to deviate from previous practice *so long as* they had no alternative plan for the water-land in question.

Three developments would change this picture: indigenous demographic decline; the exhaustion of the plunder economy that turned encomenderos into entrepreneurs who together with other settlers increased the demands on indigenous land and water as inputs; and the increase in the frequency and severity of flooding in the capital. In hydraulics, the period between the floods of 1555 and 1580 was the first stage of a transition between the pre-Hispanic past and full-fledged Hispanic colonization. During this transition period, both continuity and divergence were present. Between 1580 and 1604 the lack of serious floods in the capital allowed this transition to glide on. But that day in September 1608 when Velasco sank the blade of his hoe on the ground in Nochistongo was the definitive fork in the road. The Desagüe would become the maximum expression of the incompatibility between the material manifestations of a fluid conceptualization of what and where land and water were and the material manifestations of a worldview that saw water and land as fixed. The Desagüe would express the fact that the basin was an environment contested not just by two cultures, but by different classes and different modes of production, each with its own values for matter, energy, and life.

This contest, however, was not unmediated. Of paramount importance because of the way they interwove with hydraulic technology—in the Desagüe and before and beyond it—were the institutions by which the crown inserted itself in the relationship between colonists and Indians. The most important of these were those of the república de indios. Pueblos in the república de indios were both juridical and territorial entities. Indigenous townships or municipalities persisted, with the autonomous right to be governed by their own authorities, thanks to the fact that they retained land, water, and woodlands—and, in the case of the valley, marshlands. From soon after the conquest, these resources were communally owned and theoretically inalienable and indivisible, which meant that individuals or families had usufruct but not ownership rights.[81] By 1567, Spanish haciendas had been prohibited within 500 varas (415 meters) of Indian township boundaries, and livestock ranchos within 1,100 varas.[82] Known as the fundo legal, this endowment would be ratified, refined, and amplified by cédulas in 1687, 1695, and 1713, then encoded into the 1786 *Ordenanzas de Intendentes* that formed the backbone of the Bourbon administrative reforms.[83]

None of this made the corporate indigenous townships closed-off and static in any sense. Their dynamism and connection was not only due to the indigenes themselves being engaged with the commodity, land, water, and labor markets of the evolving Hispanic economy even as they defended their communal integrity, but also because of Hispanic landholders' appetites for land and labor. Land grants to Spaniards came in a variety of forms and sizes, were intended for agricultural, pastoral, or mixed activities, and were worked by the beginning of the Desagüe period largely with labor hired from indigenous communities (*gañanes*). Following Iberian tradition, grantees did not automatically enjoy irrigation or energy water rights, which required separate concession, although water was just as subject to usurpation as was land unless it was defended.[84]

The Desagüe introduced the city and the state behind it as powerful players in the basin's countryside. In protecting the indigenous community as an inviolable entity, the crown also (most likely unwittingly) protected the continuity of technological practices with which Indians harnessed water and the benefits of the ecosystems the water sustained as it turned land into wetlands and back again every year. That is, crown policies permitted the survival of two mutually exclusive notions of what land and water were and where they were supposed to be—the indigenous one in which water and land were mutable and interchangeable, and the Hispanic one in which they were fixed. When they fought over water and land in the lacustrine basin, these two groups were defending not only units of area and volume but also the fundamental definitions from which they followed.

The survival of indigenous communities and practices based on the conception of land-water mutability probably mitigated the effects of the four flood-inducing factors enumerated above, not because indigenes were less manipulative of land and water, but because their manipulations were kept in check by both their demographic decline and the impediments that the juridically communal regime of the townships placed on the ability of individuals to maximize resource depletion (including deforestation and inappropriate plowing) for private profit. For the Desagüe, the survival of indigenous hydraulic traditions would multiply the technological choices available to any European technician working in the drainage, whose region comprised many indigenous townships. In other words, it would make the Desagüe not quite the product of European or creole ingenuity it has been portrayed to be.

CHAPTER TWO

# Dreaming of Dry Land

It took more than four decades for the echoes of Francisco de Salazar's Latin paean to the lakes to fade to silence, but in the three short years between 1604 and 1607, the long period during which the hydraulic principles of the indigenous states were still operative finally came to an end among Spanish authorities. The digging of the Desagüe followed quickly on this sea change. The drainage system would become the culmination and maximum expression of this shift and of the supremacy of the interests of the imperial capital. More broadly, the Desagüe would be the material manifestation of the consolidation of a Hispanic state, capable of picking up where the Mexica-dominated state had left off in terms of vast infrastructural interventions upon nature but with an entirely different strategy. Each extant or rebuilt pre-Hispanic dam, causeway, or other structure would overall continue to operate as originally designed, but the fact that it was now to be hitched to a massive drainage system added a functional layer that did not exist before.

This shift happened without necessarily wiping out the continuity of commoner village usages associated with indigenous hydraulic structures. Because seasonal flooding continued to be eagerly desired by many people in the basin—as well as by the myriad other organisms that lived there—and because the Desagüe would impact the extent and locality of this flooding, commoner hydraulics at the village level would be affected by this new epoch in state-sponsored engineering. Yet indigenes would not stage a region-wide challenge to the Desagüe as a whole. This is not simply because the conquest paradoxically opened the door to the restoration or creation of local ethnic and political alignments formerly hindered by Mexica rule, but also because the logical priority of commoners was to protect their own modes of life. This explains the fact that the colonists would be able to appropriate preexisting structures such as the Cuautitlan River diversion dam into the Desagüe, but not to disassociate such structures from the underlying continuity in their usage by village indigenes.

While remaining as indifferent to what basin-wide significance the Spaniards might wish to give Desagüe structures within their boundaries as they probably were to that which the Mexica overlords had given them before, villagers would not easily accept manifest threats to the hydraulic principles *they* had embedded in these structures.

If, for villagers, hydraulic structures were supposed to support overwhelmingly communal property and production relations and the fluid interactions among water, land, and biota associated with daily sustenance, for Hispanics these structures had to be useful either to secure their capital or to ensure the daily production of things to sell. As other writers have shown, there was a philosophical dimension to these differences in valuation that underpinned the Desagüe.[1] But at the heart of this disjunction in how hydraulic structures were to be used were irreconcilable social models. In both the Old World and the New, propertied Europeans in the seventeenth century lived in the material world very differently than did communally oriented indigenes: in New Spain, Hispanics valued life, matter, and energy as things that could be measured and owned privately, separated into their components and recombined by work (understood as energy applied on matter), which was itself also subject to property relations.

Although the propertied classes in Hispanic society shared this general way of valuing the elements of existence, they had different views on how best to use living things, matter, and energy once they were measured, divided, and privately owned: as sources of rent income or as inputs in a production system governed increasingly by mercantile capital. Hispanics in the New World in general used water, land, life, and energy in both of these ways; overwhelmingly concerned with protecting built wealth, elites in the imperial City of Mexico treated them predominantly as sources of taxes and rents. Conceptualizing the Desagüe generated few quibbles among well-off Hispanics because it rested on a shared valuation and usage of water, land, biota, and energy; among elites in the capital the consensus was even stronger because they additionally agreed on the importance of preserving the built wealth and sources of rents that these four elements had combined to create. Design and money often caused discord, however, because there were different material and practical ways of expressing this consensus. In other words, among Hispanics, the strategy of desiccating the basin to protect their city was shared; the tactics of how to accomplish this were often up for grabs.

## DECIDING FOR DRAINAGE

The strategy for protecting the city from floods began to shift from containment to evacuation in 1604 when another damaging flood hit the city. The drainage idea came back with force. However, there were once again objections about the many costs of such a solution to urban flooding. The merchant guild and other important corporations and individuals agreed with the crown's audiencia member responsible for the welfare of Indians that there simply were not enough Indians left to work on such a huge undertaking.[2] Again, this proved a powerful argument against a proposal that would have required fifteen thousand Indians and almost half a million pesos, according to the plans of maestros Antonio Pérez de Toledo and Antonio Pérez Rebelto. As a result, Viceroy Montesclaros stopped further discussion of drainages and ordered instead containment works along the lines of Viceroy Luis de Velasco's.[3] A dam at Oculma was built, and the causeways of Guadalupe (Tepeyac-Mexico), San Antonio Abad (Iztapalapa-Mexico), and San Cristóbal were recuperated from the indigenous legacy or built anew, while urban canals were also cleaned to facilitate the evacuation of water from the city.

Barely three years later, a new flood swept away all hesitation about drainage. With the water invading streets and houses well into the traza with what seemed like greater tenacity, in 1607 city dwellers came to a new consensus, setting aside any concerns over the effects of a major hydraulic project on Indians or anything else. Besides all their material losses, Hispanic city dwellers continued, as in previous floods, to find themselves uncomfortably at the mercy of indigenous control of navigation: the flood inhibited canoes from entering the city acequias with provisions, and people could only leave their houses by canoe. With apprehension mounting over the apparent increasing frequency and severity of flooding, Viceroy Luis de Velasco the Younger, son of the Velasco who had ordered the 1555 containment works, easily rallied together the religious and civil authorities who had previously balked at the implications of drainage. The viceroy's position was that the causeways, dams, and embankments had been "useful for a time" in managing normal seasonal runoff "but not strong or effective enough to assure this city's safety and perpetuity by completely eliminating the risk of flooding and throwing the enemy out."[4]

Despite the bellicose metaphor, this resolve implied neither the erasure of containment measures nor a blusteringly facile endorsement of drainage. Like most officials on the city council, *contador* Diego de Ochandiano reluctantly supported the drainage "as one who lets his arm or leg be amputated in order to save his life." The maintenance of dams, embankments, and causeways would remain a high priority, even as they were now subsumed to a different overall way of dealing with flooding. This certainly provided continuity to indigenous hydraulic engineering. Most nineteenth-century chroniclers of the Desagüe would disdain this reliance on indigenous containment structures as a "regression." But in 1607 it was by no means perceived as such. Quite the contrary, colonists thought of the dam and causeway system as part of "how things are done in Genoa, Venice, and other cities in Italy and the states of Flanders for the conservation, provisioning and order of the republic," not as a separate "colonial" activity distinct from those of Europe.[5]

With this reluctant consensus, Velasco assembled weekly meetings where a commission of oidores and the royal fiscal named by him vetted proposals solicited from the public, which Velasco then passed on to the city government and a committee of experts that included master architects, friars, the master armorer Alonso Arias, and the German-born printer and royal cosmographer Enrico Martínez. In this manner, all major sectors of the elites were represented and none opposed the basic idea of drainage. They investigated eight possible projects, none of them from Indians. On October 23, 1607, the viceroy approved Enrico Martínez's proposal to drain Lake Mexico through a tunnel that went through Huehuetoca and Nochistongo, fed by a canal and supplementary structures tapping Lake San Cristóbal, the western portion of Lake Zumpango, and the Cuautitlan River.[6]

The details of how to pay for this drainage would provoke vitriolic arguments, but its exclusive aim to safeguard the city encountered no opposition. The Desagüe, designed by Enrico Martínez and reformed over time by many others, was thus in the broadest possible sense a unified Hispanic project, not an indigenous one, although it could not function without appropriating indigenous material structures, techniques, and forms of social organization of labor. This newfound urban consensus is understandable, since by then New Spain's capital boasted twenty-four convents for friars and nuns, half a dozen hospitals, a university, several schools, a cathedral, several important churches, as well as palaces for the Inquisition, the archbishop, the city council, and the viceroy, in addition to merchant warehouses and many more private and public buildings, all worth more than twenty million pesos not counting rents.[7] Perhaps this is what Viceroy Velasco the Younger was thinking about when he sank the ceremonial hoe into the ground in Huehuetoca to begin the Desagüe that famous morning of November 28, 1607.

## THE DESAGÜE IS BORN

Construction—or rather, excavation—for the Desagüe de Huehuetoca was completed eleven months later with a cumulative workforce of 60,000 Indian men and 1,664 Indian women to cook the rations of meat, corn, chilis, and salt for them. The speed with which the work was finished can mask the technical complications that it entailed, not to mention the disruption to human life and the ecosystem in the region.

Martínez only dared build a tunnel section after rejecting an open trench design for the entire work "because experience showed that tepetate is preserved and fortified by humidity, and that the sun and air break it up," which meant that exposing such terrain to the air in an open trench could be fatal to the project. Besides, he reasoned, "for each two hundred varas of open trench done with talus on fifty four varas of depth, more than three hundred thousand square varas of earth would have to be extracted," whereas with "the same two hundred varas as a tunnel of oval shape, four varas high and four wide, no more than two thousand four hundred square varas would be extracted, which is a noteworthy difference."[8] In addition, the amount of earth that would have to be extracted from an open trench with adequate slope posed a problem because unless the excavated debris was placed far away it could be swept right back into the trench by winds or runoff. As a result, Martínez concluded—presciently, as it turns out—that it would be best to build the drainage conduit as a tunnel, "with the caveat that if in the course of the works experience demonstrated the convenience of something different, the tunnel works should not impede that it should be later pursued as an open trench if necessary."

In the end, not counting the 17- to 25-kilometer-long diversion dam of the Cuautitlan River on which it depended, Martínez built a Desagüe that was 13,238 meters long (see Map 2). As he depicted it in Figure 2.1, water was conveyed from Lake Zumpango into an open trench covering 6,225 meters of ground to the mouth of the tunnel near Huehuetoca (*abd*), where it entered the 6,366 meters of *socavón* or tunnel (*cde*) to exit on the other side of the other side of the hills into another trench or canal 647 meters in length (*ef*) that took it to join the Tula River. Not marked on this image are forty-two square shafts of the underground section, that provided light, ventilation, and access to the tunnel bed to "extract the earth with many curious and fine devices and machines" that also permitted verifying whether water flowed or stalled. A structure of problematically variable dimensions, the tunnel ranged from 1.7 to 2.5 meters wide and 3 meters high.[9] It lay more than 50 meters below the ridge surface at its deepest point. (In the 1680s it would be deepened to 60 meters.[10])

The tunnel's structural variability was the result of practical adaptations that Martínez made in response to the multiple obstacles he encountered in the excavation. One of these obstacles—"the encumbrance of the great amount of water found beneath the earth"—also barred him from giving the tunnel the depth it would have required to drain Lake Mexico. Although contemporaries found a hard, clearly sedimentary thin flat rock (*laja*) and a coarse-grained granite (*piedra berroqueña*) along the line of tunnel, the overall prevalence of relatively softer materials underground generally made excavation easy but greatly complicated internal

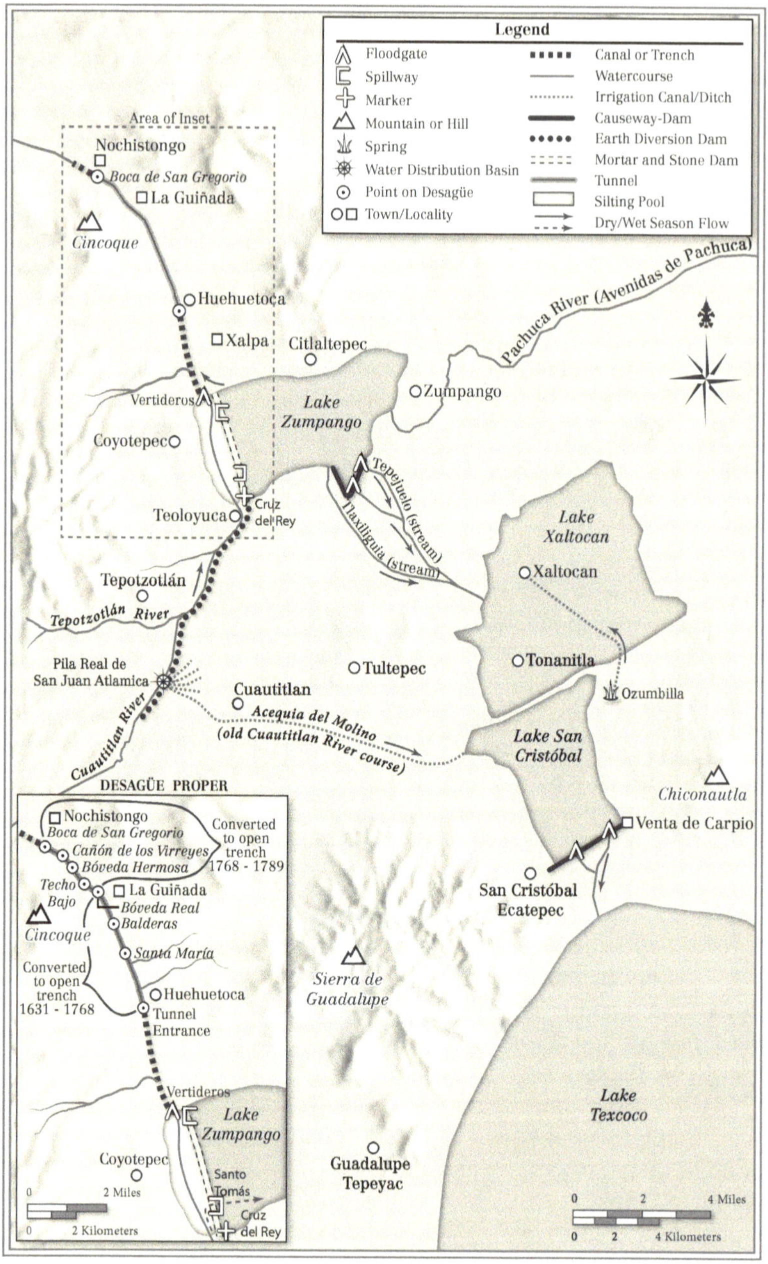


*Map* 2. Main features of the Desagüe proper and its district, 1608–1767.

SOURCES: Mexico, Junta Directiva del Desagüe del Valle de México, *Memoria histórica, técnica y administrativa de las obras del Desagüe del valle de México, 1449–1900*. 2 vols. (Mexico City: Tip. de la Oficina Impresora de estampillas, 1902), 2: "Carta hidrográfica del Valle de México"; AGN, Desagüe, vol. 7, exp. 2; AGN, Desagüe, vol. 12, exp. 5; AGN, Desagüe, vol. 29, exp. 10, ff. 29–37, 91–112; AGN, Desagüe, vol. 30, exp. 4; AGN, Desagüe, vol. 32, exp. 2, f. 329; AGN, Fomento Desagüe, vol. 1bis, ff. 213–14; AGN, Tierras, vol. 287, exp. 6, f. 30; AGN, Tierras, vol. 2028, exp. 5, f. 118.

CREDIT: Gerry Krieg.

reinforcement. The layers of schistose bedrock (*greda*), hardened sedimentary volcanic tuff (*buen tepetate* or *tepetate duro*, compressed pumice in coarse grains), clayey sand that crumbled when exposed to air (*mal tepetate* or *tepetate blando*), soapstone or steatite (*jaboncillo*), and other marls or clayey soils interspersed with lime were all friable materials.[11] Only part of the tunnel was in tepetate duro, a material that could withstand the action of water and time; the rest needed to be reinforced with mortar and stone. Only part of the reinforcement was done, and if it was not completed it was not because of negligence. It was extremely difficult to install vaulting and reinforcements on friable materials because they crumbled easily once exposed. Martínez and his sobreestantes adapted by starting with the vaulting, then constructing the walls, and only finally installing the foundations.[12]

*Figure 2.1.* Enrico Martínez's 1608 "Description of the region of Mexico and the works of the drainage of the lake," showing sections of the Desagüe. The depiction of Lakes San Cristóbal, Xaltocan, and Zumpango appears unrealistically shrunken relative to Adrian Boot's shown in Figure 2.4 and subsequent representations, suggesting the desire to impress a sense of success in draining the northwest quadrant upon viewers, royal and city authorities, and urban elites.

SOURCE: España, Ministerio de Educación, Cultura y Deporte. AGI, MP-México, 54. Reprinted by permission.

Fortunately, the challenges of other structures were easier to meet. The Cuautitlan River diversion dam was indispensible to the functioning of this main structure. It had two parts with distinct functions. The long southernmost part, going from upstream of San Juan Atlamica, near the township of Cuautitlan, to just downstream from the township of Teoloyuca, bore the brunt of the load of the diverted river with its earthen structure. The entire length of this segment was maintained by the indigenous and Hispanic riparian users of the river's water. The northernmost segment was made of stone and mortar, and although it was also load bearing and would later be reinforced with earthworks,[13] its main purpose was to create a silting pool, known as the lagoon of Coyotepec, to regulate the passage of the river's water into Lake Zumpango during the rainy season, where it was retained, or into the Desagüe during the dry season. This segment (the "albarradón del Rey" on Figure 5.2) was considered part of the Desagüe proper, since it was maintained with its personnel and monies.

What Martínez built, however, was a truncated version of the drainage that had been approved. It could not actually drain the lower lakes. The actual Desagüe was constrained to depriving Lake Texcoco of the input of northern basin waters through the regulation and drainage of Lake Zumpango and the Cuautitlan River. The reasoning was that since the Cuautitlan River was the major contributor of water to the lower lakes, directing its waters outside the enclosed basin would be sufficient to ward off most floods.

Diversion of Cuautitlan River waters outside the basin would be accompanied by a careful management of the northwestern and northern waters through the seasonal release of water accumulated in the lakes that were equipped with retention dams. Lake San Cristóbal could be released into Lake Texcoco when the latter was low enough in the dry season. Since the capital was supplied with many of its foods and materials by canoe, it was important to preserve a minimum level of water in Lake Texcoco. Lake Zumpango, which was also fed by the torrential waters from Pachuca (*avenidas de Pachuca*), was relieved preferably into the Desagüe itself, also during the late dry season. In this manner, these two vessels would be left with sufficient capacity to store rainy season runoff and the diverted Cuautitlan River as needed. The latter was made possible by a dam that turned the lowland between the township of Coyotepec and Lake Zumpango into a silting pool that people called the lagoon or lake of Coyotepec. The movement of water from all these vessels into the Desagüe feeder canal was regulated by spillways or *vertideros* equipped with sluicegates. Superintendants and guardas decided which sluicegate could be opened when to alternate the release of Cuautitlan River or Lake Zumpango waters, since it was undesirable to let them both enter the Desagüe at the same time both because its capacity could be overwhelmed and cause backups and because it was believed that the greater force (*ímpetu*) and discharge of the river impeded the entrance of any other water.

Established wisdom generally attributes the truncated design and difficulties of the seventeenth-century Desagüe to physical and financial constraints, combined with the impasse between the viceregal bodies and the city government over who would pay. Yet at the time physical limits were not perceived as insurmountable. In fact, Enrico Martínez himself had originally envisioned building a deeper and longer Desagüe capable of draining Lake Texcoco. As much as his expertise was attacked, it was less on the basis of the physical and hydrological impossibility of the project than on its potential benefit relative to its costs. Indeed, although constrained to depriving Lake Texcoco of the input of northern basin waters, Enrico Martínez's admired oeuvre worked well for a few years, encouraging authorities to rekindle the general desiccation dream. As early as October 1609 the real audiencia and the cabildo of the city were already discussing how to fix the internal weaknesses that the rush to complete the project had created—irregular vaulting, little propping, uneven dimensioning, insufficient depth, and so on—without which there could be no possibility of extending the Desagüe to Lake Texcoco. At the time, in the discussions the financial implications of the technological requirements and options that master architects and other technicians presented were not seen as an impediment. This is not to say that physical and financial factors had no impact on design, of course, but that their impact was not quite that attributed to them.

One of the physical constraints that has never been considered is the extent of cooperation from people living in the terrain traversed by the Desagüe. Both Indian and Hispanic riparian users on the Cuautitlan River had the bad habit of transgressing their rights to the river's water by making holes in the diversion dam. Or, they would drag their feet in carrying out their maintenance duties on it. Others might resist the Desagüe labor draft. The point here is that one of the most important physical constraints impacting the shape of the project was the availability of human work, as expressed in the fact that local populations had other ideas about how the terrain should look and how their manpower was best used.

A good example of this are the chinampas. Unlike the Cuautitlan River diversion dam, chinampas could not be successfully hitched to the Desagüe. Like this dam, they did embody and synthesize the substantial technological know-how in the region. Since they were integral to local indigenous economies, they also became the objects of tension between these populations and the drainage works, tensions that also encompassed the lakebed cultivation practiced by several townships in Lake Zumpango during the annual recession of water. The persistence of the diversion dam would be as crucial to the Desagüe's success as the survival of the chinampas and lakebed cultivation were obstacles to it, and the superintendants of the drainage project acted accordingly. The pushback that their efforts to destroy these agricultural practices received helped determine the limits of what was materially possible for the Desagüe.

## OF MONEY AND MEN

Because the Desagüe was a royal project for the benefit of a key node of imperial rule, the initiative for designating the sources of funding for the project came from the viceroy. So far as the crown was concerned, however, this was also a project done for the benefit of the wealthy Hispanics who lived in the capital city, and so it expected these propertied citizens (*vecinos*) to contribute as well. This might give the impression of a balanced distribution of financial obligations with regard to the drainage. This was not the case, and popular classes disproportionately paid for the welfare of their social betters through taxes. There is no reason why it should have been otherwise. But since contemporaries hinted at a certain iniquity in this situation, and because it is always important to look beyond the assumptions about who finances and who benefits from "public" works in both past and present, it is relevant to examine where the Desagüe money came from and who contributed it. In addition to determining where the funds came from, it is important to consider which costs the royal and city exchequers actually paid for and which they did not.

Viceroys assigned the Desagüe an income from two main sources—exceptional donations and assessments on property, which generally targeted property owners, and ongoing taxes (*sisas*), which the popular classes disproportionately paid. Following the 1607 flood, Viceroy Velasco instated a one-time assessment that would be the first of three occurring over the colonial period. Based on the principle of taxing the property owners who would benefit the most from the drainage, the assessment was a 1.5 percent levy on the property, rents, and goods of city owners. It yielded 304,013 pesos, including 9,375 pesos that the king had to pay for the 625,000 pesos that the mint and court jail were worth.[14] Assessments like these would be levied again in 1637 and 1767, extending to the huge holdings of the *marquesado del Valle* as well, although on the latter date city propertied classes got a break, as they had to pay only if their properties were rented out.[15]

Sisas in general were a common means for Hispanic cities to finance public works and water supply infrastructures in particular. In the City of Mexico, the sisa began in 1563 as a tax on every animal slaughtered in the jurisdiction of the city, specifically to finance aqueducts and the water distribution network. In 1571, this tax was replaced by another one of half a silver real per *cuartillo* (half-liter) on retail wine sales in the city. Although this applied to merchants of both Parras and Spanish wine, the latter received an exemption from charging and turning over taxes on wholesale wine sold in *pipas* (three-hundred-liter containers).[16] The money generated by this wine sisa was earmarked for flood prevention works, administered by the city though a *mayordomo* named from among its regidores, and kept in a three-key vault box that could only be accessed by the corregidor, the obrero mayor (who inspected the works to control the destination of expenditures), and the mayordomo, who was also in charge of warehousing all construction materials.[17] When urgent repairs on causeways, dams, and so on were needed and the *caja de la sisa* was low, the Real Hacienda provided what was missing, which by law had to be repaid by the city. In 1604, for example, the royal treasury advanced eighty thousand pesos to pay the Indians who worked on the various flood repairs, and since the sisa account only had half of that, the city paid the rest from its *propios* over ten years.[18] When the Desagüe construction started, the same principle that the city should pay through its slaughter and wine taxes applied. A royal cédula of April 23, 1616, confirmed this, and it was later included in the *Recopilación de Leyes de Indias* and made applicable to both seculars and ecclesiastics, except for the Franciscans.[19] It was not the lack of availability of financial resources, or even the difficult terrain, but the unwillingness of city, crown, and colonial elites to pay for the safeguarding of their own city that limited the Desagüe's scope.

One item whose cost could have become a significant hurdle to the project was land. The Desagüe needed space. There does not appear to have been a formal process of expropriation and compensation for private or communal lands lost to its structures. In March 1608—when Martínez was already tearing through the terrain near Huehuetoca—Marcos Zardo, "a poor, old and disabled man" married to an Indian woman from Huehuetoca, petitioned for compensation for land formally appraised at twenty-five pesos "which they have taken from us to make adits and open trenches."[20] On his land, claimed Gonzalo Martín de la Orta, also from Huehuetoca, each shaft and the earth removed to make it occupied enough space to cost him the loss of yield of one *fanega* of wheat (about fifty-five liters). In addition, Desagüe personnel had "eaten and finished all the wheat he had sown as well as the hay and [destroyed] the cisterns he had built," which he valued at three thousand pesos.[21] The lack of evidence of disbursements to compensate these or any other private or communal landholders suggests that this kind of cost was absorbed by the Desagüe's neighbors. While this may not have been excessively painful for *hacendados*, for smallholders and villages it apparently was.

The truly insurmountable and objective limit to Desagüe construction and maintenance was the supply of labor—or, more precisely, repartimiento labor. Authorities were well aware of this. Repartimiento was the system of allocation of indigenous labor whereby each township had to contribute to the Hispanic economy a percentage of its able-bodied male adults on a rotational basis. It arose to meet the labor needs of colonists and was another way in which Spaniards grafted onto pre-Hispanic structures of rule and upward siphoning of wealth. It was based on a pre-Hispanic form of labor requisitioning called *coatequitl*, the summons (*llamamiento,* or convocation)[22] of workers for temples, hydraulic engineering works, and other major "public works" that preconquest Mexica lords commanded from tributary provinces, allied lords, and even independent kingdoms.

After 1555, repartimiento even used the same jurisdictions of pre-Hispanic summons. It emerged with the 1549 abolition of *servicio personal* (labor tribute) of the encomienda system that had given privileged colonists tributes in kind from the Indian towns granted to them, and as a result of the senescence of encomienda itself and the pressure of nonencomenderos' demands for labor. Coexisting with a variety of free wage, convict, indentured (*obraje*), and unfree (enslaved war captive or African) labor forms that entailed direct employer-worker relations, repartimiento allowed the crown to encourage or block the activities and development of classes as it needed by forcing colonists to go through an official (eventually called the *juez repartidor*) to obtain labor. The system was intended to both guarantee labor for the república de españoles and ensure that this did not overtax the república de indios, driving them to extinction as it had in the Caribbean.

Despite the presence of alternative labor pools in the form of African slaves, poor Spanish immigrants, and increasing numbers of mestizo and indigenous workers detached from the pueblos de indios, the Desagüe depended overwhelmingly on repartimiento labor. Other labor forms were used in the Desagüe only for short periods and for tasks requiring skilled labor. In 1607 the Desagüe was favored with an ordinance that assigned to it people who were convicted for noncapital crimes, such as carrying weapons in the case of blacks and mulattoes or vagabondage in the case of Indians.[23] This resulted in the use of some convict labor. In 1637, for example, Superintendant Fray Luis Flores would ask for and receive two hundred Indians who had been convicted of vagrancy in various barrios of the capital.[24] Two years later, however, he refused to take any convicts at all, complaining they were more trouble than they were worth, the sobreestantes did not want to deal with them, and they tended to run away. The viceroy offered to build an on-site jail, but this was probably nothing more than the modest stocks at the Desagüe headquarters in Huehuetoca.[25] Convict labor was therefore abandoned and did not become a feature of the Desagüe again until the first decade of the nineteenth century. Wage labor was reserved for permanent skilled workers, such as smiths and carpenters, and of course sobreestantes and guardas. There is no evidence that free wage, indentured, or slave laborers were ever considered for meeting the Desagüe's general labor needs. The only discussions about the choice of labor that were held as the Desagüe took shape were not about what labor regime should prevail but over whether projects should go ahead given the demographic crisis among Indians.

There were different forms of repartimiento, as Hispanic agriculturists, Taxco silver miners, urban builders, managers of public works, salt miners, and many other urban, industrial, and rural employers were eligible to receive this drafted labor. From its inception, however, the needs of colonists in the capital city loomed large in this institution. The first repartimiento, created by Viceroy Antonio de Mendoza, was designed with the "double objective" of ensuring workers both for Hispanic agriculturalists purveying grains to the City of Mexico and for urban construction, both private and public, including the cathedral. With the flood of 1555, extraordinary urban labor demands for urgent hydraulic works, repairs, and reconstruction tested this double objective, so instead of relying on repartimiento from surrounding districts, Viceroy Velasco required the cabildos of the two halves of the city itself, Mexico and Tlatelolco, to send Indians. The two functions of repartimiento in central Mexico were separated in 1565 when purely urban Indian laborers became insufficient, and certain repartimiento districts were restored to the city public works. The remainder were assigned to serve wheat agriculturists in the city's hinterland. The specifically urban repartimiento was abolished in 1624.[26]

The agricultural version of repartimiento was the one that provided the original structure for labor recruitment in the Desagüe. This version too was proscribed in 1633 for all but the Desagüe and mining. Having already started to substitute wage workers for their diminishing and fickle supply of corvée Indians, haciendas, mines, and other mainstays of the economy would predominantly use wage labor from then on. Not so the Desagüe, where repartimiento remained in uninterrupted full vigor until the close of the colonial era.[27] We know that the first Desagüe drafts brought in more than sixty thousand Indians to build the tunnel and canals in 1607–8. Thereafter, Desagüe labor demands rose and fell with perceived flood risk; accordingly, the decade of highest labor use was after the 1629 flood (labor use would peak again in the 1760s and the 1790s).[28]

Most aspects of the Desagüe repartimiento were well regulated. Rules stipulated compensation, treatment, maximum distance to be traveled, and the percentages of the adult male population that could be drafted. Pay eventually paralleled laborer (*gañán* or *peón*) wages in haciendas, rising to two *reales* per day in 1637.[29] That year the duration of *tandas*, or shifts, was also increased from two weeks to a month, reflecting the longer distances from which many workers came.[30] At least 4 percent of the adult males in neighboring jurisdictions could be sure they would have to perform Desagüe duties any given year.[31] The overall characteristics of the system of labor recruitment, as noted by Hoberman and Gibson, held fast throughout the colonial period: the net was cast over ever-broadening jurisdictions, with those nearby targeted first when demands could be met with them alone.

Although the use of repartimiento labor provided distinct advantages to the propertied classes, the Desagüe's complete dependence on it had certain drawbacks. One disadvantage has already been alluded to: with indigenous populations in decline during the early years of the Desagüe, the availability of an adequate labor force often proved to be an objective limit on Desagüe activities. Evidence that this was a serious problem comes in the form of debates about whether to proceed and what to do in the Desagüe on the basis of what regidores thought was the most gainful use for the dwindling Indian population. At the cabildo of October 2, 1609, for example, Francisco de Trejo Carvajal and Luis Maldonado del Corral, regidores who had been commissioned to study the different ideas about how to proceed with the Desagüe, basically agreed with master architects who had recommended deepening and repairing the Desagüe right away, even at the expense of further burdening the remaining Indians, before it was too late, whatever the immediate costs:

> It would be best if said desagüe were built and finished for once and for all without all the pains of expenditures and effort, even if the Indians incur them, which would not be severe due to the care with which they are handled, because lacking Indians in the future for indeed they are becoming extinct, the difficulties we face now will become impossible obstacles for the succor of this city, which should urgently beg His Excellency not to abandon this worthy enterprise of his.[32]

While some thought it was better to use the precious labor source on necessary flood protection quickly, before it vanished altogether, others felt the opposite. In 1611, for example, regidor Francisco de Solís y Barraza believed that little tangible benefit, great losses in rents, and a dismal prognosis was all that could be expected from sending more Indian labor to the Desagüe, which caused "many serious illnesses and deaths among the indigenous." The decline in indigenous population was evident in smaller labor drafts and in reduced private and royal rents and tributes. Since "Indians are the central nerve of this kingdom, and without them everything will lack," more Desagüe drafts would result in "nothing other than their complete extinction, many and greater expenses, and no benefit whatsoever."[33]

Another economic disadvantage of using a rotational drafted workforce in the Desagüe was that each group of workers stayed for only fifteen days to a month, which made it both difficult and uneconomical to train workers for activities with which they were not previously familiar. This disadvantage, however, may have been offset by the low labor costs this form of recruitment offered. It goes without saying, of course, that for the indigenous population the Desagüe draft was not exactly a benefit either.

Working conditions in the Desagüe were unenviable, to say the least, and this was consistent throughout the history of the public work during the colonial period and very possibly beyond it as well. The workday went from sunup to sundown, and those who worked in the tunnel ate their meals and emptied their bowels there. Each man worked with iron bars or spades, as the task demanded. Under Martínez, winches seem to have been used to raise the debris from the depths of the excavation, but in the shallower digs laborers carried it out with sacks on their backs.

It is not clear what the mortality rate was in the Desagüe. Enrico Martínez claimed barely two dozen deaths during his helmsmanship. In 1848, Francisco de Garay wrote that during Enrico Martínez's time Desagüe deaths were noted in the parish records of Huehuetoca, but that with the open trench conversion the death toll mounted, so a note was inserted in the parish books stating that henceforth a separate book would be kept for Desagüe deaths. Garay claimed to have examined these separate books, where each line listed the name, the township of provenance, and the cause of death—"from the drainage." There were about fifty names to a page, "just how they must have lain on the hill, all tightly lined up." His final tally was two hundred thousand Desagüe deaths over the colonial era, but this cannot be verified.[34]

As in the rest of indigenous Mexico in the first century of contact with Europeans, epidemic disease affected the villages that sent workers to the Desagüe, but here the frequency of Desagüe assignments likely weakened their ability to fend off disease. For example, being so close to the Desagüe, Indians from Tepotzotlán were routinely sent there, until in 1635 they reported that 1,416 *tributarios* had died of *cocoliztli*, an epidemic disease that may have been hemorrhagic fever.[35] The viceroy ordered other towns in the district to replace the fifty workers Tepotzotlán would have had to send, but given the nature of epidemics nearby towns were likely equally affected.[36]

If the Black Plague had given European peasants the bargaining power against their lords to extract from them fixed rents and even land ownership, reductions or elimination of dues, and cash wage payments, in New Spain the lethal epidemics had few if any benefits for Indians as they dealt with their new lords. This is in large part because contrary to late feudal peasants, indigenes in New Spain had to be contend with a far stronger—albeit "composite"—monarchical state that tried hard to take the issue of contractual relations between Spaniards and Indians off the table altogether first with encomienda and then with repartimiento.[37]

Indians and their townships were not the only ones who paid the price of drafts for Desagüe labor or the officialdom's single-minded concern for the urban hub in hydraulic matters. Hispanic haciendas and ranchos also suffered from the increased competition for labor brought on by the drainage project, especially during the initial construction of the Desagüe and after the 1629 flood. In comparable works in Spain and other European states, agricultural labor needs took precedence over those for public works: engineers could hope to hire workers only in slow agricultural months (and this was in demographic conditions that had not seen catastrophic crises since the fourteenth-century pandemics).[38] In the Desagüe district, however, demand for labor in the drainage was often highest when Hispanic agriculturists needed the manpower too.

Both Indians and landowners had no choice but to complain to the crown's officials when faced with excessive labor demands or insufficient labor supply. For example, landowners in Tacuba joined forces to complain that the Desagüe had drained away the repartimiento labor they derived from San Cristóbal and Chiconautla, which were in the Desagüe jurisdiction and sat at either end of the dam that contained Lake San Cristóbal. This had resulted in a sharp fall in the tithe (paid in grain) from 6,000 fanegas a year before the Desagüe call-ups to 1,090 by 1631.[39] This complaint was echoed by cultivators in the district of Cuautitlan, who explained that they had formerly worked their haciendas (both irrigated and not—*de riego* and *de temporal*) with Indians from the townships of Tepotzotlán, Coyotepec, Huehuetoca, San Miguel, Chiapa de Mota, and Xilotepec. But now the Desagüe had ruined most of them, since "there is scarcely one Indian per month for each." Viceroy Pacheco eventually did share some of the Desagüe's workers with the grain producers: clearly, failed harvests were the last thing he needed in the midst of a flood crisis. Thus, during most of the rainy season (June to September) of 1632, the Desagüe's workforce was reduced to 370 laborers so that the remainder could return to the agricultural repartimiento of their jurisdictions.[40] Landowners of the valley of Atrisco took a less circuitous route, intercepting the crew of two hundred Indians leaving for the Desagüe to take seventy for their haciendas. Of course, the viceroy would have none of this unilateral piracy and immediately intervened in favor of the public work.[41]

However, as haciendas increasingly dealt directly and contractually with workers, and repartimiento died outside of the drainage, the effects of the Desagüe drafts fell overwhelmingly on the indigenous population. In addition to the effects of the hard work itself on individuals, for Indian villages Desagüe repartimiento labor was as disruptive as the mining repartimientos of Taxco described by Robert Haskett and others.[42] During the Desagüe's first three decades, repartimiento was heavy on villages near and far, as it siphoned off manpower at critical points in the agricultural cycle, even beyond the basin.[43] As it was, Desagüe draft obligations competed with essential work that needed to be done for subsistence and to garner cash for tribute, in addition to posing a very real danger of death for economically active males and increased the risk of males leaving the villages altogether during or after their tandas to escape the impositions. Indian governors typically complained that Desagüe summons left the pueblos' lands unworked, homes uncared for, and tributes in arrears. Complaints of this nature occurred during both the population trough of the 1630s and the population recovery in the second half of the eighteenth century.

All this was corrosive to the survival of indigenous townships, as belonging to one brought so many impositions that often the only solution was to abandon village life altogether. The impositions, however, cut both ways: resistance to them in the form of petitions and group defiance, in contrast to flight, served to cohere indigenous townships. As elsewhere, the tactic of first resort (but by no means the most successful one) was the petition. Indigenous governors would ask the authorities for reprieves from Desagüe drafts, citing special circumstances such as a need for laborers for their own sowing, epidemics, competing coerced obligations, flight of males to the mining districts and haciendas, and inadaptability to the cold weather of the Desagüe district, among others. Indigenes used these tactics throughout the colonial era. For example, in 1618 Indians from San Gerónimo Aculco successfully asked to be exempt from Desagüe labor for a year, arguing they needed to repair the town's convent and church.[44] In an unsuccessful petition from 1631, those from Papalotipac and Tlacuylultepeque asked for reprieve due to their loss of tributaries to epidemic and fear that sending their quota to the Desagüe would endanger the remainder, as they were *tierra caliente* people unsuited to *tierra fría*, and going there also required wading across three rivers.[45]

Although they did not work in the Desagüe themselves, indigenous authorities also bore the pressures of the public work, and this had a ripple effect on their communities. Indian *gobernadores* and *alcaldes* were responsible for selecting the individuals in their townships who would have to be sent to the Desagüe as part of repartimiento summons. In the seventeenth century, governors could be punished with imprisonment and fines of up to two hundred pesos for failing in this task.[46] Every year, when superintendants inspected structures such as the Cuautitlan River diversion dam under the care of Indian townships, ranchos, and haciendas, governors and alcaldes would be in attendance along with the mayordomos of haciendas to answer questions and to receive maintenance orders and any specific instructions about how to carry out their tasks. They were then responsible for coordinating the execution of these orders. The role of gobernadores of townships in dealings with the Desagüe was thus no different from that played by their counterparts in other interactions between the república de indios and the república de españoles: although the Indian gobernador was elected in each township and confirmed by the viceroy, his success in contesting Desagüe impositions was key to both the township's and his own welfare.

In sum, in the Desagüe and out, all forms of repartimiento shared a single fundamental characteristic: that of being a labor subsidy from the indigenous realm to the Hispanic one, which implied a net loss of human resources for the communities subject to it. The impact of this subsidy was more severe than the pre-Hispanic coatequitl on which repartimiento was modeled because of the demographic crisis context, the increased frequency with which it was required, and the fact that like monetary and goods tributes, this labor tribute did not come back to subject towns in the form of culturally coherent engineering works, administration, or ceremonies, and sometimes not even in the form of the actual human being, who might flee his township obligations altogether or die, a particularly high risk in mine work. Even when weighed against these costs, however, repartimiento as a labor subsidy was simply too attractive for city elites to pass up. The persistence of repartimiento in the Desagüe even as it ended everywhere else would have important technological and social implications.

Meanwhile, the reliance on this limited method of labor recruitment aggravated the consequences of certain other decisions in Martínez's Desagüe. Once they saw the Desagüe completed and water from Lake Zumpango and the diverted Cuautitlan successfully entering it through the access canal or trench, traversing the tunnel and pouring out the Boca de San Gregorio, as the outlet at the end of the tunnel was known, authorities immediately ordered Martínez to get ready to extend the works all the way to Lake Mexico. This meant that instead of focusing his labor allocations and attentions on reinforcing the tunnel, between the dry seasons of 1609 and 1611 a probably dazed Martínez was busy excavating a deeper tunnel. This time he started at Boca de San Gregorio and excavated under the original tunnel back toward Lake Zumpango so that gravity flow could take care of evacuating whatever underground water workers encountered (a strategy that many detractors of the first tunnel had urged).[47]

As a result of this distraction, the tunnel began to fail the season after its inauguration and worsened thereafter. Many factors contributed to the tunnel's problems: its small dimensions, excessive incline, and varying widths accelerated water flows and thus increased abrasion; the lack of revetment allowed the walls to crumble; the alternation between wet and dry conditions made the friable materials even more delicate. As a result, the interior progressively disintegrated and the resulting debris obstructed the passage of water. To make matters worse, when the Cuautitlan River water slowed down in the approach to the obstructed Desagüe tunnel, it deposited its suspended solids before entering the trench section that led to the tunnel, soon blocking the waters of Lake Zumpango that were supposed to exit to it. Meanwhile, since it was now closer to the water table, the deepening "*obra nueva*" met worse difficulties than had the original tunnel, and a checkmated Martínez opted for connecting the two tunnels instead. All this work was in vain, however, because it was not the end but the first section of the Desagüe that had insufficient depth to be eventually capable of draining Lake Mexico and Lake Texcoco.

The complaints were not long in coming from city elites, who felt the 304,013 pesos they had contributed in property assessments had been ill spent. By the end of 1611, the cabildo was writing to the king, saying that although the Desagüe was indispensible, it had thus far cost 420,000 pesos and 130,000 man-days of indigenous labor with unsatisfactory results. They asked him to send "two engineers who are architects and versed in mathematics" to fix the problems.[48]

Because it combined a tunnel with an open trench, both of large dimensions, the Desagüe presented challenges different from those faced by early modern Europeans in comparable large-scale works, such as the urban canals of Milan and their extensions, the Canal du Midi, and the drainage of the fens. Statics and dynamics of both solids and fluids, hydrometry, geology, and hydrology were involved in the Mexican project's design. Knowledge in these sciences had obviously not acquired the form it has today, but the detailed tacit and empirical knowledge of craftsmen and local inhabitants who contributed to the works, willingly or by compulsion, amply compensated. Indigenous populations of the Desagüe district were tapped for their mastery in manipulating water flows with structures made of inorganic materials reinforced with local plants. European master architects, friars, their helpers, and even rural entrepreneurs brought their cumulative experience with Mexican rocks, earths, and woods, which, depending on their qualities, were either used or avoided in the works. Friars in particular applied their expertise in directing human groups to the management of the large, multiethnic, plurilingual, and coerced workforce. Still, some critical technical problems such as the proper rate of flow relative to the gradient in the works or the relationship between soil composition and the slope of the trench walls—not to mention the various social and political problems such as the declining indigenous population and the limitations inherent in the goals of the elites—remained intractable.

Faced with contradictory reports born partly out of this confluence of varied knowledge traditions and the dangerous impasse among city and viceregal authorities, in May 1613 Philip III asked Iñigo de Cárdenas, his ambassador in France, to find him "a craftsman or engineer who is a geometer and knows how to survey elevations and waters" to send to the Desagüe.[49] Cárdenas selected a Flemish military engineer, Adrian Boot, who seemed a good choice as he was then working on the drainage of the French wetlands (*marais*). Boot was dispatched to Madrid to receive his commission and instructions. His *Diseño de la ciudad de México y del Virreynato de Nueva España desde el Mar del Norte al del Sur, para instrucción del desagüe de la Laguna de México*—the map of his destination that he drew while still in Europe, shown here in Figure 2.2—suggests that vague as Boot's familiarity with the cartography of New Spain may have been, he was aware of the lakes' enclosure. Accompanied by a translator, the artilleryman Juan Rodríguez, who as the son of a Spaniard who had settled in Antwerp knew both French and English,[50] Boot arrived in Veracruz from Seville in early October 1614.[51]

*Figure 2.2.* Adrian Boot's vision of the enclosed basin of Mexico and its viceregal capital, before embarking.

SOURCE: España, Ministerio de Educación, Cultura y Deporte. AGI, MP-México, 55. Reprinted by permission.

Although—like all other royal engineers in Spanish America before and after him—he would later work mainly in fortifications, building the fort of San Diego to protect the important Manila trade port of Acapulco, Boot's original mission in the New World was unique in that he was assigned exclusively to an urban public work. This was a measure of the importance of this particular urban center to the crown. After the engineer had reached the capital, on October 9, the viceroy commissioned senior oidor Pedro de Otalora to accompany Boot on his inspection with a notary, a bailiff (*alguacil*), and a Nahuatl interpreter.[52] On November 14, after the rains had ceased and the lakes had begun their annual recession, the party was ready to set out, with a meticulous scribe to record the proceedings.

## TOURING THE BASIN

The record of Boot's tour of the basin and the Desagüe offers an invaluable picture of how the Desagüe looked a few years after its completion, how it had affected the hydrology of the region at that point, and what the relationships among the various actors in the project were like. It also documents the beginnings of frictions over the fate of the project, not just between local and royal technicians but among the colonists themselves and between them and the crown. It is therefore worthwhile to follow some of the steps in the tour quite closely.

Authorities were extremely cordial and collaborative with the king's engineer. Otalora bent over backward to facilitate his work, and although he gave Boot specific questions he wanted answered, he studiously avoided impinging upon the engineer's activities except to ask to be kept informed "of every detail and without omissions."[53] To allow Boot as much latitude as possible and encourage his candidness, the audiencia had even taken the precaution to keep Enrico Martínez under control, ordering him to be present in Boot's inspection of his work "in the parts and occasions that seem convenient . . . but not otherwise."[54] Boot himself seemed unswayed by such courtesies and clear about his primary loyalties. His demeanor indicates that so far as he was concerned, he was there to gain an understanding of the Desagüe within the regional hydrology, to determine whether his royal employer benefited from the works being undertaken to protect the capital city as a stronghold of his rule, and how this benefit might best be secured in the future. The unflattering verdict that this single-mindedness contributed to would help garner the Fleming undying ingratitude from city elites and local technicians.

Boot understood the hydraulic component of his mission to have a twofold goal: understand both the structural aspects of the Desagüe and how the water moved through it and in the entire basin. For this, the engineer organized his tasks into five distinct, sequential activities—touring the entire basin, examining the insides of the Desagüe proper from both underground and above, observing the water flows throughout, making a careful survey with scientific instruments, and finally drafting his report and a map, possibly the one shown in Figure 2.3.

The tour began in the south of the basin. Since some of the floodwaters periodically climbing into the City of Mexico originated in the abundant springs at the foot of the mountain range arching at the south of the basin, it was important to assess how this water was managed once it entered Lakes Chalco, Mexicalzingo, and Xochimilco. Departing from Iztapalapa on foot, on horseback, and by canoe, the engineer's inspection began in the township of Chimalhuacán, where he examined the causeway-dam of Mexicalzingo, which had two sections and sluicegates that could be shut to bar the passage of southern water to Lake Mexico in case of need. Finding things generally in working order, Boot's work in the south and southeast consisted mainly of gathering survey data for an overall view of flows and terrain. Proceeding counterclockwise around the basin from there, he rounded Lake Texcoco all the way to the causeway that retained Lake San Cristóbal, seeing nothing particularly worrisome. The causeway-dam of San Cristóbal, however, had significant leaks "because it is made with false stonework and not stone and mortar." These leaks aggravated the danger created by the 2.69-meter difference in the surface of its water compared to that of Lake Mexico.

Boot and Enrico Martínez met in San Cristóbal. As Otalora introduced them, he warned Martínez to explain the Desagüe to Boot "with the peace and quiet and courtesy as it is fair to bestow upon said engineer who is the person sent by orders of His Majesty for so important a case and matter."[55] Although the two men would become bitter opponents, their relations were cordial and collaborative enough for the moment. They may have spoken in Flemish or French while they conducted their operations, as Martínez understood both.

Right away, the two technicians set off together on a canoe with Bernardino Alvarez de Rivera, the notary, to measure the depths around Lake San Cristóbal, followed on land by Otalora. Alvarez de Rivera noted their every move. The rainy season had filled the lakes, so Lake San Cristóbal communicated with Lake Xaltocan through a "river" whose depth of five varas and a half (a vara is .83 meters) decreased about a meter as it joined Lake Xaltocan. Where the river joined the lake, Boot, measuring with a long staff, noted depths equivalent to between 4.55 meters and 5.39 meters through both water and a thick layer of "*lama y lodo*" (slime of decomposed plant matter and silt). Boot asked Martínez how far the canal reached, whether it was man-made or natural, whether its depth diminished and by how much, why it had so much slime, and

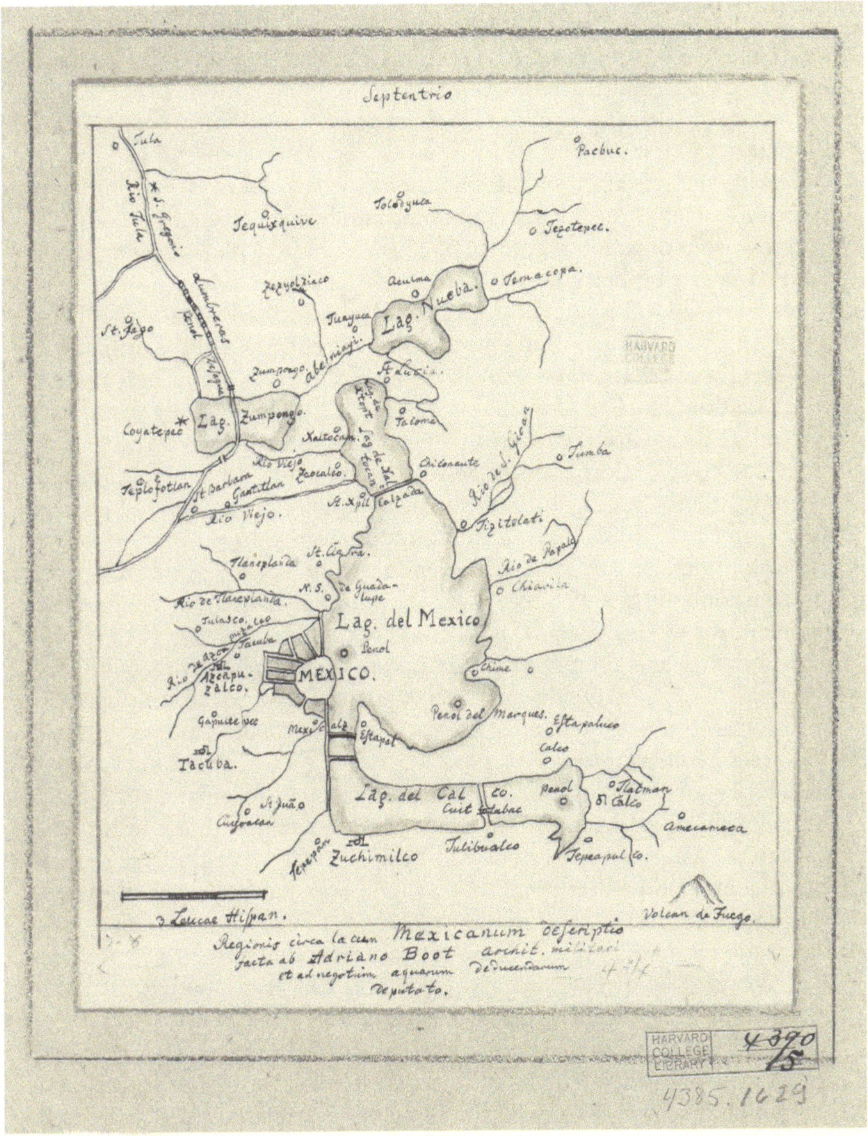


*Figure 2.3. Regionis circa lacum Mexicanus*, a map of the basin and the Desagüe, attributed to Adrian Boot, 1614. Differences in calligraphy with Figure 2.2 and written documents by Boot, as well as otherwise inexplicable inaccuracies (such as the relative locations of Lakes Xaltocan and San Cristóbal), suggest this is a copy of a lost original. The inaccuracies that appear were replicated in the 1700 edition of Gemelli Careri's *Giro del Mondo* and only corrected in later editions of the work, as other cartographical renditions of the basin began to circulate in Europe.

SOURCE: Harvard Map Collection; Giovanni Francesco Gemelli Careri, *Giro del Mondo*, 9 vols. (Naples: Stamperia di Giuseppe Roselli, 1700), 6: 57. Reprinted by permission.

whether Lake Xaltocan and Lake San Cristóbal had the same surface level. He found that the lakes had essentially joined, leaving them "at a single continual surface, so that [water] does not flow from one part to another because it is all at the same level."[56]

The shared elevation of the surface of the two lakes clearly allowed townships around the lakes to communicate by canoe, since from the island township of Xaltocan, the two men were able to reach the hamlet of San Pedro, where they disembarked to examine the "old river, where the water from Lake Zumpango used to flow to Lake San Cristóbal." They could advance no further on the "old river" toward Lake Zumpango, however, because the lake's surface was almost two and a half meters above that of Lake San Cristóbal, a difference that generated a significant current. To go against the current they would need two canoes and twenty Indians to row. Otalora immediately ordered the township of Zumpango to supply them—a clear indication that its inhabitants still found part of their sustenance in the lake. Although some eighty centimeters of water had evaporated already, all land around this site was still flooded, a boon for local communities who harnessed the seasonal fluctuations, but an impediment for the technicians, who could not advance even on horseback to reconnoiter and so ended the day in San Cristóbal. Leaving San Cristóbal the following day, they reached a canal that connected the township of Xaltocan to the lake and causeway of San Cristóbal.

To reconnoiter the area of Lake Zumpango, they boarded four canoes manned by seventeen Indians from the township of Zumpango at a bend in this canal, measuring as they went into the lake, with the notary always in the same canoe as Boot and Martínez. All around this lake they found depths no greater than 1.25 meters. Sometimes the depth was as little as 50 centimeters, and in these places they were forced to disembark and wade through dense aquatic vegetation ("*mucha yerba*"). Reaching the causeway of Lake Zumpango, Boot measured "with a cord and a level" the difference between the surfaces of the water of Lakes San Cristóbal and Zumpango, finding the latter to be three and a half varas higher. Martínez then led the engineer to see the Vertideros on the dam containing the latter lake, showing him also "the mark of the water that could be seen on a palisade of *morillos* with which said causeway of Zumpango is reinforced," which registered how high the lake rose during the rainy season. Evaporation had diminished the water level here by almost half a meter.

After sleeping in San Cristóbal, they returned to Lake Zumpango, this time embarking near the township of Citlaltepec, on the northern shore of the lake, to take soundings toward the township of Zumpango, "about a league away." On this section of the lake, the water was deeper: Boot found depths of between 3.3 and 2.5 meters of water and decomposed plant matter, which although detrimental to the Desagüe if it was disturbed and carried by the current, provided vital nutrients for the wetland agriculture of the villagers. From the center of the "lake called Citlaltepec and Zumpango because it is shared by both townships," they headed west toward the Vertideros. They were punted through a *tular* (reed patch) just beyond the town of Citlaltepec, and they found the depths slowly diminishing as they tacked along the western shore. From there, they entered "a canal where the water of said lake of Citlaltepec and Zumpango enters and reaches the beginning of the works of the Desagüe." The canal that carried the water from Lake Zumpango to the gates at Vertideros had about two and a half varas of water. Then they traversed the lagoon of Coyotepec, where a dam contained the diverted Cuautitlan River. The lagoon both created a silting pool to prevent debris from entering the Desagüe and enabled the guardas to control the Cuautitlan River's flow into the Desagüe. Probably as a result of silt deposits, the surface of this lagoon was about two meters higher than the surfaces of Lake Citlaltepec and Zumpango, creating the risk that the Cuautitlan River's waters would break into Lake Zumpango instead of flowing into the Desagüe.

Spending the night in Huehuetoca, they returned to this lagoon, which the notary now called "lagoon of Cuautitlan." They began inspecting "a trench that comes toward the township of Huehuetoca," which was none other than the access canal or trench for the Desagüe proper. Boot found its mouth to be just over two meters below the surface of Lakes Citlaltepec and Zumpango and its length toward the entrance of the tunnel full of water so deep that his 3.5 meter-long staff could not touch bottom, despite the 5.81-meter width of the trench. Along this trench, their canoe was stopped by a structure described as a "*puente de tierra*" that was not capacious enough to let them pass underneath, so they disembarked and walked to two closed sluicegates at some distance from each other whose function was to keep water out of the tunnel when necessary for inspections and repairs. They checked the diminishing water levels on the Desagüe side and proceeded until they were in front of Huehuetoca. Up to this point, Boot had seen a basin where the most challenging waters were those to the north: the level of the surfaces of the water bodies descended from the Coyotepec lagoon where the Cuautitlan River's water was controlled, to Lake Zumpango, Lake Xaltocan, and Lake San Cristóbal.

## IN THE BOWELS OF THE TROUBLED DRAINAGE

Finally, fifteen days into the tour of inspection, Martínez led Boot to his tunnel. The first leg of the inspection proceeded above ground, going from one shaft opening to the next, with Martínez explaining the underground features at each one. Understanding the bowels of the Desagüe before descending must have demanded an extraordinary effort of the imagination, since not only was "underground geometry" a field difficult to master, but the tunnel had two components that met underground. The "*obra vieja*" that was not deep enough to enable the full drainage met the incomplete "obra nueva" just under the bed of the deepest shaft in the old tunnel.[57]

Most shafts were named after people—Blas Hernández, Godina, Andrés Hernández, Torres, Herrera, Guzmán, Tejeda, Pedro Ruiz, and so on—probably the sobreestantes who headed the construction gangs. Nineteen of the shafts joined tunnel sections that were reinforced with sixty-two centimeter-thick mortar walls, and eight communicated with sections that were carved into the rock. From Shaft #15 on there were also sections supported by timbers, as in mines. Some of these shafts had a hatch and ladder that closed off access to the tunnel and facilitated the inspections from above. Descent into the tunnel was through these shafts. For this, special arrangements had to be made, since the bed of the Desagüe below was still under a meter of water. Again, canoes had to be rented from Indians of Zumpango and Citlaltepec, who were also paid for their transport and punting, while in shallower parts boards were laid out. Apparently, the prospect of going underground did not appeal to the oidor, so he swore in royal notary Jhoan Baptista Rendón to go in his place and take note of every exchange between Adrian Boot and Enrico Martínez. By mid-afternoon, they were ready, and Boot entered the tunnel for the first time.

With the light of the entrance of the tunnel farther and farther behind, the inspection proceeded by torchlight. At each shaft, Martínez explained the state of the tunnel there and how much it needed to be deepened for the drainage of both Coyotepec lagoon and Lake Zumpango, based on the assumption that it needed one vara of descent for every one thousand of distance. On this first day of two underground, they covered the distance encompassed by twelve shafts, each about 175 meters from its neighbors. They started at Shaft #3, named Tajuelo, and ended at #14. Although he sometimes followed on a canoe, Jhoan Baptista Rendón is unlikely to have been able to write down the details in the dim light: to examine things more closely, the technicians and Boot's translator rarely used the canoes. Instead, they either laid down planks across the tunnel on the timber supports, "crawling on them on our knees or on our sides because it was not possible to pass in any other way," or waded with the water sometimes up to their waists, "going through it with great difficulty and very slowly," carried only occasionally on the shoulders of the Indian bearers, and all the time translating back and forth. When they came back up after nightfall, drenched and muddy, the four men still had to "correct the numbers and additions we had memorized at each stop."[58]

On the second day underground, they went from Shaft #14 to the last one of the old tunnel, #27. Through much of this section they were under the dark thick of the hill, where the air was dank and filled with danger and echoes. Some four and a half kilometers into the tunnel, they reached what had probably been Shaft #27, but had now become a flue named Cañón de Carrillo where they had to crawl on planks:

> Passing through this whole section with great effort because the earth is very slippery and muddy, and in said tunnel we entered a place where there was a large cavity that is called the badlands [karst] where one felt great fright and apprehension as it had a great quantity of salts from the earth itself which apparently had corroded the sides and the top, and some [cavities] were as large as barrels [pipas][59] or carts and being as this location was the most dangerous in the entire drainage and the one that has impeded the works and their swift progress, we hurried through a section of sixty varas, more or less, which is about as much as the cavity of badlands and loose slabs [*lajas*] must have measured, and we passed through a narrows from which water falls from seven feet high to a guide gallery [*guía*] which was opened to reach the new tunnel, which gallery must have been some two varas high and half wide, and through it we went with the water just over one quarter vara high, thus reaching the level part of the flue they call Carrillo, under the summit of the hill.

The guide gallery for the new tunnel connected to the old tunnel under the thick of the hill. Parts of this new tunnel were finished and vaulted, although Enrico Martínez had mostly reinforced it with arches. He had also carved a series of six flues (*cañones*) that split off either side of the new tunnel and led to the surface. One of the cañones had three flights of stairs, the first made of tepetate, the second of wood, and the last of mortar and stone. The party climbed up these stairs to leave the bowels of the Desagüe.

For the following leg of the inspection, the men would access the tunnel from above, joined by Otalora, Luis Maldonado del Corral (regidor of the City of Mexico), Friars Thomás de la Cámara (a Franciscan) and Lázaro de Prado (a Dominican), and a retinue of notables. They all went down a wooden ladder in Shaft #5. All possible comforts were arranged, so the oidor and corregidor could remain dry and safe atop their hand-chairs carried by Indians through the tunnel, which wherever possible had been lit up with "a great quantity of torches and candles." Maldonado was apparently a heavy man because his Indians quickly tired out and he had to return to the surface. Despite the slime that further mired the Indians' progress, Otalora ordered the inspection to continue. With difficulty, the Indians carried him all the way to the Aguirre shaft, #17, more than two thousand meters beyond, where water and fatigue overcame them and they had to lower him. "Despite this difficulty," the notary recorded in a tone that seethes with resentment, Otalora wanted to see more of the tunnel, even though

> said engineers and said recounting notary explained to him the multiple difficulties they had experienced the day before while inspecting the tunnel. And after all this again he ordered them to go ahead to see if there was any way to lay down a floor of boards by canoe because he wished to find out by inspecting and personally reviewing said tunnel even if he experienced within it a few difficulties and it took a few days. And said engineers explained to him that he should consider the fact that he was greatly endangering his safety and health because it was in no way possible to go ahead in the manner he proposed without getting thoroughly wet, and this could provoke an illness and a threat to life.

After this concerted dissuasive effort, Otalora yielded, and the whole party resurfaced through a ladder for lunch in Huehuetoca. That same afternoon they all headed for the new tunnel (the obra nueva), where the procedure was repeated. The following day they retraced the trajectory of the tunnel on horseback from above, to assess how the Cuautitlan River water flowed between the dam that retained it in Coyotepec and the tunnel. Peering down the shallower shafts, they could all see the water rush through the old tunnel where they had struggled the previous days. At the crest of the hill, Boot and the notary listened for the rush of the water below the flue of Carrillo, but the sound was lost in this deepest adit. They were able to see water rush from the outlet at San Gregorio, to a depth that Boot estimated at 1.65 meters and followed it yet further on until Nochistongo.

At that point, the only field task that remained was the survey. Boot set up his instruments, the configuration and usage of which apparently fascinated the notary:

> Said survey was done with some instruments that had been taken for that purpose, placing on the ground a thick board and on top of it a wooden shelf, over which a metallic quadrant and a plumb with which it was kept in place were hung, and farther away another board as thick as the first one was placed and on top of it the staff of a lance whose every vara was numbered and divided up and down in one-sixteenths, and in the middle of it a white paper was placed facing said staff and behind it a black cloth so as to be able to see the white better. And said master engineer placed himself next to said quadrant and looked through it at the place where the target on the staff was located, and he had it raised and lowered until he fixed it at the point that it had to be, with myself, said notary, and Jhoan Núñez, said engineer's interpreter, holding said paper in place without moving while he came to see how much it had gone up or down[. A]nd finding it as it needed to be, he would fix the Point of said survey and Enrico Martínez, maestro mayor, with me, said notary, would go see on said staff how much the land that they were surveying had risen or descended, all three of us committing to memory the number that this translated to.[60]

The four men carried on in this manner, surveying the difference in levels along the path of the water until the rising wind made it impossible to continue, and they retired to reconcile their numbers in a little chapel where the workers usually attended mass. They worked there, under the intent eyes of Desagüe sobreestantes Antonio de Esquibel, Antonio Mañero, and Lázaro Vela, and the Desagüe carpenter, Martín Pablo, as well as Otalora, whose curiosity about the proceedings was piqued once again: the following day it was he and not Rendón who manned the staff as they continued the survey. On the following and final day of their joint inspection, the technicians presented their reconciled measurements to Otalora, and Martínez returned to the city while Boot and the oidor headed for the western side of the basin, to see Chapultepec with its aqueduct and some flooded ejidos nearer the city. Otalora asked Boot if he needed to see Lakes Xochimilco and Chalco, but Boot found it unnecessary. After twenty-four days, the tour was over, and Otalora and Boot parted ways for the moment.

The Desagüe clearly needed changes. Given contemporary ideas that the more violent the discharge the better, technicians thought that the depth of the Desagüe had to increase to give it the proper gradient. The current difference in water body altitudes and the level of the bed of the drainage's access canal and tunnel was insufficient by these standards, and Boot thought that parts of the 5,395 meter-long access trench between Vertideros and the mouth of the tunnel needed to be deepened up to 2.50 meters to effectively tap the Cuautitlan River. In addition, the tunnel was suffering from neglect in several sections over the 4,731 meters up to the junction with the new, obra nueva tunnel guide gallery, and therefore had to be redone in mortar and stone. Overall, it had to be deepened in various sections up to almost three meters in order to be level with the Cuautitlan River. If, however, Lake Zumpango and the avenidas de Pachuca it collected were to be drained, then an additional meter and half at least had to be shaved off. Boot thought the new 2,017-meter-long tunnel that connected to the old under the thick of the hill was in far better form, having some mortar and stone vaulted sections and others in a less friable type of tepetate, and needed only to be somewhat deepened in certain parts. At the end of this tunnel was an exit canal that was almost twenty-eight meters lower than Coyotepec lagoon.

Viceroy Guadalcazar read Boot's report on December 27, 1614. Although the foreign engineer concluded that the Desagüe was indeed "a grand work," he claimed that it was "useless for the purpose of freeing the City of Mexico from the risks it faces and the dangers to come," and that it could not "drain the lake of Mexico, that of San Cristóbal or that of Xaltocan" because this would require that it be connected to a very deep and capacious twenty-eight-kilometer canal, the construction of which would involve costs too "extraordinary to be sustainable."[61] He did feel the Desagüe could easily be improved and deepened to carry the Cuautitlan River floods safely out of the basin. Where the maestros and other experts in the colony had generally criticized Martínez's work by invoking his failure to adhere to Vitruvius, "prince of architecture," particularly on the matter of slope and therefore current, Boot did not rest this criticism on any authority other than his observations and calculations. Moreover, it was not the speed of the current or the slope he objected to, but the entire concept of evacuating water with enormous expense and labor for uncertain results. In this, he was a loyal servant of his master's desire to protect the treasury and the subjects that fed it.

A few days later, he proposed flood-control measures that grafted established practice in the Low Countries onto the pre-Hispanic model without draining anything, promising to "capture the water with my instruments" and to "expel it as is done in the cities of our land which are tormented by the waters of the ocean sea."[62] In addition to raising and reinforcing existing causeways and dams, his plan called for twenty sluicegates that could be shut as necessary to protect the city, and a new canal almost 11 kilometers long, 6.5 meters wide, and 3.3 meters deep arching protectively west of the city, from the causeway of San Antón in the south through the ejidos of Chapultepec and to the Guadalupe River. The purpose of this canal was not drainage, but "to gather all spring and sierra water so that it can be taken to the lake and does not fall upon the city." It would have "four canals opened in different parts to come from the large canal to the city so that said canoes of the district can navigate," with which "at all times fresh water would be made available." This part of his plan would take three years to build with a daily workforce of 150 Indians and cost almost 340,000 pesos. If Boot's proposal has been dismissed as impractical ever since the viceroy read it in January 1615, it is because Boot clearly had an understanding of the role of water for the City of Mexico that few shared then or now. In proposing cranes at the sluicegates to transfer canoes from one side of the causeways to the other when the sluices had to be kept shut, he demonstrated that he may have had something in mind like the Naviglio Interno of Milan and its additions—a multipurpose canal to go around the city, providing freshwater as well as a means of urban and urban-hinterland communications.[63]

This proposal was rejected, becoming a favorite event in the retelling of the Desagüe story by nineteenth-century liberal historiography. In the version published by José Fernando Ramírez, one of the founders of this tradition and a lawyer statesman who in 1857 headed the Junta Menor del Desagüe and later served under Emperor Maximilian, Boot the foreigner "put on airs as an engineer brought in by the world's foremost monarch," but the Mexican master architects rebuffed his "boastful and highfalutin words" and buried his proposal "in ridicule, letting him understand that they, without the airs of the engineer or any great pretensions, had done the same thing he was now proposing."[64] Despite the nationalist sieve through which the Fleming's *parecer* and proposal were poured over and over again thereafter, he did turn out to be right in the long run in one important regard: after successive additions and reforms piled onto the original Desagüe over the centuries, the City of Mexico is still not "free from the risks it faces" for the simple reason that it sinks.

Conversely, for other writers, Boot was a figure to be rescued from these ignominies. The point here is not to vituperate or vindicate Boot, Martínez, or anything or anyone else, but rather to find the historical implications of Boot's deployment to New Spain at a time when monarchs, princes, and courtly alliances vied with each other to hire Dutch and Flemish engineers to drain marshlands in territories they wished to control. What is the difference between the actions of these men and their employers in Europe and those in the New World? Is it that the former helped to consolidate nation-states while the latter deepened colonial subjection, creating a cleavage down the Atlantic? This might have been the outcome and the retrospect in the nineteenth century, but the motivations and actions of those involved, and what was achieved at the time seem to have been marked more by what class interests moved them than by their belonging to one nation or another. It would seem that the difference lay in what classes or class alliances stood behind the rulers who deployed these men throughout Atlantic early modernity, what they ordered the engineers to do, and why.

## DOUBTING

As the memory of the 1607 flood faded from the minds of the city elites, and as they heard increasingly conflicting reports about the utility of Enrico Martínez's Desagüe, their support for the project wavered, and this was reflected in how they handled labor and capital for the Desagüe. By then, repartimiento was becoming a useless institution, wage relations quickly replacing it in major economic activities, such as mining and agricultural production on the haciendas. Yet repartimiento endured in the Desagüe, mainly because of the cabildo's reluctance to let go of it and because of viceregal acquiescence, as in this fashion they both guaranteed labor supplies when and where they were needed and minimized their own expenses by shifting them to the Indian communities. Neither were city elites ever willing to invest in animal-powered machinery to minimize human energy needs. When Adrian Boot reintroduced his proposal to them in 1620, for example, while rejecting the parts of the proposal dependent on introducing expensive European technology, the regidores supported those aspects of the plan that conveniently entailed little spending, deploying them as weapons against the decision of the viceroy and audiencia to continue funding and sending labor to Enrico Martínez's works as ordered in a 1616 royal cédula.[65] As a result, Martínez himself reported that because he was receiving fewer and fewer repartimiento workers with whom to conduct necessary repairs and maintenance, the Desagüe had basically ceased to function by 1623.

In a reflection of the crown's own hesitation to continue to support the Desagüe in the face of the other option, which was to move the capital, Viceroy Gélvez in April 1623 ordered all work other than maintenance on the Desagüe to cease in order to run a risky experiment: to see what effect the rainy season would have on Cuautitlan River waters if they were barred from the Desagüe and allowed to resume their normal course. Although this measure was much maligned both by contemporary writers and later historiography, it was nonetheless reasonable for the circumstances, since it seems that what motivated it was the viceroy's weighing of the welfare of the city against the costs of its flood-control measures on Indian and non-Indian cultivators.[66] Although it is not clear that the measures contributed to the 1624 flood that triggered rioting in the city and Gélvez's own demise, they did lead to much rejoicing by rural indigenous populations, who not only would be freed from Desagüe repartimientos but also recover their traditional usage of land and water.[67] The definitive death-knell for Enrico Martínez's tunnel, however, came in the aftermath of the largest flood in the capital's history, which began on September 20, 1629. The changes engendered by the flood would ensure that when Martínez himself died in 1632, he would take an entire technological epoch to the grave with him. But all else that had calcified around the Desagüe would go on without him, misgivings or not.

The greatest continuity in the Desagüe from its very inception was who and what it was conceptualized for. Of course indigenes would suffer the consequences of its labor drafts and its effects on their access to and usage of land, water, and resources in the northwest quadrant. But even Hispanic agricultural enterprises both near to and far from the Desagüe were clearly not the priority either. During the project's first three decades, they complained about how it drained them of necessary labor, or how the river diversion works to save the city had flooded their haciendas. These complaints were dutifully recorded and then ignored. In 1632, one hacendado from Atrisco, to the southeast outside the basin, for example, complained that the Desagüe had taken repartimiento workers and thus he was losing his harvest.[68] Another one was disconsolate over the complete flooding of his lands after works supplementary to the Desagüe—the 1609 diversion away from the city of the Azcapotzalco and

Tlalnepantla Rivers—cost him not only the original ten thousand pesos of the purchase but all the improvements he had accrued thereafter.[69] Clearly, the drainage, together with its attendant flood-control measures, was conceptualized to relieve the city above all else. As a result, authorities neither encouraged nor contemplated any design for it to assist production or generate wealth, Spanish or indigenous. In this sense—and this alone—the Desagüe could be said to be "conservative" in its motivations.

CHAPTER THREE

# The Trench of Misfortunes

"Mexico is finished," wrote the Jesuit Jerónimo de Mercado to his superiors in Spain four months into the flood of 1629 and after many futile prayers for divine intervention. Everything but the Plaza Mayor, the Plaza del Volador, in the core blocks of the city, and parts of the indigenous quarter of Tlatelolco was under a man's height of water. More than fourteen thousand people had already fled—the better off to other cities and towns; the poor to higher fields beyond or the cathedral construction site, receiving handouts of bread, meat, and firewood that the religious orders bought with municipal monies. The dying received extreme unction from their priests on canoes; the dead in the coffins were covered with lime and left any place that was dry, consecrated or not. Pestilence lurked, as rotting animal carcasses bobbed back and forth in the stagnant water. Although Indians from outside the city profiteered handsomely from selling potable water and waterborne transport, their social betters faced losing it all. Commerce and even minting were at a halt.[1]

Viceroy Gélvez reported that of the imperial city's estimated worth of 125 million pesos in property and incomes from taxes and rents, six million pesos had already been lost and more than two-thirds were in danger if the water rose any further.[2] Urban elites and officials were in what was possibly their worst crisis since the New Laws of 1542. Then, the king had wanted them to give up encomiendas and other privileges of conquest. Now, he wanted them to give up their city: a royal cédula of May 19, 1631, instructed the viceroy to call on the main stake-holders in the capital to consider moving the city altogether to higher ground slightly to the west, out of reach of the lakes.

Unwilling to abandon the built wealth they had amassed, the well-heeled balked. Chastised and contrite, they went back to the Desagüe they had allowed to languish in disrepair. The problem was how to make it work. Looking everywhere for enlightenment, officials turned toward what at the time must have seemed like sunlight breaking through the stormy clouds. Or, quite concretely, sunlight breaking through to the dank and terrifying intestines of the tunnel: the idea to turn it all into an open trench. They would easily and fairly economically accomplish this in a few years, they were told, by harnessing Indian labor and the force of rushing water.

As it would turn out, converting the Desagüe into a single massive open trench would take more than a century and a half. This was not just a consequence of the difficulty of the task itself, but more so of the choices made in the formative early decades of this conversion process, and how these decisions reverberated through time. A variety of practices and understandings in the areas of construction, management, hydrometry and hydromechanics, visual representation, and work were put in place during the first three-quarters of the seventeenth century. Because these practices and understandings were consistent with the priorities of the urban elite and the social relations of power in New Spain, they were strongly reinforced at every turn. Cemented in place in this way, they became the taken-for-granted way of doing things and would prove exceedingly resistant to change.

Members of the religious orders, invited to give their opinion about the Desagüe from the very beginning, enjoyed outsized influence and power during this crucial stage of Desagüe history. Possessing convent libraries and at least some knowledge of natural philosophy and "mathematics" (in any of the multiple meanings the term had in the early seventeenth century) obtained through university or seminary studies, these lettered men of the cloth enjoyed great intellectual prestige among officials and elites. In addition, since they had built the material presence of their respective orders in the New World from scratch with Indian labor, these men were also respected as project designers and managers. This does not mean, of course, that the peculiar combination of text-based and practical knowledge about hydraulics and construction the friars possessed was technologically or objectively superior to that of other groups, only that it was perceived at the time to be so. The fatigue and confusion with all the disputes of the 1615–30 era brought these religious letrado-builders into the spotlight as men who might help the city and its elites out of their benighted situation.

It was a Carmelite who designed the open trench in 1631, convinced authorities of its advantages and feasibility, and participated together with Franciscans in the first stages of its implementation. When in 1637 the Franciscans were put in charge of the works, men of the cloth were simultaneously entrusted and empowered to radically redesign the drainage and implement their ideas and methods of work, acquiring unique control over people, nature, and matter in the Desagüe proper and the entire Desagüe complex such as no other technician after Enrico Martínez and indeed few superintendants ever enjoyed. As a group, the friars infused into the Desagüe specific understandings of matter that had an impact on the kinds of knowledge that was considered relevant to and could coalesce around this hydraulic project, and this had broader implications. Technologically and in their usage of human and nonhuman energy too their imprint was difficult to erase. Remaining at the helm of the project until 1691, these men would have ample time to entrench an enduring legacy in all aspects of the work.

## ANDRÉS DE SAN MIGUEL'S OPEN TRENCH PROPOSAL

The trench conversion's Carmelite designer was Fray Andrés de San Miguel, a native of Medina Sidonia in Andalucía. San Miguel's involvement with the desiccation went back to the days of Enrico Martínez's work on the tunnel, and it is clear he was invited to opine on the basis of his prestige as chief architect for his order's construction work in Mexico. By then, he had built or planned several convents, including the Santo Desierto de Cuajimalpa, for which he designed an acqueduct, and the Convento del Carmen in the capital, whose main building has survived earthquakes and floods, although sadly not its water supply system, now truncated by Avenida Revolución.[3]

When Martínez's tunnel began to fail due to lack of maintenance and the redirection of resources, Viceroy Luis de Velasco called a junta on the site of the Desagüe to study the situation. One of the experts invited to the meeting was San Miguel, together with other Carmelites and Dominicans.[4] Another such junta was called in 1629 by the new viceroy, the Marqués de Cerralvo, immediately after the disastrous September flood; among the twenty-six alternatives to Martínez's design presented during this meeting was San Miguel's.[5] He wrote down the details of this proposal in 1631, as a report on the state of the Desagüe which he submitted to the Padre General of his order, Fray Esteban de San José, hoping it would be forwarded to the king and then come back with royal endorsement. This is the *Relación del sitio, trabajos y estado de la ciudad de México y su remedio.*

His main foil in this text was Enrico Martínez himself, who in San Miguel's view erred in his surveys and in how he carried out the excavation. Martínez, according to San Miguel, began the tunnel "as the most barbarian men in the world might have begun it," without having first conducted test bores to ascertain the consistency of the terrain and the presence of springs. The result had been a tunnel with uneven depths, and "some shafts deeper than others, like pipes of organs."[6] Eventually, San Miguel would blame all this on the authorities who had "taken

Enrico away from his printing press and from making calendars, which was the occupation through which God wished to make him serve Him." As a result, "we ought not to wonder at the mistakes he made, but at those he did not."[7]

His condemnation of Enrico Martínez was wholesale, in other words. Given the alternative of returning the tunnel to its original functioning by measures such as reinforcing the weaker sections of its interior with the same techniques that Enrico Martínez had adapted from practices in mines and architectural vaulting, as others recommended, San Miguel chose to destroy the work of his predecessor and redesign the project entirely. Departing from Martínez's technological flood-control compromise that regulated the amount of water that could reach the city by diverting and storing the Cuautitlan River and northern catchment area waters in Lake Zumpango and evacuating this water just before the rainy season, the Carmelite returned to the old idea of a general desiccation, articulated in 1555 and 1580. This would be done by first transforming the tunnel and its access canal into a single open trench. Then, a cut would be excavated from the edge of the Lake Mexico (the part of Lake Texcoco lapping the edges of the city) to conduct these waters to the existing Vertideros (the spillway where the Cuautitlan River's diverted course and Lake Zumpango's stored water joined the Desagüe trench before the tunnel segment), thenceforth following the same path of the old Desagüe, now opened up as a trench.

The idea itself was not exactly original. Master architects had just presented a proposal for an open trench conversion that also called for raising the city streets. San Miguel represented his own proposal for a single extended open trench to desiccate Lake Texcoco through the Desagüe as an improvement on what the maestros offered, since city streets had already been raised an average of two meters over the previous twenty years, yet the flooding continued "not so much because the water has risen but because the city in its entirety is sinking, and more so with the weight of the water covering it."[8] Instead, San Miguel imagined the open trench as a hypothetical body of trapezoidal cross-sections and increasing depths. This hypothetical body had four segments: from Mexico to the Vertideros; from the Vertideros to Huehuetoca (the beginning of the tunnel); from Huehuetoca to the point at which the terrain rose the highest above the bed of the tunnel; and finally, from this point to the Boca de San Gregorio or exit. The first segment alone was more than twenty-nine kilometers long.

To calculate the total amount of earth that would have to be excavated under his own plan, San Miguel explained how for each of the four sections he computed the area of the trapezoidal cross-sections, averaged them, and finally multiplied this average by the distance, or length of the body, to obtain the volume. Following this procedure, he obtained 2,913,625 "square" varas (1,714,610 cubic meters) for the first section; 886,745 cubic meters for the second; 1,447,660 cubic meters for the third, and 636,970 cubic meters for the fourth. He added them all up in the end, obtaining a grand total of 4,685,985 cubic meters of earth that needed to be dug for an open trench from Mexico to the outlet of the Desagüe.[9] He then repeated the same procedure for the master architects' plan, arriving at a total more than *eight* times the amount of earth that would have to be removed. This huge difference and the apparently lesser cost of San Miguel's plan hinged on giving the entire trench steeper sides and greater longitudinal slope than the guild masters'.

San Miguel's presentation contained far more than a simple comparison of the economies of his plan over that of the architects. He chose to translate into inky, formal-lettered profusion the quick, penciled, and unself-conscious computations that builders routinely made on margins, scraps of paper, wood, or other surfaces. For his own project's second section, for example, he explained his procedure as follows:

> From here in the vertidero to one thousand six hundred varas of the tunnel, over a distance of eight thousand and two hundred and sixty varas, the earth rises evenly, where the earth climbs over that in the vertidero twelve varas, which with the eighteen of the vertidero are thirty, and to these correspond in the width above twelve varas, which with the four [varas] of the floor make sixteen and its half eight, which multiplied by the thirty of depth make two hundred and forty and these with the hundred and eight of the vertidero make three hundred and forty eight; half of that is sixty four. This multiplied by the eight thousand and two hundred and sixty of length add up to a million and five hundred and six thousand and forty square varas.[10]

The choice to include all this in the text may be idiosyncratic to an extent, but when taken in relationship to the image that the friar used to communicate his open trench proposal, the computational detail makes sense: it is used to support a representation of the Desagüe as a simple mathematical and geometrical abstraction. This is well conveyed in Figure 3.1, where he showed the trench in perspective, with its dimensions, but with no other indication about its materiality or emplacement on the terrain or in relation to other structures in the Desagüe. Sitting transparently on the paper as would an architectural plan or an elevation drawing, the trench is a pristine geometrical body with no humans, buildings, streets, dirt, water, or other impurities to muddy it. These features were consistent with San Miguel's effort to convince officials of the advantages of his purportedly simple plan.

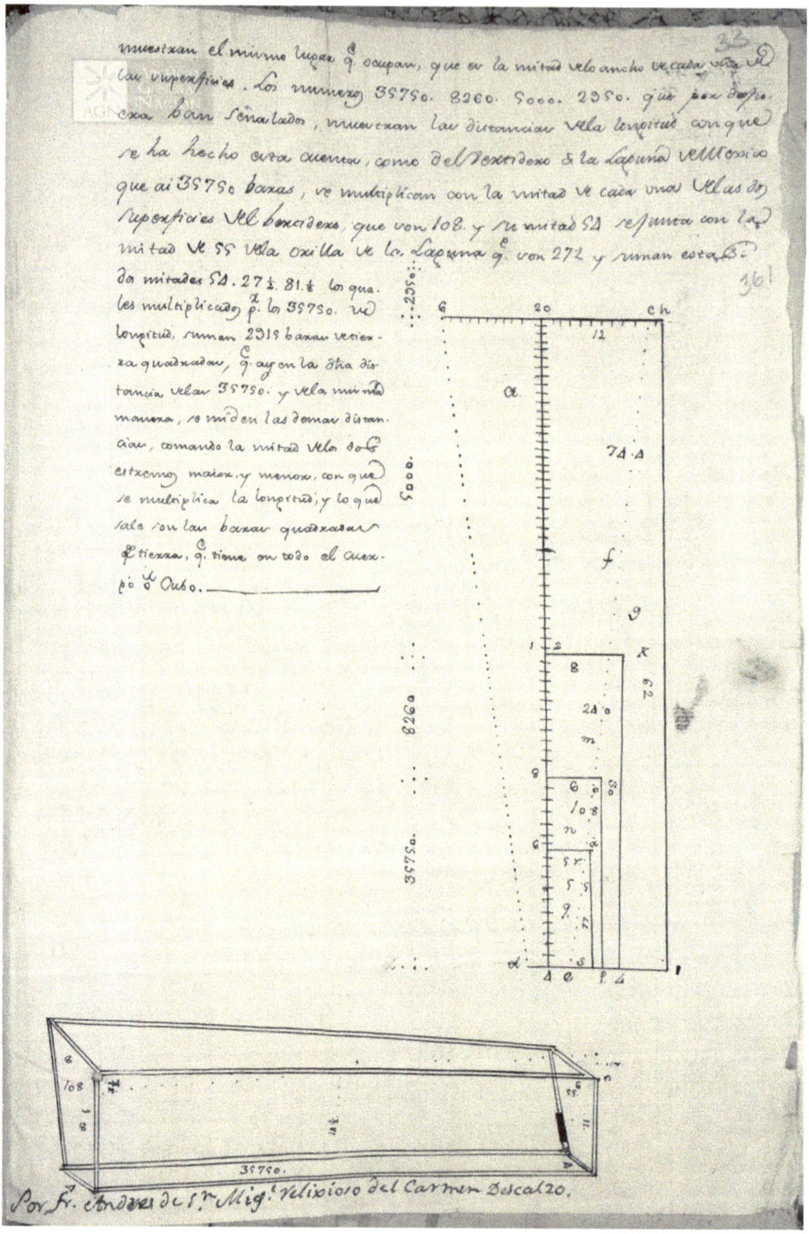
muestran el mismo lugar q. ocupan, que es la mitad de lo ancho de cada una de
las superficies. Los numeros 35750. 8260. 5000. 2350. que por despa-
cio ban señalados, muestran las distancias de la longitud con que
se ha hecho esta cuenta, como del Derridero à la Laguna de Mexico
que ai 35750 baras, se multiplican con la mitad de cada una de las dos
superficies del Derridero, que son 108. y su mitad 54 se junta con la
mitad de 55 de la orilla de la Laguna q. son 272 y suman estas dos
da mitades 54. 27½. 81.½ los qua-
les multiplicados p. los 35750. de
longitud, suman 2915 baras de tier-
ra quadradas, q. ay en la dha dis-
tancia de las 35750. y de la misma
manera, se miden las demas distan-
cias, tomando la mitad de los dos
extremos maior, y menor, con que
se multiplica la longitud; y lo que
sale son las baras quadradas
de tierra, q. tiene en todo el cuer-
po de Oubo.
Por Fr. Andres de Sn. Migl. Relixioso del Carmen Descalzo.

*Figure 3.1.* Fray Andrés de San Miguel's translucent open trench.

SOURCE: AGN, Desagüe, vol. 3, exp. 6, f. 361. Reprinted by permission.

### THE ACCEPTANCE OF SAN MIGUEL'S PROPOSALS

In the midst of the calamitous multiyear flooding that started in 1629, the idea of converting the tunnel into an open trench from the Vertideros to the Boca de San Gregorio was appealing indeed. San Miguel's proposal to extend the open trench to Lake Mexico itself was left open and hung as a backdrop in all future discussions. But his idea to destroy the tunnel and create a single massive trench to replace it gained traction. Why? It has been claimed that when the earthquake of January 22, 1637, brought down parts of Martínez's vaulting, the viceroy decided right then and there to get the trench going.[11] This serendipitous coincidence could not but help in prodding authorities along. But it does not explain what made them receptive in the first place.

With the city literally mired in floodwaters and after reviewing a multitude of contradictory proposals about what to do about the situation and the Desagüe, confusion spread among royal and city officials. San Miguel's idea must have seemed like a simple and definitive end to their hydraulic nightmares and, most importantly, would use methods of work that translated into urban and imperial savings both on repartimiento labor and on funds. Because it addressed none of the complications that would soon ensue from the populations, activities, geographic accidents, and other obstacles that lay in its path, the Carmelite's design allowed the Desagüe to be reimagined, after two decades of headaches, as a comparatively clean cut in the terrain. The evidence corroborates the later opinion of Francisco de Garay, the engineer who in 1856 redesigned the drainage route through Tequixquiac and supervised the construction of the Desagüe del Valle de México the following decade. Garay stated that the open trench was certainly an engineering mistake but must have made rational sense to the colonists:

> To make a tunnel with shafts, an up-front expenditure in currency was required to pay for the masonry work, the machines, the animals. In the trench everything could be provided by muscle power and this is the reason why this is the option that was approved; not because it was good or scientific, nor even economical, but because under the circumstances it was regarded as the only one possible. It was not even humane; on the contrary, it was cruel.[12]

Subsequent writers tended to accept the open trench redesign and its implementation by human muscle power alone as the only possible course of action, approaching the issue as a moral problem that measures the universal good of the nation and progress toward capitalist development against the regrettable but ultimately necessary and unavoidable evil side-effects of advancing both. However, it turns out that the people who chose to implement Fray Andrés de San Miguel's proposal did not have universal welfare in mind: like all individuals, they belonged to social classes and groups, and chose at least in part based on what seemed best for their class, while loading further costs upon subject populations. The social, economic, and ecological costs of the open trench decision and the mechanisms by which they were shifted from one group to another will be examined later. For the moment, it is important to note that the effects of these measures on those who bore their costs were both significant and evident to all. By contrast, although these decisions also had significant long-term negative impact on the Desagüe and on the elites themselves, this could not possibly be foreseen at the time.

In formulating his proposal, San Miguel recognized that a key problem in opening up the tunnel would be what to do with the earth removed from over it, especially as the excavation deepened. The amount of energy that would be needed to haul the earth out of the excavation was of such a magnitude that it had to be addressed if the project was to be considered at all. It should be recalled that this was the trough of population loss in New Spain, so the availability of human labor as a source of this energy was a factor in technological choices. To reduce the dependence on human energy for the construction of the tunnel, Martínez had used animal-powered hoists at each vertical shaft or *lumbrera* to extract debris, in the manner of the mining industry, as shown in Figure 3.2.[13] This type of hoist could be adapted for a variety of uses, including drainage, lowering wood beams, and raising workers and debris, with animals providing the energy. But the friars who succeeded Martínez do not seem to have prioritized alleviating the demand for human labor with more animals, hoists, or other simple machinery.

Beasts of burden do not appear to have been considered as the primary means to either power hoists or drag the debris out on ramps. This is intriguing: technicians and authorities knew that mules were five times more efficient at moving loads than humans.[14] In its recommendations for the pursuance of the open trench project, for instance, the junta of 1630 called for twenty-eight hundred oxen and six thousand mules to supplement the work of three thousand Indians.[15] When asked in 1648 to raise the causeway of San Cristóbal using tepetate as a filler, maestro Juan Serrano proposed using mules to haul it to the site "to save His Majesty the outlay of Indians and money and because the work of each mule is equal to that of six Indians."[16] The rare references to animals in the Desagüe record permit only an approximate explanation for their dearth and of their use. Cost must have been the foremost consideration: while the price of mules had been declining steadily since the 1530s, it was still an up-front full payment for the animal's potential future labor.[17] Additionally, in the Desagüe district grasses on *tierras realengas* or *tierras baldías* were reserved for repairs of earthworks in the public work, and

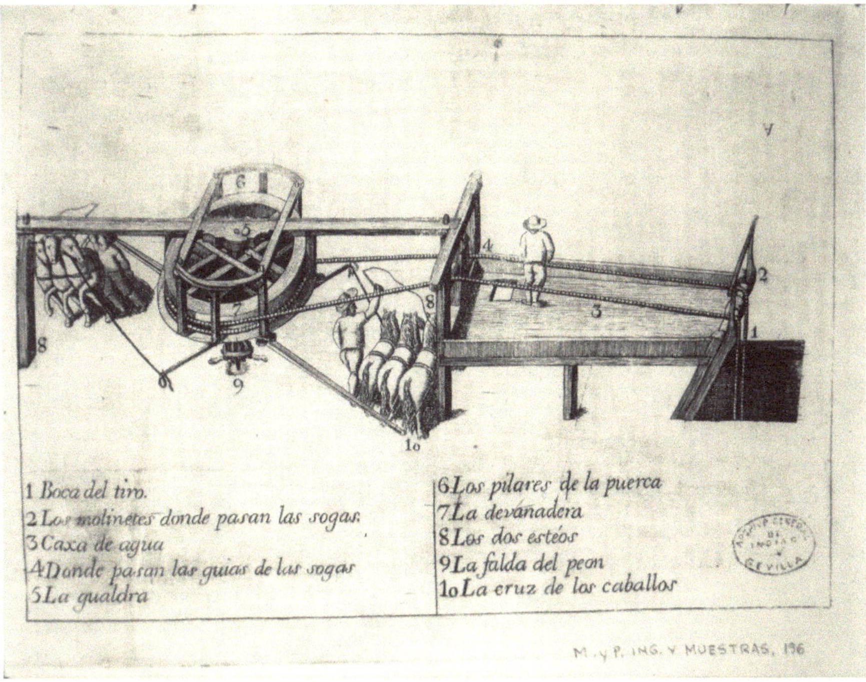


*Figure 3.2.* Multiuse hoist, probably similar to those used under Enrico Martínez.
SOURCE: España, Ministerio de Educación, Cultura y Deporte. AGI, MP-Ingenios, 196. Reprinted by permission.

those belonging to indigenous townships were at least in theory protected against outside livestock, so access to pastures would have been limited and the price of forage high, especially given the drier and less fertile pasturelands of the district. In other words, a mule was more expensive to acquire and maintain than a human because although the Desagüe provided the latter with rations, Indians supplemented their caloric needs at their own expense and only their travel time and actual labor on the Desagüe was paid, and minimally at that. This preference for human labor had a history and a logic: in the very beginning of Spanish settlement, entrepreneurs chose Indian porters (*tlamemes*) over beasts of burden for transport because thanks to coercion and lax enforcement of pay regulations they cost less per load unit over distance.[18] In all enterprises since then, the Desagüe included, the effect of this type of choice was an indirect energy subsidy of the Spanish economy by the Indian one.

Whatever the case, San Miguel proposed to solve the problem of the considerable energy needs of the project by combining human energy from repartimiento labor with mechanical energy from rushing water.

Indians would begin by digging up the earth overlying the tunnel following the line of the shafts and letting the debris fall down each shaft; as they turned this excavation into a trench, they would pry the rock and earth off the sides, also letting it all fall. Crews along the bottom of the tunnel and trench would break up these materials as they piled up. Rushing water would perform the rest of the work by sweeping everything away.[19] Superintendant Juan de Cevicos had this method of work tested shortly after San Miguel proposed it. For four days earth was excavated and thrown down the shafts at the treacherous La Guiñada segment of the tunnel, the highest part of the hill and thus the most difficult portion of the open trench conversion, and water was released to sweep it away. At the end of the experiment, the foremen went into the tunnel to see whether the water had been able to carry away the dirt: they found it clean. The success of the test, boasted San Miguel, brought the viceroy and his eminent entourage to the Desagüe to delight in performing the experiment themselves.[20] This work procedure was adopted together with the open trench redesign.

But San Miguel's proposed method had objective problems. He seemed not to remember that in 1611 Enrico Martínez had tried using the mechanical energy of water to sweep away debris from the interior of the tunnel as he was deepening it. At the time, this had been an expedient that Martínez had resorted to with great regret, due to the ill effects it caused on the works, only after his detractors had blocked his repartimiento labor allotments.[21] Besides the inevitable erosion of the faces of the trench this method provoked as the water rushed through, it had other negative implications.

Differences in the composition of the soil and rock would become an obstacle for excavation of the open trench just as they had been for Martínez in 1608. In Enrico Martínez's time, two types of problems had afflicted the works—those specific to the tunnel and those specific to the canals or open trenches. San Miguel understood these problems: the tunnel was weakened by the fact that the ceiling was generally not reinforced by a mortar barrel vault but by arches placed at intervals. These sustained utterly different kinds of loads, from the weight of the hill at La Guiñada to that of only a couple of meters of earth closer to mouth of the Desagüe. The tunnel was therefore structurally weak, and the unvaulted and unpropped ceiling areas crumbled easily. In canal portions, even in Martínez's time, the problem was the erosion and crumbling of the talus on the sides which silted and diminished the carrying capacity of the canal.[22] Nor was opening up the tunnel as a trench a matter of merely digging: excavations in sections of softer rock could cause the tunnel to cave in, leading to dangerous stoppages along the watercourse. Lateral supports had to be secured while earth was removed by hand or by water and extant tunnel parts had to be protected with masonry to withstand the greater "volume" (discharge) of water from sections already converted to trench as it crashed against the constricted tunnel entrances and accelerated within.[23]

At the time he wrote the open trench proposal, however, San Miguel considered none of the foreseeable problems in its execution. The proposal did not suffer for it in large part because while it was under consideration the Carmelite was asked to visit the works, allowing him to improve on his ideas. In the dry season of 1635, for example, he was sent with two maestros and the elderly Franciscan Sebastián de Garibay to assist Superintendant Juan de Cervantes Casaus in inspecting the newly built vertidero for Lake Zumpango, the existing trench portions, and the tunnel vaulting and the tunnel side supports at La Guiñada and El Tunal. Again, in the dry season of 1636, he accompanied the superintendant, Adrian Boot, and three maestros into the tunnel to recommend further repairs and wood supports for the weak tepetate portions.[24] This practical and regular onsite intervention certainly lent San Miguel increasing credibility as a man intimate with the Desagüe's interior. It also allowed him to make timely amendments to his open trench solution without compromising the essential features of the plan.

In his 1636–37 report, San Miguel stated that not everything that was pried away during the excavation could be thrown down the shafts because the ground was "varied and of many mixes, because it has some banks of black earth, others of clay, others of hard tepetate, others of soft tepetate and others of soapstone, and other differences." For the water to be able to sweep the debris, "earth, clay, soft tepetate and the like" could be thrown in its path even in large chunks, but hard tepetate had to be broken up into smaller pieces.[25] It is difficult to reconstruct how San Miguel imagined the workers would proceed, as there is no explanation in the text. We know that the Indians had no other equipment than ropes suspending them from the ledge above and iron bars to both pry and strike the walls of the widening trench, but how once thus suspended without scaffolding the workers solved the problem of using the bars as levers without being able to use their own weight, or how they struck without bouncing away from the wall remains unclear.

Without addressing the problem of a lack of fixed pivot or support for each suspended worker, as well as the difficulties of varying hardness, San Miguel offered the solution that once the open trench was put into practice, diggers hitting layers of hard tepetate between two shafts could simply work their way around and beneath the hard material, leaving the decision of whether to remove the offending layers for an undetermined point thereafter. Thus all the weak terrain would be excavated and dropped onto the path of the water in the old tunnel below, with the hardest materials possibly being pried off in the dry season, when workers would descend to break it all up and deepen the bed as much as possible.[26] This was the most important modification that San Miguel made to his original 1631 proposal, and while it certainly helped its approval along it did not alter the fundamental principles behind it. After a swift series of consultations, on July 17, 1637, a junta general resolved to put the open trench proposal to execution.[27] This decision seems momentous in retrospect, but it seems to have been taken comparatively quickly and easily.

There are several factors that make this favorable decision rather surprising. Technically speaking, for all Andrés de San Miguel's number dexterity and capacity for geometrical abstraction, his proposal was in no way an improvement on Martínez's. Quite the contrary: if we were to agree with a liberal view of technology as something that should "advance," then the friar's design and method of work was a technological retreat relative to that of his predecessors. Having pushed tunnel and adit construction from mining technology into a new realm, combining it with architecture,[28] Martínez was technologically "progressive" in this sense, regardless of whether his choice was dictated by labor availability amid plummeting indigenous populations not yet compensated for by migration or natural reproduction among other groups or by his own considerations of efficiency: he had also used machinery to assist in the construction process, in effect replacing direct human or animal energy with mechanical multipliers of work.

In fact, contemporary builders and engineers actually believed that more machines and capital-intensive methods were preferable to labor-intensive methods. Military engineer Adrian Boot's 1615 proposal articulated his opposition to Martínez's Desagüe not by rejecting capital-intensive methods but by drawing from a different technological tradition (that of the Low Countries' polders) and redeploying it to manage, rather than eliminate, the water. Boot's proposal had suggested the use of cranes to lift canoe traffic over closed sluices, pumps to eliminate water from seepage or springs beneath the dams, iron gratings to keep the urban canals clean, and other unspecified diverse "instruments" of iron and wood.[29] Masters of architecture who inspected the Desagüe when Martínez was excavating the tunnel felt that "insofar as removing the earth, the method seen was very good and for the sake of abbreviating the work an additional machine [a hoist] could be placed in each shaft, moved by beasts, which saved much on cost and the employment of people."[30] Animal- or water-powered labor-saving machines and energy-multiplying devices much like those used by Juan Herrera in the construction of the Escorial were ubiquitous in construction sites throughout the viceroyalty, as can be seen in paintings of the period such as Antonio Ramírez de Montúfar's 1678 *Construcción de la Catedral de Guatemala*.[31]

What recommended San Miguel's proposal in the eyes of contemporaries, therefore, was not any objective technological "superiority." Does this mean that authorities acted irrationally or impulsively in favoring the Carmelite's ideas? If, following the same progressive view of technology, efficiency, technical merit, and performance were objective factors inherent to the things humans make to transform or move matter, then this technological choice might appear irrational because none of these three elements are optimized in San Miguel's proposal. But since these three factors and rationality itself are culturally and socially contingent, then the answer is "no." The open trench design, being "inefficient" technologically speaking (it demanded more energy, time, and matter than required), prevailed over one which might seem superior to us because was it was perfectly consonant socially and culturally on three counts. It was consistent with the imperial and urban-centered conceptualization of the Desagüe that letrados helped articulate, with what all these sectors considered a priority at the time, and even with the demographic and economic context of the works. It also helped that the proposal came from a man who personified the union of two exemplary groups of loyal servants of the urban elites: San Miguel presented himself and indeed was both an architect and a letrado. For this reason, it is useful to explore this friar's cognitive universe.

## SAN MIGUEL'S GEOMETRIC CONCEPTUAL UNIVERSE

San Miguel was not the only lettered man of the cloth to serve in the Desagüe. Carmelite, Franciscan, Jesuit, or Mercedarian—men from all of these orders would at some point or another play a role in it as well. One of the things they all had in common was their usage of geometry as a tool in their New World ventures. With geometry they structured spaces and rituals for the conversion of Indians, organized convent life, and created architectural styles that would help the faithful distinguish one order's church from another's.[32] Geometry was thus a habit of mind for these men, a way of thinking about worldly matters that had a deep impact on the manner in which they intervened in the Desagüe.

Among these orders there were also important differences in the meaning and use of geometrical conceptions. Unlike their fellow religious consultants in the Desagüe, the Carmelites were involved neither in the Christianization of Indians nor in the higher education of the elite youths. Their focus was on the moral regeneration of those already confirmed in the faith in urban centers and on the moral rigor, austerity, mortification, and self-abnegation of their own friars. For them, geometry helped sustain the strict and austere spiritual and physical principles of the order as expressed in the Constitutions of Santa Teresa. These constitutions stipulated that the convents of the order had to be designed by Carmelite draftsmen-architects, so the order encouraged the formation of draftsmen and builders among its ranks.[33] San Miguel became one of these men. As the foremost architect for his order as it expanded in New Spain in the first decades of the seventeenth century, San Miguel advocated a strict adherence to the use of straight lines and a stark geometry that reverberates in his Desagüe ideas.[34]

For Fray Andrés de San Miguel, God's cosmic order was reflected in geometric proportion and mathematical law. He expressed this belief through his activity as his order's draftsman-architect and as an architectural treatise writer, but these notions were actually embedded in the architecture of all religious orders, which linked the practical and the ideal through the forms and relationships of geometry, mathematical proportion, and arithmetic. Religious architecture could only deepen the connection between religion and dimensions of mathematics that was already embedded in Christian belief and its interpretation of a cosmic order, and indeed this very idea was expressed in several writings in New Spain at the time.[35] San Miguel's radical proposals for the Desagüe must therefore be seen as deeply rooted in this conceptual universe. But the actual implementation of these proposals came about in part because this universe and the habit of resorting to geometrical abstraction to solve problems were not unique to him. They were in fact shared by both the religious and the secular men who influenced the Desagüe with their opinions and decisions.

In contrast with other orders who worked more freely with architects trained in the guild traditions, however, the Carmelites had a self-enclosed practice in which the only outsiders were the skilled and unskilled laborers (*oficiales* and *peones*). The Carmelite construction sites directed by San Miguel operated in this way. To erect his order's college at Tenanitla in 1615, for instance, he hired no outside architects and closely supervised carpenters, masons, and other artisans from outside the order as they carried out his design.[36] Since collaboration with secular architects was discouraged by the rules of the order, Carmelite architects like San Miguel learned primarily from their predecessors in the order or from architectural treatises, and indeed, scholars have identified several textual influences in San Miguel's written work, including many of the treatises circulating in the Iberian realm at the time.[37]

San Miguel left three writings that express his thoughts on architecture, perspective, geometry, and arithmetic. Two of them demonstrate how he directed these thoughts to the problems of the drainage, but they were also clear attempts at self-promotion. In the *Relación del sitio, trabajos y estado de la ciudad de México y su remedio*, written right after the 1629 flood, he proposed the open trench solution, while the 1637 *Informe dado al Virrey Marqués de Cadereyta acerca del Desagüe de Huehuetoca* was a progress report of sorts. The third text is a well-known illustrated manuscript on a variety of topics in architecture. Possibly begun before his intervention in the Desagüe and continued thereafter, during his career as his order's architect, San Miguel's treatise illuminates his conceptual universe and how water, earth, and the material world fit into it. It helps explain how he looked at the Desagüe overall and his and others' role in it.

In the treatise, San Miguel's view of himself as an architect depends heavily on the Vitruvian differentiation between the "mechanic," or skilled manual laborer who executed plans (the *oficial* in New Spain) and the artist architect (the *maestro*). As expressed by one of the Carmelite's sources, Diego de Sagredo, in this hierarchy stonecutters, silversmiths, carpenters, bell makers, and others who worked with their hands and ingenuity (*ingenio*) were "*oficiales mecánicos*" ("mechanics" in an English-language equivalent), but those who worked "only with the spirit and ingenuity" were the liberals, as were the men who worked with grammar, logic, rhetoric, music, arithmetic, geometry, and astrology, such as painters and sculptors.[38] Architects, according to Vitruvius, were "obliged to be masters in the sciences of philosophy and the liberal arts." Lacking these attributes, they "cannot be perfect architects: whose tools are the hands of the mechanics." Lastly, the good architect above all must be armed with geometry: his models having begun their works with Euclidean geometry, so did San Miguel, including the basics of planes, the properties of points, lines, and angles, the methods for computation of volumes of different bodies, and how to draw them correctly, measure, divide, and calculate their dimensions according to the rules of perspective.[39]

For all its imitation, San Miguel's text differed from his European models in ways consistent with its author's identity and aims. This coherence is exemplified in a section that to modern eyes might seem "delirious."[40] In it, San Miguel praised the Mother of God by calculating how many cabbage seeds would fit into the terrestrial sphere as compared to the exact number of graces Mary possessed, on the basis of an arithmetical progression:

> In the first act [she] already earned as much grace as that of two of the highest seraphim, and with the second act [she] earned as much grace as that of four seraphim and in the third act as much as that of eight seraphs and in the fourth as much as sixteen seraphim and deserving always with each one of the acts as much grace as she already possessed and this is something like one thousand and twenty-four seraphim.[41]

And so on, until "with the last of the first hundred acts she earned as much grace as one million and 957 millions of millions and 488 thousand 248 million and 937 thousand 424 million and 983 thousand 768 seraphim."[42]

This arithmetical juggling illuminates the logic behind San Miguel's choices in his open trench proposal. While his largest contribution to his order was as a practical builder, by penning a treatise San Miguel could partake in the privileges of the lettered city, augmenting his qualifications with the precious disciplines of a Vitruvian artist. In this light, his decision to demonstrate his confidence in arithmetic and geometry in both open trench proposal and treatise seem strategic: they could elevate him high above the oficiales mecánicos and both justify and strengthen his authority on important matters in the kingdom such as the Desagüe. San Miguel managed to convince contemporaries that the open trench and the means of executing it were viable in large part thanks to his prestige as an architect-letrado, a man who built material structures with celestial principles transmitted through a long textual tradition that began with the Bible and went through the treatise writers of the Renaissance.

## IMPLEMENTING SAN MIGUEL'S PLAN: FRIARS AS DESAGÜE SUPERINTENDANTS

The crisis caused by the Desagüe's failure amid the 1629 flood that had opened the door to Andrés de San Miguel's radical redesign of the drainage also ushered in a technological and managerial reorganization. When the drainage began, like other projects where the crown was involved, an oidor was assigned to oversee things. All technicians, from Enrico Martínez and Adrian Boot to friar consultants and maestros, were subject to the authority of these commissioned oidores. However, there was no "executive" director to speak of with specific guidelines and clear powers, and in fact it appears that the viceroy simply sought the ideal individual among either members of the real audiencia or of the city council. The viceroy often chose these men to ensure that the execution of any major decisions about the drainage was made, as was customary, in consultation among the audiencia, the city government, and often too the major religious orders, the church, and the consulado. In this manner, it was not a single person that looked after the Desagüe over time. The Desagüe was run in this consultative and somewhat decentralized manner from its inception until 1631.[43] But in 1630, Viceroy Rodrigo Pacheco de Osorio, Marqués de Cerralvo, decided to give the Desagüe a superexecutive—a superintendancy.

This made sense. Precisely because of the fluidity of water, hydraulic projects such as the Desagüe were entirely unlike other public works in that they affected and were affected by the hydrology of rivers and water bodies that connected over hundreds of miles. To run projects of the complexity of the Desagüe, such as the Dutch and Italian water-control systems, required extensive powers, both territorially and jurisdictionally. European princes otherwise quite zealously protective of their own domains could concede overarching regulatory, judicial, and executive powers either collectively to *waterschappen* (water boards) or individually to *sovrintendenti alle acque* (water superintendants).[44] The jurisdiction of the Desagüe's superintendant transgressed district boundaries and its administration straddled royal and city prerogatives, albeit constrained from time to time by the consultative traditions of the Spanish monarchy.[45]

In the seventeenth century, a superintendancy was considered less a specific position than a "manner of exerting authority," so that officials occupying the post oversaw several activities and subordinates in a given geographic area, where they also enjoyed the right to administer justice.[46] Like a mini-viceroy, a Desagüe superintendant could order whatever was needed for completion of the works. With viceregal approval, he had the power to order a magistrate (*ministro con vara*) to summon repartimiento labor from Indian pueblos. Technicians reported to him. So did civil authorities of the jurisdictions with Indian towns subject to the labor draft. The superintendant's jurisdiction even replaced that of the local judges (*justicias ordinarias*), who were henceforth legally inhibited from meddling in Desagüe affairs (although they would continue to do so whenever local interests and power arrangements were touched). The superintendants thus acquired ex-officio jurisdiction over any civil or criminal cases that might arise between parties pertaining to or involving ministers, officials, or other people in Desagüe works. For all this, they got two hundred pesos a month to cover their expenses.[47]

It also made sense for an oidor to occupy this post, given these audiencia members' broad powers and territorial jurisdiction. And that is what happened between 1631 and 1637, when oidores Juan de Villanova Zubiaurre, Juan de Cevicos, and Juan de Cervantes served in short successive order. They were assisted by a *guarda mayor* (general warden) who resided on-site in the Desagüe headquarters in Huehuetoca and ran the daily affairs of the works assisted by three *guardas menores* who lived next to the structure each was assigned to supervise. A nonresident *maestro mayor* and a notary completed the team.

But this was apparently just a temporary expedient and not what the viceroy had in mind. When he created the superintendancy, Cerralvo actually indicated that the post should be occupied by someone from a religious order. His successor, the Marqués de Cadereyta, fulfilled this expectation on July 20, 1637, bestowing the post upon Fray Luis Flores, a member of one of the most powerful religious orders, the Franciscans. This was just in time for Flores to supervise the beginning of San Miguel's open trench conversion.

As a Spanish Franciscan who had arrived in New Spain in 1633 and became commissary general of the order's Provincia del Santo Evangelio in central Mexico, Flores obviously had administrative and leadership capabilities.[48] But so did the oidores, who after all helped run an entire territory. So why this turn-around? It was not only a reversal of the practice between 1631 and 1637, but, more seriously, it was at odds with longstanding and stated royal intentions to rein in the power of the church and most particularly that of its wealthy New World regular clergy, a campaign that went back to the 1574 *Ordenanzas de Patronazgo*. In an effort to undermine the religious orders who had accumulated land, wealth, Indians, and privileges in the New World, these ordinances had reduced the tax exemptions of the regular clergy (their vast properties paid no tithe, for example, despite the fact that the crown used this income to support the clergy) and begun to dislodge the orders from indigenous parishes, which could instead give employment as priests to droves of secular letrados that the universities continued to supply in excess of actual demand.[49]

Cerralvo's and Cadereyta's logic seems to have been grounded firmly in the healthy traditions of "*obedezco pero no cumplo*," as they felt a religious superintendant would assure the "greater fidelity, lesser cost and better treatment for the Indians."[50] Cadereyta's choice of not just a friar but a Franciscan—a member of the order specializing in Indian conversion and rule in the Mexican heartland—can thus be understood as a choice for perceived efficiency, cost-effectiveness, organizational ability, and incorruptibility, which, combined with the perceived mathematical and technical expertise of an Andrés de San Miguel, seemed to all point toward men of the cloth. This was all much more than a matter of perception, as the religious orders that these men belonged to had an established reputation for the hydraulic engineering skills accumulated over centuries of building projects on both sides of the Atlantic to supply their large monastic establishments with water for drinking, mill operation, fishing ponds, waterborne transport of stones for construction, waste disposal, and washing, as well as to control runoff and drainage.[51] Friar superintendants could adapt all this technology for the Desagüe, often quite ingeniously. Take monastic hygiene, for example, where control of water permitted facilities such as latrine blocks (*reredorters*) that were swept clean by water flushes and basins (*lavatoria*) for washing.[52] This notion of flushing with water, applied in Carmelite monasteries in New Spain, would make its way to the heart of the friars' redesign of the drainage.

Much as they had done in colonizing Islamic Iberia, as they boarded ships to their New World destinations, members of religious orders thus carried far more than their breviaries, expectations, and textual training. As lettered men with scholarly training and access to libraries, the engineering friars in both the Old and New Worlds were in the unique position of being intermediaries between the learned realm of texts and the practical world. They linked the material realm and the ideal realm of reflection in the landscapes they transformed through their engineering and among the various populations who originated or developed the technologies they needed, and they did so less through formal writings than by transmitting and implementing tacit knowledge and reshaping matter.

By the time they were tapped for Desagüe service, in terms of their ability to synthesize and deploy these different types of knowledge, all religious men had the benefit not just of the collective legacy of their orders in Europe, but also of having had to erect the material foundation of Christianity at the same time as the spiritual, and to think hard about how to arrange and order space, how to build, in which shape, with what materials and at what cost, and so on, all to suit the evangelical mission.[53] As a result, when they chose men they considered to be experts in geometrical and mathematical operations to run the drainage, authorities were unwittingly also turning the project over to a specific manner of thinking about space and matter that would deeply affect the design of the Desagüe.

## SAVING MONEY, LABOR, AND SOULS

There were more pedestrian reasons for calling upon the religious orders and, specifically, putting Franciscans in the powerful position of superintendants. These reasons had to do with the steep financial demands the Desagüe had already made—and threatened to further make—and the managerial challenges of the huge project. The friars would become the pin that tied all cost-saving and managerial measures together.

Cost-saving measures were articulated by an assembly of notables (*junta de notables*) of the provincials of the orders, the rector of the university, the prior of the consulado, ten maestros, and consultants assembled in 1630 to evaluate the Desagüe. This junta sent the viceroy four groups of proposals to reduce the cost of the Desagüe for the vecinos of the City of Mexico and the royal exchequer by spreading them out beyond the boundaries of the city and bringing in the religious orders' resources. The proposals were not enacted all at once, but rather adopted over time. While the specific provisions in each of these four categories of cost-saving measures varied, the principles remained the same, as did the drive of the vecinos of the City of Mexico to minimize their share of the costs. The burden of the twenty-five pesos in taxes that since 1629 merchants had to pay on every pipa of wine they disembarked at Veracruz, for example, was actually passed on to the entire population of the viceroyalty in the form of higher prices.[54] The weight of labor costs, in turn, was to be placed on the backs of Indians, who would have to contribute in-kind and personal service. That of materials, on Spanish agricultural enterprises. Estates in Chapultepec, San Cristóbal, Chalco, and Tepotzotlán, for instance, were asked to donate the work of sixteen hundred oxen harnessed for plows and driven by hired Indians, while those distant Oaxaca, Atlixco, Tlaxcala, Cholula, and Veracruz would send two hundred equipped carts—and all equipment would be returned in whatever state it was after Desagüe usage. Corn to feed the workers was also "requested" in the bishoprics of Tlaxcala, Michoacán, and Oaxaca.[55] Small-scale subcontracting or job-work (*trabajo a destajo*) would also serve to lower costs. This was common in public works in the early modern period, including the Canal de Castilla.[56] Although contracting out by auction or as job-work did not exist within the most technologically challenging features of Desagüe proper, it was tolerated in structures that demanded digging only.

Finally, the junta's last group of proposals called for replacing the paid secular supervisors of the Desagüe, who received very respectable compensation for their commissioned work on the project, with members of the religious orders, who would serve for free, out of duty.[57] As if the experience from the hydraulic works of friars during the aftermath of the 1555 flood had not been enough to prove just how much cheaper their management was, the cost savings were thoroughly confirmed when in 1635 the cabildo was faced with the need to clean out the urban canals filled with debris of all sorts left behind by the 1629 flood. The job was put up for auction (*a remate por pregón*) to private contractors, none of whom bid for under 140,000 pesos. Viceroy Cadereyta intervened, naming Fray Luis Flores to supervise the work, with the senior oidor Juan Alvarez Serrano to oversee things. Flores mobilized another twenty-three members of his order to assist him without pay in directing Indians, who were organized by town of origin and assigned canal sections to dig out the muck *a destajo*. The reduction in cost to the exchequer (although certainly not to the Indians) was significant: twenty thousand Indians worked at two reales per day to clean more than twenty-five kilometers of *acequias* for a total cost of thirty-four thousand pesos.[58] In this manner authorities confirmed what they suspected all along—that one of the most effective and sustainable ways of saving money and yet remaining efficient was to replace civil authorities on salaries with members of the religious orders as directors of public work. This allowed them to even save themselves the time and effort (hence money) of coordinating all three of the other measures proposed by the 1630 junta, since the friars would deal with subcontracting for portions of the work,[59] Indian labor, and the reception of donations and contributions directly on the site.

Resorting to men of the cloth, moreover, could also help in the acquisition and management of labor for the Desagüe works, problematic matters in large labor-intensive public works that depended on the indigenous labor drafts (*repartimiento*) that were then being phased out. Orders were expected to harness the labor of Indians from their parishes (*doctrinas*) for some of the projects, as they already did for their own huge construction projects. Workers obtained through the normal channels of repartimiento requisitioning of submitting requests to the draft magistrate (*juez repartidor*) had to be paid—minimally, but still paid. This is not to say that the Desagüe henceforth relied on unpaid labor from parishes under regular clergy in its vicinity, as this would be insufficient and unwise; rather, it points to the fact that in the minds of those making the decisions, labor procurement and management and friar supervision were intertwined. When in 1637, Viceroy Cadereyta asked the corregidores and alcaldes mayores of jurisdictions liable for Desagüe labor to promptly comply with repartimiento summons, he also called on all regular and secular clergy in the corresponding parishes to collaborate. Moreover, Fray Flores was to draw up a plan for repartimiento drafts by consulting with the heads of his order's convents throughout the kingdom for available sources.[60]

The numerous and constantly changing workforce for the Desagüe was made up of both Nahuatl and Otomí speakers. As a member of a nonmendicant order (the Carmelites did not have Indian doctrinas) San Miguel had nothing to offer in terms of savvy management of and access to vast numbers of Indian laborers for his project, much as his other talents were appreciated. The Franciscans, in contrast, had plenty of both in the region and beyond, as they had convents in San Cristóbal Ecatepec and Cuautitlan, both critical locations in the drainage works.[61] This did not mean it had to be a member of that particular convent that became superintendant, as other qualities were necessary, but that proximity to the works gave the order a specific advantage in administrative and "ethnographic" terms. It is clear, then, that their access to numerous and geographically spread out Indian populations in the region cast the Franciscans as better directors of public works than other orders and secular authorities, at least for the time being.

For laborers in the Desagüe, a repartimiento summons was a curse to be avoided at all costs, regardless of who supervised them. It was dangerous work: in La Guiñada, the deepest and most deadly segment of the open trench, locations were even nicknamed "La Boca del Infierno" (Hell's Mouth), "El Purgatorio" (Purgatory), "El Credo en la Boca" (The Credo on the Lips), and "El Mal País" (The Badlands). Workers referred to the open trench as a whole as "El Tajo de las Desagracias" (The Trench of Misfortunes).[62] The Desagüe provided rations of maize, salt, chilis, and lard,[63] but Indians often spent their meager repartimiento wage on other victuals and *pulque* sold illegally by overseers. Nor did it help the reputation of Desagüe labor among Indians that from 1607 until the late 1630s convicts and vagabonds of all ethnicities were sent there as forced penal labor.[64] The friars' organizational and motivational skills thus helped bring, keep, and encourage a labor force under circumstances that were difficult to say the least.

Summoning a labor force suitable for the enormous tasks of Desagüe work required either wages competitive enough to outweigh these infamous working conditions, or coercion. These narrow options resulted from the combination of the dramatically reduced indigenous populations, increasing demand in mining, haciendas, and obrajes for the labor of those who remained, and the growing role of the market in the distribution of these workers. Because of their increasing reliance on privately hired Indian gañanes rather than repartimiento, hacendados and other labradores in the Valley of Mexico were serious competitors for labor precisely at the time the Desagüe needed it most.[65] So obtaining a labor force as large as the public work required was not easy. The number of workers varied with the urgency and size of the works, but in 1616 there were in all ten thousand laborers and the daily workforce by the beginning of the open trench conversion could reach nine hundred people.[66] The vecinos of the City of Mexico agreed that they did not desire to tax themselves any further in order to secure the funds for a budget large enough to pay acceptably high wages to draw such huge numbers of workers.[67] Coercion, then, was the only solution the authorities were willing and able to contemplate.

It was the need for coercion that made friars even more attractive superintendants. Savvy management of vast, diverse, unhappy, and relocated populations was a sine qua non in the Desagüe, and only members of religious communities had proven their skill at this on such a large scale. Whether with cacique donations of labor from their communities or labor drafts, all religious buildings in the valley had been erected with Indian labor. In the aftermath of the flood of 1555, when repartimiento was beginning for works such as the reconstruction of the pre-Hispanic dike of Nezahualcoyotl, Viceroy Antonio de Mendoza chose the Franciscans to supervise this and other flood-control projects. The choice was obvious, given that the Franciscans controlled the most populous Indian towns. Fray Torquemada, guardian of the monastery at Tlatelolco, supervised fifteen hundred to two thousand Indians daily on the reconstruction of the Mexico-Tepeyac causeway (renamed Guadalupe). His tasks included coordinating the transport of earth and stone needed for filling in the causeway to the site by canoe, and he finished the job in five months. Meanwhile, Fray Gerónimo de Zárate, guardian of Cuauhnahuac (Cuernavaca), supervised equivalent efforts to build another causeway between San Cristóbal and Chiconautla (named San

Cristóbal) to prevent water from the north and northwest from reaching Lake Texcoco. Later, these men were tapped to direct the restoration of the causeways of San Antón (Iztapalapan) and Chapultepec respectively.[68]

So when Viceroy Cadereyta named the first Franciscan superintendant, he spoke convincingly on all the benefits of a friar superintendancy with regard to the correct treatment and payment of Indians and how the friars could ensure the supply of repartimiento tandas. Friars provided savings on other fronts too: they ran the hospital in Huehuetoca, which not only tended to those who became ill in the Desagüe but also translated into another saving for the exchequer because it was the attending brothers' order who paid for it.[69] Even their spiritual services were part of the economy, as they probably helped reduce tensions and induced docility in the workforce in ways inaccessible to secular administration. For this activity, they eventually built a portable chapel, equipped with its bell and adornments and mounted on iron-belted wheels, "which is very useful to the project because having the mass so close to the works themselves we avoid wasting time to go for mass in the town . . . and even more so when one sees the consolation the chapel offers to these people who are devoted to San Antonio," coincidentally the saint of the chapel.[70] Fray Luis Flores immodestly assessed in his first days in office that "the Indians were delighted, and went happily to work."[71] Improbable as this sounds, it nevertheless suggests that the presence of the Franciscans helped ameliorate unbearable conditions. It was they, after all, who in 1555 complained that Indian workers were being brought in from as far as twenty leagues away with no pay at all (which probably overtaxed Franciscan coffers), thus obtaining from the city council some food and stingy assistance to pay for wages, due upon completion of the works.[72]

Whatever the objective truth about Franciscan beneficence, what is important here is that authorities felt Franciscans made labor management easier. As Viceroy Conde de Baños would explain later when he named Fray Bernardino de la Concepción, guardian of the Texcoco convent, to become superintendant after Fray Luis Flores retired, "it is of the utmost importance that the person who takes over this superintendancy be a member of this order, as the Indians are very well accustomed to work in this project with the religious of Our Father San Francisco."[73] And work they did, destroying the tunnel their forefathers had built generations earlier.

## SAN MIGUEL'S PLAN IN PRACTICE

With Fray Luis Flores ensconced as superintendant, in the late summer of 1637, when the authorities resolved that the Desagüe should henceforth proceed by open trench, San Miguel's script finally found its producer.

Like the secular superintendants before him, Fray Luis was given the power to order workers and supplies from both civil district magistrates and spiritual guardians (*ministros de doctrina*) of Indian parishes within reach of the Desagüe. Along with the appointment came the first order for draft labor for his term, to be fulfilled in month-long tandas or shifts. Indians, 630 in all, came from towns within the jurisdiction, like Teoloyuca, but also from as far away as Puebla.[74] The viceroy also sent maestro mayor Juan Gómez de Trasmonte, Fray Andrés de San Miguel, and the Mercedarian savant Fray Diego Rodríguez to "stake out on the surface of the earth the width and length that the trench shall have,"[75] and in December Flores began destroying the tunnel in a euphoric atmosphere. As in the 1607 groundbreaking, officials celebrated. "I verified with my experience that which I thought impossible," the fiscal of the Audiencia of Guadalajara who witnessed the event wrote to the viceroy, "for the great amount of earth that the current takes away is a miraculous effect." This, of course, went without minimizing the felicitous effects of Fray Luis Flores's "personal attendance, the affection with which he treats the Indians, and the punctuality in payments."[76]

The problem with this miracle was that it was quite seasonal: water was available in abundance only a few months out of the year. Hence, soon after the digging began, it became clear that reliance on substantial flows of water would create complications. Most of the water for debris-sweeping came from the Cuautitlan River and was stored in Lake Zumpango with water from other sources in the regional catchment. While technically perennial, this river carried much less water in the dry season, from October to May. During the dry season it was partially dammed up to provide water for irrigation for Indian townships and for private agriculturists, upstream at the water-allocating tower of the Pila Real in San Juan de Atlamica. In practice this meant that the Desagüe would have enough water for debris cleaning only from June to September, so the rest of the year superintendants would have either to change the work methods, which they never did, or resign themselves to a desultory pace in the open trench conversion—unless the audiencia intervened to reallocate water use in the region.

Even much later, when in 1743 the Desagüe secured a larger proportion of the Cuautitlan River's water,[77] most progress on the excavation of the canal was restricted to the rainy season. Only then did abundant water from both the river and runoff collected in Lake Zumpango create a strong enough current in the tunnel and beyond to carry away the debris. Unswept debris could create a blockage that could force the Cuautitlan River to back up in the rainy season, burst over and through its embankments and diversion dam, and resume its natural course down through the low area of Ecatepec-Chiconautla and onto the vulnerable city. On the other hand, securing a greater share of water for the project caused friction with and even sabotage from its agricultural neighbors.

To address the problem of debris removal in dry spells, friars dammed up what water they did have access to, releasing it all at once with force. This was known as "working by sluice-bursts" (*trabajar a comportadas*).[78] It was tried for the first time in the dry season of 1637 using water released from Lake Zumpango. In addition, Fray Luis Flores also collected water in two cisterns (*jagüeyes*) from the two rainfalls that occurred that December. Sluices at Vertideros and along the trench retained this water until Indians had loosened the clumps of earth and soft rock on the bed. When the sluice gates were opened, the water was released to sweep the debris away. "On the day it was released, the water carried away more earth than could be thrown down by two thousand Indians," Flores reported triumphantly, "with the two times that the sluices were opened the tunnel was left practically clean."[79]

But San Miguel himself gives us a hint at how disproportionate to the immensity of the whole task the comportadas were. The figures in his proposal indicated that more than four million cubic meters of dirt would have to be removed—over half of the volume of earth that would be removed to create the monumental Canal du Midi in France. Fray Luis Flores's *Memorial* of 1653 throws light on just how laborious the excavation of all this material was, and not just because of its volume. As of that date, in the open trench, just under 9 linear kilometers lay "in all perfection," although later reports would show just how imperfect this perfection was. That left only 3.5 kilometers of tunnel to open up, of which 1,874 meters were rough-hewn into the rock and the rest vaulted. The problem was that this was the deepest section, and workers would have to dig down 56 meters to reach the tunnel bed. After that, if the water in Lake Texcoco was ever to drain through the Desagüe, more than 12 kilometers of trench floor would still need to be deepened 3.32 meters.[80] Furthermore, it had to be done without obstructing the tunnel bed or the trench with debris while excavations above went on.

This was hard to avoid, given the varying cohesion of the terrain ranging from friable tufa (tepetate) to volcanic rock. Neither San Miguel's design nor the Franciscans' implementation of the open trench excavation used an adequate angle of repose—the angle of slope under which particles on the sides of the trench would tend to remain in place. It is true that it would not be until the eighteenth century that the resting point of different materials would be systematically studied and codified by engineers such as Jean-Rodolphe Perronet, who served as the bridge expert to Louis XVI.[81] In the meantime, however, builders of all sorts had substantial tacit knowledge at their disposal, as evidenced in the wider angles in the master architects' open trench proposal that San Miguel had refuted, as well as in the gentle sides of myriad irrigation ditches and indigenous dams in the region, including the Cuautitlan River's diversion structure. Too steep, the sides the friars carved eroded easily and clogged everything. The friar technicians' difficulties in approaching slopes in dynamic terms, relative to the forces that would act upon them and the granularity of the materials involved, was visible in other structures in the Desagüe as well. In building the spectacular 5,460-meter-long and 11-meter-wide mortar-and-stone dam that separated Lake San Cristóbal from Lake Texcoco, for example, the Franciscan Fray Gerónimo de Zárate unfortunately gave barely any slope at all to either talus. This meant that the water thrown against its base by the wind quickly destabilized the structure.[82] The reasons for these problems with inadequate slopes are not fully clear. But their effects were quite visible, both aggravating and worsened by the water-conveyance method.

Even by 1770, nothing had succeeded in making the method of water conveyance reliable "because one way or another the work depended on abundant rainfall." Despite San Miguel's prognostications about speedy completion, this dependence meant that after a century and a half of the practice it would still be impossible to "give a prudent estimate of what remains for the ultimate perfection of the works."[83]

In time, it also became evident that so long as the water conveyance method remained in operation, increasing the number of laborers on the tunnel demolition did not necessarily hasten the completion of the project. This was because dirt that was tossed into the tunnel risked staying there if there was not enough water available to sweep it away, and the amount of water was not unlimited even in the rainy season. In the early eighteenth century, decades after the open trench was supposed to have been completed, master architects still plodding ahead on it recommended "making the number of Indians proportionate to the amount of water that [the trench] carries so that they will not throw down so much earth that they silt it up."[84] In short, the open trench conversion moved ahead only when there was water. Under the friars, regular maintenance on the bed of the tunnel and competed trench portions was also dependent on the same water source.

Despite these difficulties, nothing changed in the conceptualization or in the design of the works. This was in part due to the fact that as a letrado, Fray Luis Flores—and his successors, for that matter—shared San Miguel's outlook on the Desagüe. He too deferred to Vitruvius as his authority on water carriage and on the problem of how much slope along the axis of the tunnel and trench to give to the Desagüe. Like San Miguel, Flores tried his utmost to observe what this "Prince of Architecture" stipulated as gradients for conducted waters of "one half foot for every one hundred."[85] This gradient was excessive for the Desagüe.[86]

The Franciscans managed other problems that emerged in the implementation of the open trench in a similar way, applying the principles derived from architectural or Aristotelian authority, particularly in the realm of mechanical work, understood as an amount of energy transferred onto excavated earth to move it over a distance—down, up, or out the Desagüe. This energy could be transferred to the earth by water, a human, an animal, or a machine, or some combination of these. Because in the Desagüe they would automatically be allotted repartimiento labor and were allowed to use the water of the Cuautitlan River, Fray Luis and his successors had little incentive to think hard about the latter two methods of performing work. They were biased in favor of techniques and technologies that utilized very cheap, coerced workers, and they lacked motivation to think hard about technology that might save on labor or water. The friars left no evidence of having contemplated the use of hoists or escarpments for human or animal removal of excavated debris. The combination of repartimiento labor and the seasonal water method remained in use throughout the colonial period because it was believed to be cheaper than investing in machines or animal energy to carry away the debris. This had long-term consequences and wound up being far from inexpensive.

## FRIAR SUPERINTENDANTS AFTER FLORES

Flores left the Desagüe in 1658, three years before his death of old age. His fellow Franciscan Bernardino de la Concepción was asked to step in. Seeing little aptitude in this man, Viceroy Mancera asked Fray Diego Zapata, Superior of the Franciscans, to recommend a replacement.[87] Upon Zapata's and Mancera's recommendation, Fray Manuel Cabrera was named superintendant of the Desagüe by a royal provision in December 1664. Like San Miguel an Andalucian by birth, Cabrera had completed his theological training in Puebla in 1652. By the time he was named Desagüe superintendant, he had accumulated exceptional managerial experience as the head of Franciscan convents in Santiago Tlatelolco, Coatinchan, San Juan de la Penitencia, Huexotzingo, and Cuautitlan, in the Desagüe district. He brought to the Desagüe not only his knowledge of Nahuatl and possibly Otomí—the two indigenous languages prevalent among drainage draftees—but also a familiarity with the hydrological conditions of the Cuautitlan River basin.[88] Serving between 1665 and 1675 and then again from 1687 to 1691, he could be said to be the culmination of Franciscan supervision of the Desagüe not only chronologically but also cognitively and managerially because during his term a series of conflicts and confluences over technological choices ripened, as did the scientific stakes that derived from them.

Ten years after Cabrera's instatement, however, the audiencia grew impatient with the slow pace of the trench conversion. Claiming that it could be finished in a year if the budget was doubled and more Indians were assigned to it, crown attorney (fiscal) Martín de Solís and his colleagues removed Cabrera and took over the Desagüe.[89] The interregnum of audiencia rule in the Desagüe lasted between 1675 and 1687. Solís consulted with royal engineer Francisco Pozuelo Espinosa, who had been assigned to fortifications projects in the viceroyalty. The result of this collaboration was that Pozuelo Espinosa demolished ten different segments of the tunnel even where it was unadvisable to do so given the friability of the tepetate. According to a March 1687 inspection report by a more experienced military engineer, Jaime Franck, Pozuelo Espinosa had in addition to this not given the segments of trench he opened faces with gradients adequate enough to prevent crumbling. The results of these actions remained hidden from view because the interior of the tunnel was not subsequently inspected for a period of four years. The city—and the fiscal and his engineer—were therefore lucky that it had not rained excessively during those years, as a flood would most certainly have ensued.[90]

Solís inflicted several other wounds on the open trench. In order to have enough water to aid his efforts to accelerate the trench, he diverted several streams into it. In doing so, he undid one of Cabrera's important modifications in the open trench conversion, which was to divert the streams whose water dropped into the open trench's northeast face from around the township of Huehuetoca, "eating and throwing its walls or sides into its bed." With "dams and levees of strong masonry," Cabrera had led the streams over "mortar and stone bridges [aqueducts, in this context] so that crossing over the Desagüe to the other side or the west they may end up incorporating themselves into the waters of Coyotepec, and from there return to flow through the Desagüe without any risk whatsoever through the Vertideros."[91] Since all the sections Solís opened up had steep slopes, turning the streams back into the trench only aggravated the problem of erosion and crumbling that obstructed its channel. In addition, Cabrera would later claim, during the peak season the streams formed small falls as they dropped into the trench, impeding the flow from the Cuautitlan River, putting the entire functioning of the Desagüe at risk.

When all this came to light after Franck's report, Cabrera was reinstated by a junta general and was back on site by June 1687.[92] Two years later, he published a vitriolic denunciation of Solís and Pozuelo under the title of *La verdad aclarada y desvanecidas imposturas con que lo ardiente de una pluma poderosa en esta Nueva España en un dictamen mal instruido, quiso persuadir haberse acabado y perfeccionado el año de 1675 la fábrica del real desagüe de la insigne Ciudad de México.*

Historians have generally accepted Cabrera's version of the events—that it was a simple matter of Solís's ineptitude, and by implication, his audiencia colleagues' cover-up of it. But the reasons why the oidores withdrew from the hands-on direction of the works and the friars were returned to it in the face of royal concerns about the powers of the clergy need to be explained, not assumed. Financial considerations and the ability of the friars to summon collective knowledge from different sources in the service of the project seem to have been as important considerations in Cabrera's return as his technical skill.

Solís had spent more than normal, and this was a problem. In order to deliver on the promise to finish the trench in two months, he bought equipment for debris removal that had not been utilized under the friars. An inventory taken in 1683 shows that in addition to iron bars, spades, shovels, and other tools that friars customarily used in the Desagüe, Solís had bought six small rammers (*almadanetas*), four pulleys (*roldanas*), 253 leather sacks (*huacales*), and ten coils of good quality *agave heterorocanta* rope (*maromas de lechuguilla*). The sacks and pulleys are interesting because they suggest that the oidor and his engineer were trying to extract the debris from the trench. This was obviously a more expensive way to conduct the open trench conversion than the friars' method of water-sweeping during the wet season only. Also listed are four counterpoise lifts (*bimbaletes*), or shadufs, as they are also known. Traditionally an irrigation tool that diffused from the Muslim world to Iberia, the bimbalete had become common in Mexico as well.[93] Solís used these bimbaletes in the repair of a breach in the Cuautitlan River diversion dam during the rainy season. The rushing water made it impossible for the Indians to reach the gap with materials to hold the water back and permit the repairs. After failing to get horses to brave the current and deliver the workers where they were needed, Solís had finally ordered this device to enable the Indians to reach the gap through the air.[94] In the open trench conversion, the fiscal probably used the four bimbaletes to assist in the extraction of debris. In other words, it would appear that Pozuelo Espinosa and Solís were breaking out of the paradigm established by the friars by working on the open trench year-round. This cost more in labor and materials, and therefore would have added to the displeasure with their performance.

Solís and Pozuelo did apparently do irreversible damage to stretches of vaulting, however. In his own work of uncovering of tunnel sections, Cabrera had left intact whatever resistant vaulting he found to help prop the sections of the tunnel that were carved into weak tepetate and other friable rock, while also allowing the water to flow both under and over these vaulted sections at peak season.[95] Solís and Pozuelo, by contrast, had barged ahead and demolished everything as they moved further into the tunnel. Naturally, this created numerous problems. The new opening to the tunnel they thus created had not been designed to bear "the entire load of the water" pressing against it as it tried to enter the tunnel, lacking "the great strength of the mortar and stone walls that with all his expertise Enrico Martínez, the eminent master of these works, put [in the former mouth]." In addition, this opening was not only small, it was located under the thick of the hill, with more than fifty meters of terrain still covering it. The result was that "the water cannot flow above it and being as the façade of this small vault is of bad quality earth, it is inevitable that it will be eaten away by the eddying that the water that cannot fit in its cavity will create precisely at this location."[96]

## FROM GEOMETRY TO A SCIENCE OF WATER?

How did all this dedication to the hydraulic project affect how the friars—and all other letrados, technicians, and savants, for that matter—understood what they were doing with water, matter, and forces in the Desagüe? Cabrera's critique of his enemies' work confirms his intuitive mastery of the hydrology and hydromechanics of the Desagüe. Yet he never questioned the nature of any of these helpful or disruptive phenomena. Decades ago, historian Cyril Stanley Smith suggested that the traditional view that technology was applied science and thus came after it was wrong and pointed out how in the early modern period science followed technology, explaining how and why techniques and tools worked after they had been empirically developed.[97] Recent scholarship has confirmed and developed this insight, showing the far-from-linear connections among craft, practical knowledge, and scientific knowledge.[98] This more sinuous path is indeed that which Galileo Galilei, Bendetto Castelli, and their followers took when they used geometrical language to dissect and explain what they saw in the motion of water through engineered as well as natural flows. Dealing with hydraulic challenges in Italian projects comparable in size to Mexico's drainage, Galileo and Castelli noted, systematized, and codified connections among hydraulic design and the behavior of water and earth, time, and forces. Hydrodynamic and hydrostatic phenomena started to be explored by means of experimentation and observation, sometimes yielding testable scientific conclusions about the behavior of fluids.

The Desagüe offered the same possibilities to the friars, as they became intimately familiar with its functioning and problems. Did they use their observations of the behavior of water and matter, formulating hypothesis that could be generalized beyond the Desagüe, and then testing them in the Desagüe or elsewhere? In other words, did their reputed familiarity with mathematics and their vast experience using geometry to rearrange space and matter enable them to understand the behavior of water in the

Desagüe in a manner autonomous from, yet parallel to, the new scientific directions emerging in Europe?

In short, they did not, mainly because their ontological and epistemological orientations did not predispose them to this kind of scientific interrogation and because they felt no pressure from their social surroundings to do or think differently. For Cabrera as for the rest of the friars, the knowledge that mattered in the Desagüe had already been set forth in either classical treatises or in natural philosophy. At the same time, whatever was not to be found in the texts came from personal experience, "because knowledge in these works is an experimental science which is not to be studied in books but in the announcements and accidents of time."[99] Despite this statement, there is no evidence that Cabrera used the Desagüe as a site of experimentation in the Galilean sense. Rather, he seems to have regarded it as a place where knowledge could be accumulated from observation, not produced through the testing of hypotheses. There simply was no perceived need to question established truths, and his intuitions served him well in practical matters.

They may not have sought new knowledge from the hydraulic beast under their care, but the friars certainly mobilized what they did know to explain its behavior. San Miguel and the Franciscan superintendants believed that the root of all problems in the Desagüe was a failure to surrender to the natural order of things and their elemental constitutions and places, understood in Aristotelian terms. Every person and thing had its place in a well-designed universe, and to act in a way that disregarded this order was to court trouble. The Desagüe's tunnel had not worked because it was put in the hands of Martínez, whom God had meant to remain among his books; the masters and oidores who "lacked knowledge of geometry, weights, and measures" had proceeded blindly because they were essentially usurping a position that was not for them.[100] More importantly, water rebelled and continued to cause havoc in the Desagüe because its elemental qualities were ignored. To correct this, the implication was, nature need only be heeded and harnessed. San Miguel believed that "things which human effort cannot bend can usually be bent by industry when it finds the possibility of using one element against another." Earth, the solid obstacle to his open trench and also the "lowest" and "least active" of all elements was susceptible to water, which "unifies it and gives it the capacity to bear fruit, runs over it and tears it up and takes from one part to another." Even a small stream could "tear away a great piece of a mountain and transport it a great way, with its trees upright, with greater speed than a ship with full sails."[101]

In this view of the elements and the natural order of things the universe was a static thing, whose physical phenomena could be observed and noted but did not require speculation about qualities or behaviors not already known or new questions about nature. Fray Andrés never wondered what made the empirical knowledge he acquired about matter, water, and phenomena earlier and elsewhere useful and applicable to the Desagüe. What mathematical and geometric knowledge he possessed he applied only to the two extremes—the divine and the practical—before his eyes.

These friars' intellectual acquiescence to higher authority—Vitruvius, Aristotle, God—meant that unlike some of their brethren in the trans-Atlantic realm of European culture, who were asking what physical and dynamic factors influenced the velocity of water, or indeed, what this velocity was and how it related to flow, in New Spain's cultural heart the behavior of water provoked no hypothesizing with experimentation. This deference to established authorities was upheld in New Spain just at the time when turmoil was destabilizing established scientific paradigms in many places in Europe. New questions emerged as refined instruments such as telescopes and microscopes yielded insights into nature and the cosmos, and the qualities of motion tore through the universal order.

Motion had not been absent from thinking about water. However, although discharge as the product of cross-section area times velocity had been defined in antiquity by Hero of Alexandria in his *Dioptrics*, in the Spanish and European realm in the Middle Ages and Renaissance, water in motion continued to be measured by the cross section it wetted, and its velocity not at all.[102] It must have been difficult indeed to interrogate the correlation between slope and the movement of water when San Miguel and other technical advisors and superintendants of the Desagüe remained faithful to the authority of classical architectural treatises such as the Roman Marcus Vitruvius Pollio's *De arquitectura*, written in the first century BCE but published, translated, and digested into further architectural treatises during the Renaissance. In Spain, too, masterful practical hydraulicians as well as authors from the early seventeenth century, such as the royal cosmographer Andrés García de Céspedes, continued to defer to Vitruvius on all things related to water.[103]

Yet there is no reason why Andrés de San Miguel should not have found, like Benedetto Castelli, that "where the river hath lesse velocity, it shall be of greater measure, and in those parts in which it hath greater velocity, it shall be of lesse measure; and in sum, the velocity of several parts of the said River, shall have eternally reciprocal and like proportion with their measures." Observing the behavior of water in the Desagüe and the Cuautitlan River with his keen powers of observation, his memory, and his Euclidean toolbox, he did not reach such a conclusion. It was instead Castelli in northern Italy who reintroduced this correlation between "measures" (cross sections or area) and velocity in his 1628 *Della misura dell' acque correnti*, leading to the notion that discharge could be expressed as the product of these two dimensions ($Q=AV$).[104]

Back in New Spain, San Miguel and his cohort were applying the recommendations and formulas stipulated by Vitruvius, who called for gradients in gravity-flow water systems of no less than 1:200. The Roman's recommendations, however, had been made for aqueducts and other water-supply and sewer systems built out of stone, masonry, or even brick, where they apparently worked well, given the stability of materials and the regularity of dimensions and shape. The Desagüe, by contrast, was carved in very friable earth and compressed volcanic ash; its tunnel was only partially protected and the trench not at all; and throughout the trench and the tunnel cross sections varied in shape and dimension and changed all the time. With the Vitruvian slopes, friars and other Desagüe technicians sought to increase the velocity of the water, but the irregularity of the water's path caused further and uncontrolled acceleration, which compounded the erosion caused by the abrasive debris in the current, attacking the sides of the tunnel, causing crumbling and cave-ins and generally further aggravating problems of design. None of this, however, triggered doubts about the scientific premises of Aristotelian ideas about how the elements water and earth behaved and interacted.

To the friars who implemented the trench conversion with its water-sweeping method of debris removal, what was most important about the movement of water in the Desagüe was that it flow in violent torrents. In other words, the Desagüe they were reshaping had to convey water to the other side of the basin and do so in a way that permitted the open trench excavation to proceed with the least cost possible, replacing scaffoldings, winches, carts, animal traction, and to an extent also human energy with the energy of rushing water. This likely reinforced their weak incentive to adopt strategies to measure the volume of water or the discharge and gave far greater importance to "elemental" qualities and "forces" that contributed to rushing water, as San Miguel expressed:

> As water is by nature slippery and in the exit of the desagüe there is such a sizable fall and current, at the same pace at which it comes out from there it summons other water. Since the body of water in the ditch is so elevated, it shall run with such velocity toward where it is being summoned, pulling so as to snatch the gaze and all that one might put in front of it.[105]

This "summons" (*llamamiento*) was a term used among craftsmen as well as military engineers to describe the behavior of running water. It was involved in explaining the velocity acquired by water flows under different conditions and therefore was subject to investigation and generalization under both an Aristotelian and a new "scientific" paradigm. For San Miguel, however, it was not general principles about this phenomenon that justified its harnessing in the Desagüe, but his own direct personal experience with it. If Seville's Guadalquivir River, "when it flows with much water the force of its torrent is such that it upsets and wrecks the ships that cross it; then how much greater would its current be if it had a free exit and as much slope as the Desagüe possesses?," he asked, concluding that "where the body [of water] is raised more and the exit is free so much greater will be its torrent."[106]

Another phenomenon that San Miguel noticed was that in the Desagüe water accelerated around and over obstacles in its course and eroded the spots where it impacted after acceleration, be it in the bed or the sides of rivers and canals. San Miguel did not formulate it thus, but we can see that he intuited it in his recommendation for a very careful leveling of the bottom of the tunnel, to prevent the water from carving cavities into it or into the sides with a motion he called "resurgence" (*agua resurtida*). In fact, he claimed, it was this phenomenon of resurgence and not the friability of the soil that caused the tunnel to cave in: the water ate away at the side supports of the tunnel, widening it so much that its walls could no longer bear the load of the vaulting (in itself an important observation for statics).[107] In the basin of Mexico and the Iberian world at large, empirically derived knowledge of the behavior of water was quite elaborate and widespread among craftspeople and builders who worked with it. In 1676, for example, master architect Rodrigo Díaz de Aguilera expressed concern that the water flowing from San Cristóbal to Zumpango lagoons en route to the Desagüe would retreat toward the dam that separated the lagoons and that the very turbulence caused by this motion would erode the foundations of the dam. His solution, based on custom rather than principle, was to provide the course of the water with "slopes and cutwaters" (*taludes y cortaaguas*) to force it to flow in a straight line toward the sluices.[108]

All these observations are in fact addressing hydrodynamic problems. For example, how "free" and "unfree" flows of water behave: water flowing from a long and narrow opening in a vessel full of water was free, but that flowing through a cross-section cut in a river was not, as it is subject to pressure and friction from surrounding water. The difference in the characteristics of the flow of free and unfree water was actually a critical problem in the Desagüe, as San Miguel intuited, since it was directly related to the problems that the confluence of flows of water created in locations such as the conjunction of the Cuautitlan River with the Desagüe at Vertideros and the transition of water from extended lagoons into narrow canals, among others.

Some of the difficulties in finding solutions to these hydromechanical problems in the Desagüe had to do with the way the volume and flow of water were measured—or rather, were not measured. Suitable units for the measurement of fluids were necessary to properly design, redesign, and operate the Desagüe and probably even to think about its hydrodynamic challenges. In 1804, as he toured the Desagüe with Viceroy Iturrigaray on the eve of the independence insurrections, the Prussian savant Alexander von Humboldt would remark how despite the need to have determined "exactly" how much water the lakes and the Cuautitlan River carried at peak season the tunnel and trench were supposed to accommodate, he had found no such calculations in the Desagüe reports and records written by Fernando Cepeda and Fray Manuel Cabrera, in the seventeenth century, or Ignacio Castera and Joaquín Velázquez de León in the eighteenth. Thus none of the Desagüe's structures had been dimensioned relative to the water volume, let alone the discharge. Enrico Martínez's tunnel was too constricted for all the water it was supposed to carry; Andrés de San Miguel's open trench was too ample and became more so as Flores, Cabrera, and all who succeeded them deepened and extended it and as erosive forces acted over time.[109] This is a real puzzle, not an anachronistic expectation of Humboldt's.

In the Iberian realm, people measured water either by the surface area it wetted, as was done for water allocations in agriculture with units such as the *buey*, *surco*, *naranja*, *paja* and so on, or by container measures like the cuartillo and the pipa that were used in the freighting of wines and other fluid products.[110] These measures were sufficient for the uses for which they were designed. Consistent with this tradition, friar San Miguel had always used units of area and never once used "cubic varas" (*varas cúbicas*) when referring to water, nor did Luis Flores and Manuel Cabrera after him. Why not?

All groups of technicians and officials who intervened in the Desagüe knew cubic units. Andrés de San Miguel clearly knew how to calculate cubic volumes, as he showed in his treatise. Fray Luis Flores too had seemed comfortable using these units to subcontract a segment of open trench excavation to a Spaniard who in 1639 agreed to do it for "half a real per cubic vara."[111] Even officials with no technical training were using units of volume to measure solids. In 1637, for instance, members of the Mexico and Guadalajara audiencias visiting the open trench works reported the progress in cubic varas.[112] Master architects routinely used cubic varas to provide estimates for any digging to be done and considered the ability to "reduce, square and cubicate," as their draft guild ordinances read in 1736, as essential to their craft as literacy, numeracy, and geometry.[113]

What is telling here is that this comfort among these technicians applied to dry materials, not to liquids. Thus, they applied to matter that was understood to be in stasis, not subject to motion. Yet there were people in New Spain using units of volume for water. A foreign military engineer sent to inspect the Desagüe in 1647 explained that before giving his opinion on its tunnel, trench, dams, and other structures, he needed "to very carefully measure all of these lands, rivers and streams that come into these lakes, and how much higher one is from the other, and what is the width of each one of them, in order to determine how much water there is in each receptacle of these lakes, rivers, and streams to see how they are to be defended from."[114] What units the royal engineer planned on using for such measurement is hard to tell, but the intent to determine the *volume* of water seems clear.

This points to an awareness among some of the technicians operating in New Spain during the tenure of the friars about the usefulness of cubic units for both a more precise assessment of where the dangers of flooding were coming from, and a more appropriate dimensioning of structures relative to the volumes or discharges of water they were supposed to manage. Putting aside the technical advantages of cubic units for a moment, it is possible, as Cabrera's statement intimates, that shifting to these units in the Desagüe could have carried the downside of disrupting communication with foremen and wardens who might not have been used to them. Whatever the case, the truth remains that without them it was difficult to address many problems of dimensioning and flow in the Desagüe and in the basin as a whole. The reasons for the persistence of linear and square units for the treatment of solids and water indistinctly in the Desagüe even in the eighteenth century are not obvious, but we can discard ignorance. What seems to have been at work was choice on the part of different groups of technicians for the approach and tools best suited to their social and cultural alignments and priorities.

For the friars water had elemental qualities and was not subject to the new laws of a universe in motion. From the mid-seventeenth century on, people contaminated with post-Galilean paradigms began showing up at the Desagüe and using different tools and understandings. The application of cubic units of volume to water as well as dry materials was essential to this subjection, for without these, it was difficult to use the new mathematical tools that were emerging in the last third of the seventeenth century, such as logarithms and calculus, to understand the behavior of water in stasis and motion that so affected the success of the Desagüe. Yet in the face of these challenges, the friars held firm on insisting on the use of linear measures alone.

What made the rich empirical tradition of these friars immune to scientific interrogation was their deference to Aristotelian certainties about the qualities of matter. And this is why more precise measurements and instrumentation were not just irrelevant but also unacceptable: "The use of instruments and the belief in the evidence they could provide," historian Mario Biagioli has stated about Galileo's disputes with natural philosophers of his time, "were as alien to Aristotelian philosophy as the use of mathematics to explain physical phenomena."[115] The significance of the friars' refusal to switch to cubic measurements for water despite Madrid's request that they do so suggests not ignorance or conservatism—oft-used stereotypes about Spanish culture—but the protection of a corporate identity invested in an elemental natural order. The friars defended their Aristotelian worldview much like the university-trained academicians Galileo debated in Biagioli's work defended theirs. As in Biagioli's study, what was at stake in the prestige of the friars' identity was patronage, which the superintendancy of the Desagüe certainly was. The Franciscan who occupied the position got a modest salary from the Desagüe, while the same junta general of July 17, 1637 that approved the open trench conversion and the naming of Fray Luis Flores as superintendant also removed the secular priest of the township of Huehuetoca and turned over the curacy—one of the forms of patronage to the clergy—to the Franciscans (it was not returned to the secular clergy until the 1710s).[116]

This is not to say these men of the cloth were unscientific in an absolute sense, only that the kind of science they invoked to solve the difficulties posed by the desiccation project was pre-Galilean, and they did not use it to formulate testable hypotheses. This scientific tradition allowed them to legitimize their proposals scholastically, by reference to scripture, Euclidean geometry, and classical and neoclassical architectural treatises, all of which were consistent with an Aristotelian understanding of the physical and social world. That is, they sought to authorize observations derived from experience and anecdote by reminding their listeners or readers that they were only confirming what wise men before had said. They neither interrogated the natural order nor experimented to reach answers and did not venture new hypotheses beyond the knowledge they were familiar with. As a result, neither did they formulate new general principles (new knowledge) from any of these two activities.

This confirms Elías Trabulse's observation about many scientific minds of the period in New Spain, including people like the polymath Carlos de Sigüenza y Góngora, who also had brief dealings with the Desagüe: they excelled at incorporating the knowledge of their age—even that contained in officially banned books—and demonstrating their mastery of it, but not at elaborating or augmenting it with autonomous investigation. The reasons why these friars behaved this way cognitively may have something to do with their rather undisruptive role in colonial society, where they benefited from a hierarchical order that was in their interest to believe immutable.

If they did not reach for "scientific" ways to understand nature in the Desagüe that by then were becoming current in the realm of hydromechanics, friars—or at least Cabrera—did reach for indigenous knowledge. In August of 1688, Indians from Teoloyuca submitted to him a plan to redesign a segment of the Cuautitlan River diversion dam, which to them was simply in the wrong place. The diverted river itself, they claimed, was demonstrating this by breaking through its dam in the rainy season to carve its own path along lower ground between the township of San Lorenzo (located half way between Cuautitlan and Teoloyuca) and the Desagüe, basically trying to recover its old course (the "Rio Viejo," now the central canal fanning eastward from the Pila Real de San Juan Atlamica). For the Teoloyucans, the problem with this was that the water frequently flooded their fields. They proposed acquiescing to the behavior of the river by shifting the diversion dam and the new bed of the river eastward to guide it into the seasonal lagoon of Coyotepec (between the township of Coyotepec and Lake Zumpango). This would give the diverted Cuautitlan River a straighter course and make it less dangerous "because as it will flow straight to the Desagüe with no meanders at all, it will deposit no silt in the center of its riverbed and the latter shall always maintain whatever depth it might be given because of the violence with which the waters will flow." This beneficent effect "does not happen in the riverbed that it currently has because since the water pools there the silt settles and the river starts losing its bed." The deposition of silt in the channel of the river resulted "in the water pressing against the causeway and by necessity overflowing it or breaking it," the Teoloyucans observed, but this could be "totally avoided by making the new cut."[117]

Cabrera consulted with the guarda mayor and with guarda de Vertideros Alonso de Quesada, and all three agreed that it would be "patently useful" to the Desagüe, although it did "not seem possible given the meager numbers of these Indians." Oidor Labastida in turn approved the plan and provided for materials and extra hands, but it is not known whether it was carried out. Still, the dynamic reaction of the indigenes to the new conditions created by the Desagüe was matched by Cabrera's own ability to identify indigenous technology that was useful to the Hispanic project under his supervision and to then implement its appropriation and redeployment. If the colonial project depended on such appropriations, it also needed peculiar, practical letrados like the friars to assess the value of skills among indigenes. Nobody better than the Franciscans, and a man like Cabrera, head of several convents which themselves would not have existed without such appropriations of skills, labor, and wealth.

All this knowledge is to be found not in treatises drafted to secure patronage or for the restricted market of literary production, but in reports and the daily record of the works, which circulated because it was the stuff of decision making over work, not leisure. Indeed, under the Franciscans all of these cognitive matters were inseparable from the very tangible and muddy world of work. The manner in which friars went back and forth between their ideas about geometry, building, forces, and matter and the Desagüe had repercussions in all dimensions of the public work, from the connection between tinkering with its material structures and the evolution of hydraulics and hydromechanics as realms of activity and study to the effect of their design decisions on the laborers and neighbors in the Desagüe district. As resident superintendants of the Desagüe, these men infused into it managerial and social arrangements that were both inseparable from their redesign of the project and indelible from its fate.

## AN ENDURING LEGACY

By the time men of the cloth left the administration of Desagüe affairs in 1691, the Desagüe was in no better shape than it was under Martínez—and from some perspectives it was arguably worse. The tunnel had been partly destroyed, but the hydrodynamic problems that had plagued the original designer were still there. The new mouth of the tunnel was still too small to accommodate the rush of floodwater in the rainy season, and the water stalled and backed up onto the Coyotepec, Zumpango, and even San Cristóbal reservoirs, when not breaking out of the Cuautitlan River channel itself. This particular problem would not be "solved" until the consulado completed the destruction of Martínez's work.[118] However, even this "solution" was purely empirical and led to no scientific formulation of principles that could be invoked later or elsewhere. This demeanor toward speculation about possible *new* general principles explaining the particularities of the Desagüe was in part a legacy of the friars. During their long superintendancy, they had made little effort to link their understanding of water to other realms of knowledge, some scientific, contained in the libraries of their convents. Despite this, however, when they were ultimately phased out of the Desagüe superintendancy in 1691, it was not because of perceived deficiencies in their philosophy or approach to science.

Over the course of the seventeenth century secular expertise had increased. Rising numbers of master architects, the arrival of royal military engineers, and the emergence of domestic scientific intellectuals such as Carlos de Sigüenza y Góngora, who became a royal cosmographer and also opined on the Desagüe, diversified the resources authorities could tap whenever they needed specialized knowledge. The ultimate reason why friars returned to being consultants rather than project directors is that the crown's political desire to recuperate powers that had been dispersed earlier in colonial rule at all levels was now solidly matched by its ability to do so.

As the open trench proceeded under the friars and beyond, the distant echo of Martínez's use of machines would become inaudible. Neither would the plans of the military engineer Adrian Boot be mentioned again, even though, expensive and possibly impractical as his proposal may have been, its individual components could be appropriated and implemented independently. In ignoring Adrian Boot's reiteration of water-management practices and technologies that pre-Hispanic and contemporary indigenous populations used in agriculture and transport, they helped remove the possibility of imagining the public work in an economically generative context. In destroying in it the features that it had in common with mining, Fray San Miguel and the Franciscans who implemented and consolidated his recommendations drove a wedge between this most capital intensive and dynamic sector of the economy of New Spain and the Desagüe.

This severance meant, concretely, that all manifestations of mining technology—the adit-like tunnel, the use of animal-driven hoists (*malacates*) installed in vertical shafts to drain water seepage in the tunnel bed and assist in debris removal, the excavation of shafts to provide ventilation, and other creative redeployments of mining technology—were abandoned in favor of a "simpler" technology. The simplicity of the open trench is deceiving, however, because the trench design combined with the method used for its execution turned the Desagüe from a capital-intensive project into a labor-intensive and costly "nature- and time-intensive" project—to the extent that human energy was itself replaced by seasonally available flowing water. It could be argued that these were adaptations to depopulation, but in the long run the open trench turned out to be ill conceived and its reliance on seasonal water flows highly inefficient, since lengthening the time it took to complete the project introduced other variables such as erosion, a changed terrain, and altered water flows. This prolongation had implications for the Desagüe's relationship to its neighbors and to indigenes beyond as well.

In terms of technological choices, the question here is not why these friars "failed" to come up with solutions that might seem evident to us but were conceptually and materially inaccessible and unavailable at the time. The question, rather, is why they discarded technologies that were in use or proposed before their administration, and why they favored the path of digging up the open trench with minimal labor-saving technologies or animals and extensive reliance on human energy. The answer is that they responded to the challenges they faced in a manner that was consistent with their geometrically ordered worldview and with their letrado status. This, in turn, was both permitted and perpetuated by the availability of repartimiento labor. But the friars were not alone in these choices. They had help cementing this in the Desagüe proper from the guild master architects, who worked with them and also diffused throughout the Desagüe district the singular marriage between geometrical abstractions and an apparently cheap method of execution using human calories and the mechanical energy of rushing water. Both groups of technicians enriched this subsidy of energy and matter that the indigenous realm involuntarily offered the Desagüe with techniques all their own.

CHAPTER FOUR

# To Serve the City in Desagüe Country

Making the open trench idea a reality depended on more than the skills of the friars. The reconceptualization of the Desagüe and the methods of work and practices it was articulated to gelled through the cooperation of other groups. One of these groups was the city's guild master architects (*maestros de arquitectura*). These men played a key role in ensuring that the structures and the methods of work in the Desagüe district, but outside the trench and tunnel, dovetailed with the designs and methods of the friars in the Desagüe proper. Friars and maestros had consulted each other in the Desagüe before Martínez's death, of course. Between 1630 and 1691, their mutual cooperation took on greater specific weight, in that it was responsible for the consolidation of the open trench and all attendant methods. Working with the maestros and friars was another group of men whose lack of a unified or corporate status and self-conscious identity makes it easy to overlook. These men were the general wardens (guardas mayores), the subordinate wardens (guardas) and the foremen (sobreestantes) of the Desagüe, who lived on-site and in whose hands the day-to-day functioning of the Desagüe actually rested. They were essential to the works.

Unlike the friars, maestros and the on-site men participated in the project over the entire colonial period. As a result, together they were one of the main agents of continuity in practices and arrangements established in the first half of the seventeenth century. Friars, military engineers, and academicians came and went, but the maestros, like the guardas and sobreestantes, were there to stay, infusing their particular ways of working, knowing, and communicating into the project, along with their urban loyalties.

The maestros' specific job was to translate into the tangible structures of the Desagüe the desires of the growing imperial city. This reflected the special functions that these men had as non-letrado servants of the capital and its rulers, on whose patronage and employment they depended. Above and beyond the Desagüe, the fundamental job of maestros was to give material meaning to the rather abstract act of founding cities, by which the Spanish colonized territories. They were the ones who designed and coordinated the execution of urban infrastructure over the course of the sixteenth century, built and expanded water supply systems, public fountains, markets and squares, streets, acequias, bridges, government offices, and so on. Much like their European counterparts, these men described their work in these projects as part of their role as servants of the city and its rulers, which obliged them to use their art for "public utility," where the public was that which was civilized, or urban. Yet the largely experimental nature of Spanish urbanism in the New World forced these men to be far from servile reproducers of established building and aesthetic formulas.

To deal with the lacustrine region of the basin of Mexico with its resilient clayey soils, wild seasonal fluctuations in rainfall, and brackish lakes communicating with fresh-water ones, Mexican architects used bodies of knowledge found sooner among their indigenous work gangs, oral transmission, and the reverse engineering of built or carved things than in formal texts or images of European treatises. Keen-eyed and experienced practitioners who were also heirs to the multiple hydraulic traditions of Iberia, they identified and appropriated key indigenous hydraulic technologies and knowledge about materials and the processes that were linked to them, injecting them into the city's public work. They became intermediaries between the indigenous and Hispanic hydraulic cultures and among various technological actors within Hispanic society who decided what to incorporate and how in the Desagüe, and this made their practices quite inventive, albeit in a way that should not be understood as automatically synonymous with beneficent.

There was a logic to the way maestros synthesized American matter and labor with European ideals and craft traditions in the single most important structure that urban life depended on—the Desagüe. To better explain the elements involved in that logic, it is first necessary to examine the maestros' historical evolution and loyalties and the means through which they acquired their knowledge and skills. On this foundation, analytical questions can be explored. What enabled the maestros to infuse their peculiar forms of working and knowing into the drainage? How did maestros de arquitectura combine elements from the hydraulic traditions they carried from the Old World (particularly from the Iberian synthesis of Islamic and various pre-Islamic technologies, which constituted the main trunk of their knowledge) with those they encountered in the New World, and why did they choose one practice, species, material, or process over another? Why and within what constraints? These choices mattered very much at the time, and they should also matter to historians, particularly since it was largely the expertise of maestros that permitted and facilitated the urban core's drive to subject the hinterland as well as its producers and inhabitants to its needs through the Desagüe. The choices the maestros made in the drainage helped freeze into place the specific usages and social arrangements established under the friars. The guardas and sobreestantes then took care to ensure their daily enforcement under instruction from the technicians and superintendants. These usages and arrangements—burdensome to the indigenous rural population—would prove incredibly difficult to extricate from all future technological decisions, even during the enlightened reformism that swept into all sorts of physical projects from the mid-eighteenth century on, pushing along the exploitation of the indigenous peasantry while also guaranteeing its survival as a class.

## GUILDED URBAN LOYALTIES

Through their guild, the maestros were the only technical group in the Desagüe with explicit, structured, and exclusive loyalties to the city and its elites, whether the latter were imperial or creole. This was in large part a function of their multiple roles, essentially akin to those filled nowadays separately by public health officials, planning commissions, urban planners, and assessors, making early modern architects very different creatures from modern architects. The guild and the rules of the trade had to be sanctioned by the city government, who in turn guaranteed the members of the corporation a monopoly over all public works financed out of city coffers and the right to regulate who could enter the profession. This relationship was rooted in a European Renaissance legacy: the ideal of the master as champion of the civilizing mission of the city in the hinterland.

When they drafted their first guild ordinances in 1599, Mexican architects enshrined this ideal in them, while also regulating the many practical matters of the trade and their internal governance by elected *veedores* (senior architects serving as examiners and executives of the guild) who served as the liaison with the city council.[1] As in the classical tradition, architects were supposed to know "where to build for the salubrity of human habitation" and how to erect edifices of all sorts to reinforce the city's supremacy.[2] Masters in the basin of Mexico who built the public infrastructure of urban life remained true to the legacy of this ideal even as they adapted to and appropriated American realities. The relationship between architect and city was conceived as a symbiotic one by both parties—the master served the lords of the city; the city "lords" (the municipal council made up of vecinos) protected their servants by approving their ordinances, making the mechanisms the guild established to control access to the profession and customers legally enforceable.

The maestros' intimate, structured, and exclusive ties to the city as a node in the organization of the Spanish empire was manifest in a number of ways. Although the blanket prohibition for nonexamined builders to take on construction projects was routinely ignored by most, in works commissioned and paid for by the city, and some of the crown's, the monopoly of the guild's master architects seems to have been firm. This was certainly the case in the projects of longest duration: the viceregal palace, the cathedral, water supply, and the Desagüe.[3] Also, architects worked intimately with the municipal *juzgado de policía* of the City of Mexico, the commission that regulated the material aspects of urban living. The 1640 *Ordenanzas de Policía* stipulated that in addition to the *juez de policía*, this commission had to include a senior master architect to collect and review the drawings for all works conducted in the city. All estimates of construction works were to be submitted to this officer.[4] In time, the maestros serving on this commission were expected to act executively "without the intervention of the juez de policía" to punish those who had been remiss in their duties to "maintain [the city] with the necessary luster, adornment and symmetry, without allowing the squares and streets to fill up with filth causing notorious indecency."[5]

In addition, master architects enjoyed privileged control of a key financial aspect of urban life that was also central to the Desagüe. This was the business of appraisals for the real estate market in the City of Mexico. For the one-time 1 percent tax to fund Enrico Martínez's Desagüe, it was a maestro—Andrés de la Concha—who provided the estimate of the value of buildings liable to the tax. When over the course of 1631—with the water from the 1629 flood still high—the crown insisted that officials and notables in the city seriously consider moving it to a less vulnerable location, it was again master architects who provided their employers the most important weapon to justify their rejection of the royal proposal: by then, these values had climbed from twenty to fifty million pesos.[6] In addition, because loans issued by convents or merchants (the only sources of credit in the absence of banks) needed collateral, real estate became a lubricant for credit. The amount of the loan or mortgage sought would hinge on the value of the estate (*finca*) or other property used as backing, and only the master architects' notarized and itemized appraisals were acceptable.[7] By 1736, master architects would be fulfilling this function with even greater precision relative to the market, calibrating their appraisals to a map of the city that indicated the value of each lot according to its location. Each master had to possess his own certified copy of the map, and failure to assess properties "with the required fidelity and purity, without damaging either party with over- or undervaluations" was punishable with a fine of a hundred gold pesos and a year's suspension from practice.[8]

Finally, from the beginning of the drainage, maestros were also instrumental in the management of the hinterland for the good of the city. They did this in a number of ways both inside and outside the Desagüe. Through surveying terrains to decide where to place dams, aqueducts, bridges, and other structures, they gathered information about land, water, and property and production relations. In building such structures, they effectively placed objects in the hinterland that were designed to fulfill one or more of three functions—convey hinterland resources to the city, deflect unwanted things away from it (water, people), or spatially organize or coordinate the movement of goods and people between city and country.

Unsurprisingly, none of the master architects' reports or proposals about the Desagüe or water management deviated from the notion that the lacustrine region had to be disciplined to suit the needs of the City of Mexico, regardless of the fact that adjoining areas were also densely populated in countless townships and hamlets and had organic links to the lakes and surrounding fields. The linkage between architects and urban life was thus both broad and exclusive from the inception of the Spanish city to the end of colonial rule, with the civil engineer elbowing a space into this relationship in the national period. This link helped determine the parameters of the possible, the conceptual range within which maestros made choices for the water management of the region and in specific the Desagüe.

By contrast, although apparently no less loyal to the city, the guardas and sobreestantes had no social organizations, codes, or training guiding their conduct in the Desagüe. The only visible qualification they seem to have possessed in both the seventeenth and eighteenth centuries is that guardas mayores in particular tended to be men of rank and property. Alonso de Quesada (1670s on), for example, a guarda de vertideros, was a *licenciado*;[9] guardas mayores Fernando de Chirino Villalobos (already working 1682), Baltasar del Vivero Quevedo (named in 1687), Juan de Espínola (named in 1689), and Francisco Martínez (named in 1693 after the death of his predecessor, a sargento mayor) were all captains. Joseph de Valladolid (named in 1710) was a Caballero de la Orden de Santiago and owned an hacienda near the Salto de Tula;[10] Francisco Enciso de Moctezuma, guarda mayor in the 1680s, may have been in the family of the counts of Moctezuma.[11]

Rank may have been regarded as a more important qualification for their job than technical fitness at the time of hire: it was perceived as central to claims to and adjudication of honor; in turn, to have honor meant having moral integrity.[12] Rank—hence, honor—would have been considered an important qualification for a position responsible for receiving and disbursing payments, materials, and labor and which carried broad jurisdictional powers in the Desagüe.[13] Subordinate wardens and certainly the foremen were not men of rank, and neither were the architects, for that matter, although they aspired to be.

Guardas menores and sobreestantes got their jobs more by family ties or their proximity to the Desagüe, where they often lived or held other office. At all levels of guardas or sobreestantes, some of these men founded entire lineages of Desagüe attendants, like maestros and engineers did in their own professions. The Quesadas provided two *guardas de vertideros y calzadas*—Alonso, who died in 1708 after a lifetime in the Desagüe, and Miguel, his son, who got the post in 1709;[14] the Moctezumas, three guardas mayores (Diego de Audelo, Francisco Enciso, and Pedro Francisco); the Valladolids, two, with a third member of the family providing local expertise, since he was a priest in Zumpango (Joseph Alfonso).[15] Among the sobreestantes, the Tapias gave three—Joseph, Nicolás, and Juan; the Porras, Pedro and Andrés (1670–80s).[16] Like their social betters, some of these men profited from their positions by using the Desagüe's stores, workers, and even lands for their own gain. The Tapias, for example, colluded to illegally sell pulque to the Indians who came in the repartimientos.[17] While the maestros built and repaired things, the guardas were in charge of the daily regulation and functioning of the Desagüe as a hydraulic mechanism. The number of guardas who resided on-site varied, but there were always at least three to maintain a close eye on the work's central features. The guarda de vertideros kept watch over the sluice and spillway that regulated the release of water retained in Lake Zumpango into the Desagüe (although some guardas de vertideros were also *de calzadas*);[18] the guarda de calzadas made sure that the most critical causeways and dams were not damaged and that they performed as expected, and for that purpose had a lodge or *garita* on some of the causeways, such as that of Lake San Cristóbal; and finally the guarda mayor had general oversight over the entire Desagüe, collected information from the other guardas, corresponded by messenger with the superintendant in the City of Mexico, and ensured the daily execution of orders. It seems that new guarda posts were created as new structures were created or old ones decayed, since at least as of 1677, a *guarda del tajo* was hired to carry out the guarda mayor's instructions in the open trench conversion project and to alert him in timely fashion of blockages and other emergencies.[19]

As a result of their constant presence at the works, guardas and sobreestantes played roles nobody else could and that were logistically and cognitively essential to the technicians and the superintendants. Knowing the social and physical terrain as well as established practice in the works, guardas and sobreestantes helped initiate incoming technicians and officials, such as royal military engineers who were sent to the Desagüe for brief periods to assess the works or incoming viceroys who were expected to become quickly familiar with the Desagüe on which their capital depended. In addition, they steered the works during the rudderless periods. Finally, they were the ones who trained and oversaw the repartimiento crews.

During periods when disputes erupted in the Desagüe among those in positions of authority, guardas and sobreestantes could also wind up providing the only fairly disinterested information on the true state of the works. This was most evident during the fractious period of 1675–87, and the fierce struggle between Fray Manuel Cabrera and fiscal Martín Solís. In 1675 Solís slated eight sections of the extant tunnel between La Guiñada and Huehuetoca for demolition—or rather, digging out the terrain above the vaulting and leaving only the latter in place. The first segment Solís tackled was in La Guiñada itself, the most problematic sector of the tunnel because it lay at the deepest point beneath the surface. If the tunnel here was obstructed in any way, water released into it could not surmount the vaulting and flow above it as it did in other sectors, where the open trench work had left little or no earth above the tunnel, and could instead back up dangerously. Solís gave his orders to proceed with the excavation, but instead of obeying the order the sobreestantes debated it, forming two camps. Those who worked aboveground felt that the tunnel sector was wide and clear enough to convey any rush of water unobstructed, and so opposed the order. Thomás de León, Andrés de Porras, and others who worked underground, however, knew that in some segments this sector was too narrow. Their impasse forced Solís, military engineer Pozuelo Espinoza, oidor Gonzalo Suárez de San Martín, and a royal notary to follow sobreestante Francisco Rodríguez down into the tunnel, where they found that indeed, the underground crew was right. The order was then carried out.[20]

An example of these men as gatherers of information was a sobreestante from the Porras family who worked under both Cabrera and Solís and saw the effects of the demolition on a site he knew well. Pedro de Porras submitted a plan to alleviate some of the problems caused by the retreating water during the rainy season. By 1676, he had observed how, in flowing back into Lake Zumpango, the water formed eddies that stressed the foundations of the spillway of this lake, while also depositing its sediments right in front of its floodgates, hindering their functioning as well. If this continued, Lake Zumpango was liable to fail as a regulator in the Desagüe. Porras therefore proposed a masonry dam to lead the water from the Cuautitlan River and other sources smoothly toward the floodgates that released it into the Desagüe and to prevent them from eddying back. After surveying relative water levels and the terrain, maestro mayor Rodrigo de Aguilera, who was at the time in charge of the cathedral construction, recommended that the sobreestante's plan be carried out.[21] Thereafter, maestros kept an eye on this site, alerting authorities in 1677 and 1678 that tunnel obstructions were causing waters to "pile up" at the entrance to the Desagüe, again generating dangerous eddying there and upstream.[22]

Porras was surely unaware of the representational choices he made in drafting the image, shown here in Figure 4.1, to convey the details of his

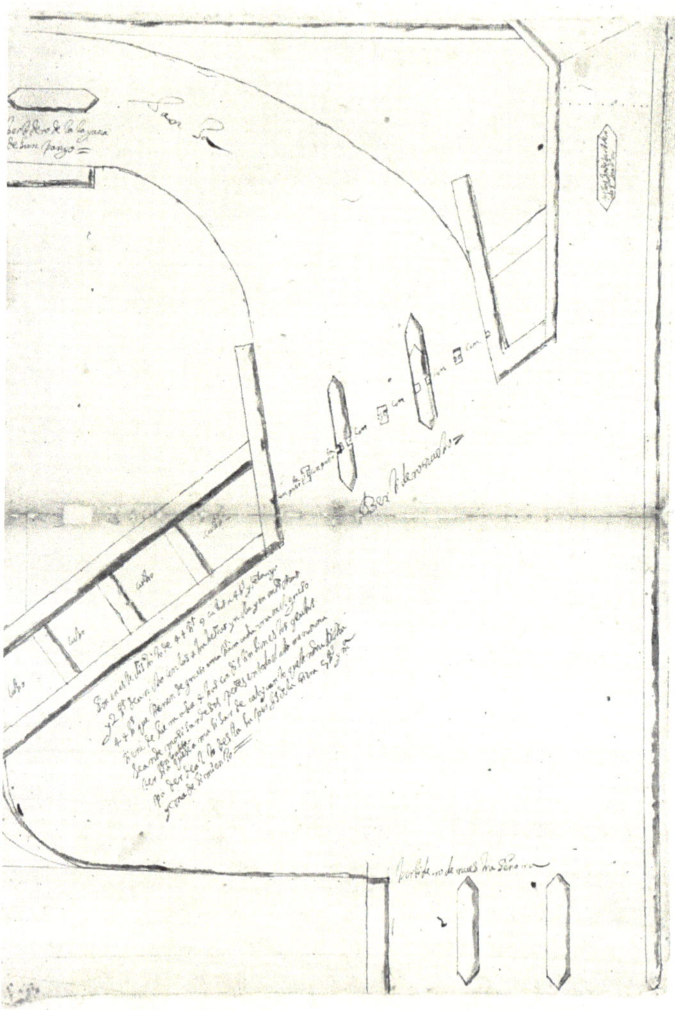

*Figure 4.1.* Sobreestante Pedro Porras's 1677 plan for the dam at the Coyotepec silting pool.

SOURCE: AGN, Desagüe, vol. 7, exp. 1, f. 5. Reprinted by permission.

proposed dam, but its style contrasts with San Miguel's careful, lettered, and authorially self-aware depiction of his open trench idea. The roughness of the lines and the lack of measurements of any sort suggest a sketch, not deliberate draftsmanship. Why would he visually convey ideas like this and not like the friar? Probably because Porras's concerns were overwhelmingly practical and immediate, decoupled from expectations of patronage or reward—he was already gainfully employed, after all, and the possibility of promotion from sobreestante to guarda or maestro was foreclosed by guild and status considerations. If correct, this interpretation would suggest that sobreestantes in fact worked hard at being gatherers of information and experts on local conditions—and at profiting from this in rather pedestrian and also local ways, such as usurping land to grow crops—and were unperturbed by the lettered aspirations and resentments endured by others in the Desagüe. It seems to have paid off: clearly a valued sobreestante, Porras was still in the Desagüe decade later, supervising repairs in the house for the guarda de vertideros.[23]

It was the sobreestantes' specific job to get work crews of diverse and ever-changing composition to carry out their work. Although they often used violence or the threat of it to accelerate the pace, this could not replace teaching each new shift of repartimiento Indians how to clean up the interior of the tunnel, for example, a task that few if any Indians would have had experience with. Military engineer Jaime Franck was impressed by Desagüe sobreestantes and guardas, particularly guarda Alonso de Quesada, who by the time he met him in 1688 had spent forty years in the works. Sobreestante Fernando Rico too stood out as a man "who knows his letters and numbers . . . an upright man of very good will," whom he observed organizing the difficult work of removing obstacles in the tunnel with Nicolás de Tapia. Foremen of underground crews, these men "fulfilled their obligations very well," thought Franck, "not wasting time at work and the Indians captains going forth with their sticks to awaken and encourage the lazy ones. When the sun rises they go in to work, and they do not come out until sunset, so that they eat their lunch underground in the tunnel itself."[24]

It is rare to find even the acknowledgment of the existence of sobreestante and guarda knowledge in modern narratives about the Desagüe. Yet the valuable store of knowledge of these non-letrado plebeians was both noticed and valued among their lettered, expert superiors, much like French Canal du Midi engineers sought the expertise of laundresses and Venetian *provveditori sopra boschi* eagerly absorbed the knowledge of their social inferiors.[25] Maestros, guardas, and sobreestantes each brought somewhat different perspectives into the project, but nonetheless guaranteed basic continuities in design, practices and arrangements, and the empirical knowledge of guardas and sobreestantes and even Indians.

## WAYS OF KNOWING

The different perspectives these kinds of men injected into the works were shaped by other cultural factors, not just social ones. The conceptual range within which the maestros in particular made choices and operated was also cognitive. Maestros handled the practical information the on-site men gave them in ways that makes it necessary to understand how the maestros learned their craft and from what sources. The maestros' way of knowing relied heavily on experience and precedent, both in the practical and the philosophical sense.

Their senior colleagues in the guild gave them their experiential learning. An aspiring architect became a maestro after long years of apprenticeship by passing the guild's exam, but this did not mean that he mastered every dimension of his trade. As stipulated by guild ordinances in 1599 and 1643, the exam was mostly practical, although infused with some theoretical content about stylistic matters and the proper location for cities and different kinds of buildings.[26] In practice, the exam was limited to only some of the subjects listed in the ordinances, and required the candidate "to know of compass and ruler"—little more than the rudiments of measuring, surveying, and drawing. One architect quipped that all masters would have to relinquish their examination certificates if the ordinance of comprehensive examination were enforced rigidly.[27] Thus, the range of expertise gained through apprenticeship, the first source of knowledge, could vary greatly in practice.

A second source of knowledge was lettered and ranged from specialized books to texts of ordinances and laws. It is likely that the guild's maestros owned at least one architectural treatise or manual, and it is possible that they used them to supplement their experiential learning. Luis Gómez de Trasmonte, who worked in the Desagüe under Fray Manuel Cabrera in the 1670s, inherited from his father Juan, also a maestro, at least part of a Spanish translation of Sebastian Serlio's *Sette libri dell'architettura*; Luis's colleague, Rodrigo Díaz de Aguilera, owned a Latin *De architectura* by Vitruvius, which he annotated.[28] Only rarely would a maestro possess a proper library. Melchor Pérez de Soto, who worked in the Desagüe in the 1650s and owned some fifteen hundred books, was probably the only one in Habsburg New Spain with such an extensive collection, which included seven editions of Vitruvius's work, four of Alberti's, several other architectural, fortifications, and mathematics works by Vignola, Palladio, Sagredo, Arfe y Villafañe, López de Arenas, Albrecht Dürer, Cristóbal de Rojas, and Andrés García de Céspedes, among others, and more problematically for himself Copernicus's *De revolutionibus orbium caelestium*, as well as texts on astrology, alchemy, and hermetism. This impressive library seems to have done more for its owner's pursuits outside architecture than for his practical work building things: Pérez de Soto had astrological interests that led to his participation as a "perito cosmógrafo" on explorations of Lower California under Pedro Porte Casanate, and eventually to his death in the Inquisition jail.[29] The fact is that even this presumably well-read maestro left no evidence of linking the knowledge in his books to hydraulic or any other problem directly pertinent to the Desagüe.

Even when maestros owned architectural treatises, then, there is no certainty as to how they used them, or, in the case of the Desagüe, whether they were useful at all, as the architectural treatises did not discuss hydromechanics in sufficient detail. Possession of architectural treatises thus does not translate easily into a distinct manner of learning, especially since texts—both as treatise and legal norm—collected and codified existing knowledge that more often passed along through work as one person learned from another in an apprenticeship setting. In any case, not much could be learned from the treatises about how to build cities on lakebeds and wetlands such as those under and around the indigenous City of Mexico.

Nor did royal regulations in the *Nuevas ordenanzas de descubrimiento, población y pacificación de las Indias* of 1573 provide maestros much usable guidance. These ordinances also assumed that cities needed dry land as the basis for *policía* or its geometric representation in the grid: it was impossible to lay out and assign regular property lots and streets on marshlands. To make matters harder for the maestros, the ordinances also stipulated that cities should not be built in places either too exposed to hostile assaults by water or at excessive elevations.[30] Mexico thus added its highland setting and marshlands to the list of its unfortunate features for which texts provided too little guidance—and too late, given that the first serious flood after the founding of the Muy Noble y Leal Ciudad de México came eighteen years before the *Nuevas Ordenanzas*. As codifications of settlement that had already taken place, moreover, ordinances, legislation, and treatises about city emplacement are unlikely to have been consulted much when concrete problems in nascent cities arose.[31]

A third way of knowing is harder to trace: not just maestros, but all builders, as in the case of friars, learned by observing physical structures built by their predecessors.[32] Practicing architects and treatise authors such as Leone Battista Alberti, for instance, admitted to "having carefully and diligently gathered [knowledge] from the [built] work of the ancients, from which I confess having learnt more than from the writers."[33] In New Spain, these ancients happened to be indigenes whose major built works were destroyed during the conquest and the subsequent recycling of materials for the Hispanic building boom before they could find their Albertis to write about them in detail. On the whole, indigenous hydraulic structures fared relatively better and thus could be interrogated to reveal their secrets in much the same manner as Renaissance writers and architects examined ruins if imperial Rome's acqueducts and other artifacts to solve urban hydraulic problems of their day.[34]

Thus, while Hispanic chroniclers culled information on indigenous construction materials, processes, and techniques from indigenous informants, Hispanic maestros learned it from directly examining extant pre-Hispanic construction, particularly hydraulic structures, and observing Indian workers and artisans on their crews. The Indian usage of *tezontle*, the light and porous reddish volcanic rock of the basin, is one example. Although not considered a noble stone by Europeans, maestros took note of its pre-Hispanic use and how their workers treated it and redeployed the material into distinctly European shapes such as vaults and arches. This was particularly the case after the 1542 earthquake painfully demonstrated to them the risk of using denser and heavier stone on the clayey soils that covered much of the bottom of the basin. Architects learned to carve and place tezontle to lessen the load of the monumental structures of civic and religious colonization that they built on the none-too-solid ground of the city without sacrificing strength.[35] As importantly in a lacustrine region, maestros mixed ground tezontle with lime and wax or tar to seal foundations before building on them to prevent humidity from creeping up by capillarity.[36] In 1628, they used these techniques along with earth and grasses to increase the height of the causeways of Mexicalzingo, Calvario, San Antonio, Tacuba, and Atzcapotzalco in the city environs and the dam of Lakes Zumpango in the Desagüe district.[37]

These were innovative reworkings of the unfamiliar into familiar principles, born out of tinkering with materials from the basin and their behaviors. Clearly, they were not simple applications of technological packages from Spain, where people tended not to erect palatial buildings on marshlands. Likewise, observing Indian usages taught the maestros about the properties, locations, and processing of other materials, from timber to sand, stone for making lime to pigments for paint. This kind of observational learning about what methods and materials worked in New Spain began to be expressed in lettered form—the guild ordinances—in the early seventeenth century. Maestros who did not abide by this codified observation did so at their own peril—and that of the people for whom they built.

By 1735, when master architects codified standards for the various materials and usages that their trade had been adding to its corpus of building practices since 1599, appropriations of indigenous knowledge such as these would be complete. Now building rules stipulated various qualities of materials and their proper treatment: the proportions of length, width, and thickness for various types of planks and beams hewn out of named Mexican tree species; the types and measurements of blocks of local stone, which were always to be of good quality "and never tepetate, with which is involved grave deceit"; the quality and origin of sand for construction to avoid its adulteration with mud or *tequesquite* (sodium carbonate compounds); and finally the dimensions of bricks, only two fingers thick to allow for proper, even drying of the specific clays found on the basin floor.[38]

As "organic" architects of the land, they had advantages over foreign technicians in the Desagüe, from royal military engineer Adrian Boot to his colleagues from later periods, for maestros could assess the viability of repairs and works based in part on this knowledge of local materials and conditions. This need to be dexterous at handling local materials equipped the maestros with a sensitivity to indigenous technique that would play an important role in the Desagüe.

## EUROPEAN KNOWLEDGE IN PRACTICE

Spanish colonists who settled in the countryside of the basin of Mexico from the sixteenth century on were familiar with the sight and purpose of hydraulic structures of different types and legacies. Not all devices used to manipulate water in Spain required the intervention of a specifically trained craftsman—obviously, as many were built according to local tradition in agricultural practice. Using easily found and inexpensive materials like branches, rubble, stones, mud, and dirt, rural producers had over time become masters at building weirs, irrigation and drainage ditches, diversion and storage dams. These were very common types of flow-control devices in the Iberian countryside that shared the principle and many of their features with those that indigenous communities had implemented in the northwest quadrant of the basin, especially on the Cuautitlan River.

In both New and Old Spain, and before and after 1521, these types of structures were often collective in manufacture, ownership, and maintenance: all the labor, materials, and skills involved in these tasks were provided by peasant producers. This means that a broad swath of the Hispanic population had sufficient exposure to hydraulic devices and practices in the Old World to allow them to identify, analyze, and understand their counterparts in the new. Thus, potentially useful indigenous hydraulic practices could be identified not just by chroniclers, missionaries, and other letrados but even by those illiterate colonists who had managed to elbow themselves into a caballería or peonía or two of land in the City of Mexico's hinterland. It was the latter sector of the population that seems to have provided the subordinate guardas and sobreestantes for the Desagüe.

At the same time, more specialized knowledge was also necessary in other Desagüe features—the tunnel, the canals, and the causeway-dams—as well as the great variety of ancillary artifacts—floodgates, spillways, smaller masonry retention walls and structures, metal tools, ropes, measuring devices. In part, it was the complex technical requirements of these things and their greater cost that demanded specialized craftsmen who could build them well, durably, and without wasting valued materials, such as iron, which was imported. But specialized craftsmen in this new colonial setting served an additional function, which was to tap sometimes unfamiliar materials, techniques, work practices, and even designs to create these objects. Thus, if the guardas and sobreestantes could convey the knowledge of local terrain, materials, and usages that they derived through observation or from the indigenes, the people actually choosing and combining this knowledge were the technicians.

The maestros were the best prepared to choose and recombine technologies, materials, and processes available in the Desagüe district. This was not just because of what they knew and learned, but also because their undivided loyalty to the city directed them in the process of selection. Of course "choice" does not mean that everything the maestros did was about consciously selecting materials, techniques, or processes. Like all humans, these technicians often did things without reflection, as they had always been done or as they were done in adjacent sections of any given structure. The deployment of particular materials and techniques was thus a matter of both established norms and conventions *and* situationally contingent decision making. As a result, this act of choosing was more complex than might appear at first glance, complicated by the fact that the technologies, materials, and processes were of different provenances. The floodgates used in the Desagüe were of European provenance, for example, as were the masonry techniques and structures. But earthworks were not, as they tended to be done in the indigenous fashion and by indigenes themselves. How and why did the maestros actually choose from among what was available, then? We begin with choices made from European knowledge.

In the seventeenth century—and indeed, throughout the basin during the colonial period—the kind of floodgates that Enrico Martínez and his maestro successors emplaced to regulate the movement of water throughout the Desagüe were European in design and utilization. Enrico Martínez and all the maestros that followed him up to and including Ildefonso Iniesta Vejarano, maestro mayor of the Desagüe in the 1750s and 1760s, used two types of water-level control gates in the Desagüe: stoplog gates (*compuertas de trabas* or *de vigas*) and screw or spindle gates (*compuertas de husillo*) like the one shown in Figure 4.2. Both types were typically made of beams one-third of a vara thick that came from increasingly rare "aged cedar wood," or from fir.[39] A modular type of gate made with regular quadrangular cross-sectioned beams that slid vertically into grooves in the masonry reinforcement piers on the both ends of the channel opening whose course was to be controlled. Their biggest advantage was durability, some lasting up to eighteen years.[40]

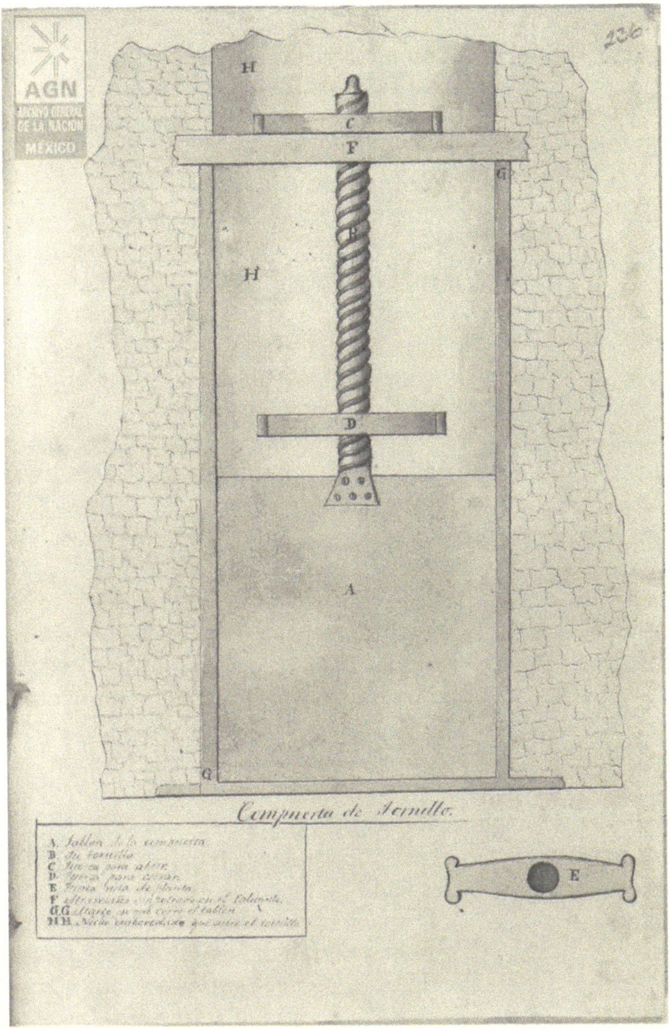


*Figure* 4.2. This screw or spindle gate (compuerta de husillo) used in the Desagüe in the 1850s probably differed little from those used during the colonial era.

SOURCE: AGN, Fomento Desagüe, vol. 3, f. 236. Reprinted by permission.

In contrast, the screw type was encased in iron (although lead was preferred because of its greater malleability and its resistance to corrosion) and raised by turning a crank with a bolt that pulled a threaded drive fitted to the gate, allowing the water to flow out the bottom. The dam-causeway that separated Lake San Cristóbal from Lake Texcoco had two screw gates. Rot and leaks in one of the gates required that it "be lined anew inside and outside with cedar planks and bound with two iron loops on its two heads and tarred all over." The other one was destroyed beyond recognition, so a completely new one had to be built, also "entirely built of cedar, with ten-vara beams, boards of the same wood inside and out, completely tarred, with its two screws, bolts, loops and all the remaining ironwork and nails all done anew."[41]

Screw gates cost four hundred pesos apiece, much dearer than the stoplogs. This was one reason why given a choice authorities preferred the latter, so that "except for the floodgates of San Cristóbal, all the others of the remaining lakes and those that the [Cuautitlan] River are and have been stoplogs." They may have been cheaper, but so long as the wood was healthy, they seem to have been quite impervious to water seepage because their "lapped beams fitting one on the other" sealed snugly both with each other once water expanded them and into the "strong masonry groove" where they were slid.[42] So long as they had not rotted or become too waterlogged, the stoplog's modularity allowed operators to remove sections manually from above using a hook and pothook (*garfio y garabato*) to allow the gate to operate like a weir, with water flowing over the remaining beams.

Certain locations demanded specific gates. After it was decided that the open trench conversion would be done with a water conveyance method, maestros always used stoplogs for the Vertideros spillway. Stoplogs were simpler for guardas and sobreestantes to operate than screw gates when it came time to release the water through vertideros into the Desagüe with *comportadas:* "It is not the same to remove a whole floodgate than to remove it in parts." This was because after six months of staying closed and in contact with water, the wood expanded and the sludge accumulated, which made it very difficult to loosen a complete gate from its crib. Because the Vertideros gates had "no machine for this purpose, such as there is in the [floodgates] of San Cristóbal, one risks destroying [a screwgate] in order to lift it," whereas rotten or broken beams in a stoplog could be replaced individually from Desagüe stores without disabling the entire device.[43] Given the frequent release of water into the trench, be it for debris conveyance or for regular Desagüe operation in this nodal point, these features made it a better choice than a screw gate, whose oak screw bore the combined weakening effects of torque and alternating exposure to brackish water and dry conditions. Maestros worried that any failure in a screw gate "could do great damage to the Desagüe." In addition to a variety of hand tools, ropes, and other equipment for the Indian trench laborers and the smith shop, under some superintendants the Desagüe stores kept spare screw gates, complete with their hardwood spindles, iron or lead loops, hooks, and other iron floodgate operating equipment and wood beams and planks for stoplogs.[44] But in an emergency of too much water pressure building up on the Vertideros wall, the time wasted with such replacement, if it could even be safely done at all, could be disastrous.[45]

In addition to usage and materials, maestros chose stoplogs on the basis of their empirical experience with their hydromechanical effects, which deserve a full quote:

> The advantage and benefit is that in floodgates the desired amount of water can be released, and tempered as necessary removing one, two or more beams. With these beams the water comes out above, and therefore gently and noiselessly; with the whole floodgates none of this can be achieved, because even if the gate is not raised entirely (which in the ones that have been emplaced I do not consider to be an easy task) the water comes out the bottom, and therefore with greater force and violence, pressed by the weight of that which occupies the part above. The violence will perforce cause very harmful erosion in that terrain. Perhaps for this reason the masters, experts and attendants the Desagüe has had have not used any other kind of floodgate in all these years than those made with beams. None but these has been used in the bridge where single piece ones have [now] been placed, nor in the other two water intakes that there are on the [Cuautitlan] River. The floodgate of the vast Lake Zumpango is made with beams. Same thing with the one in Acolman, admitting that only on the causeway of San Cristóbal are there sluices of a single piece, which must be attributed to the fact that the latter was built by the consulado, which made the cost of building a lodge where it could be raised by machines acceptable; nor does the water of San Cristóbal have to be distributed to cultivators like that of the River, since it pours immediately into Lake Texcoco.[46]

Because all of these devices derived from the European technological tradition would be deployed on a landscape in the Desagüe district still greatly shaped by indigenous engineering devices and structures, maestros had to figure out how to integrate these objects with very different purposes and, if possible, how to harness the indigenous objects and the skills that went into them into the drainage.

## MAKING INDIGENOUS TECHNOLOGY USEFUL TO HISPANICS

A Hispanic project, serving only a slice of that population, the Desagüe was nonetheless not the product of only European or creole technology that it appears to be in the literature about it. In fact, it incorporated important components that were indigenous, such as the crucial Cuautitlan River diversion dam. Because of the ways their background and work sensitized them to the materials, techniques, and processes involved in the construction and maintenance of such structures, the people most directly involved in appropriating such components into the Desagüe as a whole were the maestros. However, this does not mean that the Desagüe's maestros conveyed the social, economic, or ecological values that the indigenous creators of these techniques invested in their original usages. Quite the contrary, they approached and incorporated indigenous technologies into the Desagüe acting exclusively as urban agents looking for effective solutions to practical problems. The result was not an example of mestizaje. Rather, it illustrates a process of technological "transfer" and "diffusion" whereby the flow of that transfer was multidirectional (as opposed to merely from Europe to the colonies, and from settlers to natives, as is often the case in the literature). While it is of course true that European technique and technology traveled to the Indies through texts, objects, people, and practices, there was a simultaneous current of technological appropriation by colonists of indigenous technology. Without it, colonization was impossible.

Such appropriation had been going on for a while. Angel Palerm showed how after the 1604 flood, for example, Spanish authorities supervised the restoration of the pre-Hispanic causeway-dams of Nezahualcóyotl and Tepeyac radiating to the west and north from the city using not just indigenous labor and techniques combining stone-and-mud walls with an external stake support but also Moctezuma's system for their appropriation.[47] However, because the survival of indigenous technology depended largely on the survival of its bearers, the Spaniards' ability to learn about the techniques in the hydraulic technology of the Indians was restricted by drastic depopulation.[48] The city was in this sense lucky to have technicians who, thanks to backgrounds infused with the Iberian Islamic legacy, allowed it to profit from—and thus sustain—extant indigenous tacit knowledge. Fortunately for the maestros and the city, however, the Hispanics in the Desagüe district from among whom the subordinate guardas and sobreestantes were selected had begun to accumulate this knowledge long before the indigenous demographic trough, so they may well have been the first line of technological appropriation. Because scholars have not interrogated or explained the logic of appropriation, it is hard to tell exactly how it happened. This is what this section aims to reconstruct.

So far as hydraulics is concerned, there are several sites and practices where this dynamic of appropriation is evident, but nowhere more than in the Cuautitlan River diversion dam. This earth-and-grass dam dated back to the early fifteenth century. It drove the river into the western portion of Lake Zumpango, called Citlaltepec, extending right into the lake itself, keeping it flowing along a new northward channel that had been built to divert the river away from its original eastward flow. It had three main purposes—to deliver extra water to Lake Citlaltepec to support raised field agriculture there; to feed a network of irrigation canals fanning eastward from the new channel, and finally, to diminish flood risks in the town of Cuautitlan. However much the indigenous *planning* of the Cuautitlan River's reengineering may have depended on "cosmopolitan scientists who had at their disposal a great deal of accumulated information,"[49] the actual construction and subsequent maintenance and modification were done by local indigenes. Hence, while perhaps lacking the scientific apparatus of their lords' civil engineers, the commoners of the Cuautitlan River basin certainly distilled and passed down substantial practical and tacit knowledge from the whole process.

When the Desagüe was built, this originally indigenous structure was incorporated into it lock, stock, and barrel in terms of materials, workmanship, function, and placement. But not in terms of purpose—in the hands of Hispanic technicians, this diversion's new purpose was to allow the city to deprive of water the lower-elevation lakes that flooded it. When the open trench began, the diversion dam's additional function was to supply water for the excavation and removal of debris. By the middle of the eighteenth century, the segment of this dam that projected into Lake Zumpango was known as *La Albarrada del Rey*, or the king's dam. It went from a location known as Cruz del Rey to Huehuetoca.[50] Made out of mortar and stone, it was maintained out of the royal coffers. The part upriver, by contrast, remained a grass-and-earth dam maintained by local Indians and hacendados at their own expense.[51]

But there was a problem with this dam. Although cheaper to build than structures of stone, rubble, and mortar, earthworks were vulnerable, and large ones like this required significant and constant investments of labor. Guardas and sobreestantes knew this well and informed the maestros and other technicians of the specific regional conditions that made it so. In their opinion, all earth dams and causeways in the Desagüe district suffered from the debilitating effects of the fact that the grasses for repairs often came from brackish waters, and that the region was infested with gophers (*tusas*) and snakes, which burrowed into all freshly repaired or built earthworks.[52] In addition, the alternation of wet and dry seasons debilitated them, and in the case of the Cuautitlan River dam in particular, the body of water controlled by the dam was liable to break through during peak rain events. On the Cuautitlan River dam, human mischief also took its toll: cultivators opened clandestine outlets (*ladrones*) through its earthen structure to irrigate their fields on the sly, even under threat of severe punishment (by the eighteenth century, Indian violators could receive two hundred lashes and six years in an obraje, and Spanish ones sixteen years of overseas exile).[53]

Because of these maintenance problems, the earthwork dams containing lakes Zumpango and San Cristóbal were eventually redone in part or whole using permanent materials.[54] Enrico Martínez himself had actually begun to convert the Cuautitlan River diversion dam as well into a mortar-and-stone structure, but his strong belief that this dam in particular deserved more permanent materials hit a wall when Viceroy Gélvez put a temporary freeze on further Desagüe investment in 1623.[55] When the money started to flow again, the retrofit of this whole dam did not recommence at any time for the rest of the colonial period, despite repeated observations about its fragility as an earth-and-grass structure. Why not, important to the safety of the city as this dam was?

The answer has to do with calculations of costs and benefits. Oidor Juan Francisco de Montemayor y Cuenca expressed this calculation clearly in a 1673 report on the Desagüe to the viceroy. Whether the Cuautitlan River diversion dam was to be converted to permanent materials "has to be studied further," he explained, "reckoning whether it shall be convenient to make this dam with [mortar and stone for] greater strength to prevent the anticipated risks with regard to whether the costs of expenditures of such a precaution will exceed the damages which are likely to occur [in case of a breach]."[56]

To understand what this reckoning involved, we must first take a closer look at the constructive features of this structure as well as to how the masters worked on the diversion dam and similar artifacts. As a load-bearing structure, this dam was not a simple thing to emplace or build. Careful survey of the terrain was critical to its success in redirecting flows and resisting the force of water acting against it. In its construction, it was probably similar to the causeway-dams in the city: two parallel rows of wooden stakes driven into the ground to serve as a caisson for earth and grass placed both inside and out, and finally protected from erosion not with stones as was done in city causeways but with living grasses, and later, sometimes, trees. It is not entirely clear how the grasses and earth were applied in this particular dam. Although not devised for Hispanic society or for the study of technology, adapting James Lockhart's "new philological" method of tracing the movement of Spanish lexicon and grammar into Nahuatl to study the process of indigenous social change after the conquest helps shed some light on the technique using earth and grass.[57] Perhaps they were laid within the anchoring structure of stakes or piles as "clods of earth extracted [turned over?] from the surface terrain together with its rooting vegetation."[58] This is suggested by the philological associations of the Nahuatl nouns for grasses (*cueptli*, *çacatzontetl*, *tlachcuitl*), for reeds or weeds (*çacatl*), and *vitzoctli*—"a pointy oak stick to pull up grasses and open the earth"—with the verb forms for "to turn," "to turn over," and "to break turf" (*cuepa* and *tlachqua*, *ni*).[59] In any event, when he worked in the Desagüe in the 1630s under Fray Luis Flores, master architect Juan Serrano (better known to scholars for his later participation in the construction of the cathedral and the university in the capital), explained the materials, process, and timing involved in the maintenance of the Cuautitlan diversion dam:

> It is customary to extract the grasses with which we build this dam during the rainy season, because if we did it during the dry season in the time that it would take us to build one vara we could build ten in the rainy season, and the reason for this is that when the soil is damp the grasses can be extracted with less expense and more comfortably, and thanks to this the little roots are able to grow and bind themselves, because they are still green and tender and stamping them very moderately, after two days under the sun they become as hard as a wall and with the humidity therein the grasses sprout and the little roots become entangled one with the other.[60]

Just how "customary" was this, and what did it mean?

Serrano's is merely the most detailed exposition in a plethora of references to the use of grasses, whether as sods or plants, in earthen dams, often layered with tepetate in the Desagüe record.[61] It was indeed a widespread practice. Sods of earth with live plants were common construction materials in rural Europe as well, of course, but everywhere it was important to know what kinds of plants were binding the turf together. Choosing the right kind of turf for each usage, knowing how to extract it and when, and then how to layer it with other materials required a certain degree of—well, rootedness, so to speak. Only those who had worked in Mexico for some time, observing where, when, and how such structures were built and maintained there, were able to distill a clear understanding of what constituted appropriate processes, sequences, and species for Desagüe uses and what were counterproductive ones. Serrano's usage of materials, processes, and seasons suggests that he relied on knowledge and practices distilled—frequently through the sobreestantes and guardas—from local populations, particularly those of indigenous townships with a history of hydraulic and agricultural manipulation of the terrain.

For earth in dam construction Hispanics used the terms "tierra" and "tepetate" interchangeably. For grasses and grassy earth sods, on the other hand, Serrano used the Spanish term "césped" as well as the Hispanized Nahuatl name "zacate," but he had different species in mind for different usages. For example, maestros learned from the sobreestantes to avoid zacate that had been extracted near brackish water, probably because as each season's rains dissolved the salts the sods lost mass.[62] By contrast, the grasses had to be hardy like *Hilaria cenchroides*, a species endemic to the Huehuetoca and Tepotzotlán area identified by Aimé

Bonpland and Alexander von Humboldt in their travels through Mexico. This versatile plant had many uses in the region. While the stalks and leaves were used as fodder for livestock, the plant's Desagüe utility came from the spreading rhizomes (underground stems) detailed by Bonpland, and shown in Figure 4.3, which consolidated the soil and likely made the plant a good choice for earth dams like the Cuautitlan River's. Better still, it was well adapted to the salinity and drought characteristic of the northwest quadrant of the basin.[63]

*Figure 4.3.* Plants useful in earthworks: Hilaria cenchroides, as depicted by Aimé de Bonpland.

SOURCE: Aimé Bonpland et al., *Nova genera et species plantarum*, 7 vols. (Lutetiae Parisiorum: sumtibus Librariae Graeco-Latino-Germanico, 1815–25), I: 37, I: 96, and I: 117.

For the earthen part of the dam-causeway retaining Lake Zumpango, however, Serrano used a plant he called *césped de tule*. Tule referred to a number of wetlands rush species, most likely *Typha latifolia*, *Typha angustifolia*, *Scirpus californicus*, and *Scirpus paludosus*. Indians had a huge variety of uses for the reed of tule, most of which involved cutting it above the root and weaving or bundling its fibers or stems in a variety of objects for daily use and for sale. Upright bundles, for example, were sunk about a meter from lakeshores to create nurseries for *axayacatl* eggs, which when harvested and dried were used as human food of high protein content.[64] Was Serrano's use for tule different from that for grasses or grass sods? It seems so: one of his sobreestantes was to use carts—a Hispanic means of transport—to bring grasses from farther away to reinforce the top and dry side of the dam, while another oversaw the Indians on "canoes who bring the *césped de tule* to protect the dam from the waves that the northerner whips up, because otherwise the water would lop off the top of the dam."[65] Since the beginning of the Desagüe it had been the practice to use sods on the outside of retention earth dams while the inside was reveted with "mud, sods and reeds."[66] As Serrano explained, both types of living products were not cut, but extracted with their roots. This allowed a solid matt to consolidate on the dry side of the dams, while on the wetted side it is possible that the reeds were placed not on the face of the structure but at its foot, where their growth would temper the waves. In other words, different kinds of grasses, sods, and reeds were clearly not interchangeable, and only a local sobreestante, guarda, or technician could tell the usable differences among species (as opposed to genera) and both extract and use them properly.

Emergencies illustrate the logic of technological appropriation and when and where it did not work. In the midst of the rainy season of 1653, master architect Melchor Pérez de Soto was urgently summoned to the Desagüe to repair a gaping hole more than twenty-one meters long in the Cuautitlan's diversion dam. The swollen river had broken through just in front of the church of Teoloyuca, threatening to create an even bigger breach. To repair the dam while the river was still in full force, rather than at the ebb of the dry season, the maestro had to first create dry enough conditions, and this meant forcing the water back onto the river channel. Only after this was accomplished could he begin to rebuild the dam section. Both steps needed to be accomplished quickly to protect the unaffected dam portions and preempt a major flood if the water rushed toward the city. More than two hundred Indians mobilized to extract and haul grass sods and tepetate and to ram them into hastily constructed cribs anchored along the desired riverbank. But this was not enough, and Pérez de Soto had to spend 1,969 pesos on materials for a structure that would normally be maintained at Teoloyucan expense.[67] His bill eventually included nails of imported Basque iron—three hundred one-quarter vara long, four hundred one-half vara, and four hundred of a type known as *clavos de barrote*—and several varieties of milled wood—sixteen beams *del colegio*, twenty-three planks from Río Frío, one hundred boards of *xalocotle*, and 196 *morillos*.[68] The morillos, which in more expensive urban construction were typically made of cedar so they could bear loads in foundations, walls, and even low albarradones to protect buildings from water, were most likely used as stakes to which the boards would be nailed.[69] The wood alone required sixteen cartloads for its transportation; the grasses and tepetate, brought in from farther beyond, eighteen mules and horses as well as two pairs of oxen.[70] Working "from sunup to sundown" for two days, the maestro and the Indians managed to stop the breach and begin repairs with

> two cribs in place of the broken caisson and embankment [*tixera y tajamar*] of said breach, which are on the other side of the river bed, filling them in with very well rammed tepetate and grasses so as to begin to overcome the water, and likewise we strengthened the front and staking of said cribs, giving them some buttresses . . . leaving a section six varas long and one wide on the side of the palisade of stakes facing the river bed, and the two cribs of the caisson and embankment on the outside, filling the stakes with well-rammed grass and tepetate, with which said breach and the body of water that flowed out through it were entirely sealed.[71]

Although on the one hand grasses and tepetate were used, iron nails and hewn wood products were hardly the materials of indigenous hydraulic construction. On the other hand, Indians used their own energy and not draft animals to carry materials to their normally allotted dam maintenance sections. Had perhaps the Indian technological package been discarded in the few years between Juan Serrano's work on the dam and Melchor Pérez de Soto's? Conversely, is *this* perhaps an instance of technological mestizaje?

The answer is "no" on both counts: Desagüe technicians and personnel continued to appropriate indigenous technology according to a specific logic, not by indeterminate mixing. While they knew that varying degrees of compulsion were necessary, in terms of capability and skill, authorities trusted indigenes to perform their hydraulic duties well. Under ordinary conditions, it was best to rely fully on indigenes for the Cuautitlan's diversion dam, particularly those of Teoloyuca, "who will do it as it ought to be done because of their knowledge and experience," senior oidor Francisco Calderón y Romero reported to the viceroy after his Desagüe inspection in 1665. But the problem is that the whole indigenous technological package for earth dam construction and repair entailed multi-community coordination of tasks that needed to be performed at specific times of the year. Under the circumstances of the 1653 Cuautitlan River breach, including the many other stresses that bore on local communities, it seems unlikely that this coordinated and autonomous mobilization would have happened swiftly enough for the capital.

More probably, then, Pérez de Soto sifted through the indigenous and Iberian techniques with which he was familiar, selecting and combining the most appropriate components of each quickly but with care and deliberation for the flood conditions before him. The spectrum of materials and practices on dams that maestros used ranged from indigenous to Hispanic. But the logic of their choices was determined by the function of each structure, by their culture of origin and extant practices, and of course by their social loyalties.

If the maintenance of the Cuautitlan River dam under normal conditions lies on the indigenous end of that spectrum, all practices and materials used for the dam that contained Lake San Cristóbal were on the Hispanic end. Erected in 1604 as part of postconquest flood-control measures, there was neither an indigenous dam precedent to go on in that site nor the habits associated with such structures among the population, at least not comparable to those found in townships tied to the Lake Zumpango-Cuautitlan River watershed. Maestro Luis Gómez de Trasmonte noted in October 1674 that the dam was of such good quality and strong masonry that it had survived the forty-odd years of neglect since its last repair. On the side of the water, he found that the scarp sat solidly on foundations of stone comparable to granite in hardness. Although full of trees that had sprouted and cracked it in some places, the causeway had only minor leaks that did not threaten the city.[72]

So the answer to why the Cuautitlan River diversion dam remained a structure of impermanent materials throughout its colonial history is that this was how authorities in the city passed on the cost of building and maintaining it to communities who had provided the technology, the labor, and the materials in the first place, and because they had the technicians to facilitate and organize this subsidy from the indigenous realm to the Hispanic one. This subsidy took intangible and tangible form. The intangible was the knowledge itself, which although vested collectively may have been concentrated and coordinated by a specific official, at least in some of the townships. At least from the early eighteenth century, when it is first recorded, the Teoloyucans, for instance, elected their own "juez de albarradas," or dam-causeway magistrate, who as the person responsible for construction and repair of these structures would have been a human repository of the relevant tacit expertise.[73]

The tangible subsidy is measurable, and not negligible at all. After a century of compliance with this obligation, the Teoloyucans calculated that if the king paid for their share of maintaining the Cuautitlan dam, which went all the way from the limits of the township of Cuautitlan "to the end of the trees that are right by the spillway of Coyotepec,"[74] and cleaning the river channel, it would cost him up to 8,000 pesos every year.[75] Prorating this figure to the entire length of the diversion dam, it turns out the capital was saving its elites and the crown the tidy sum of 23,111 pesos every year by shifting to the hinterland the cost of maintaining just this particular structure.[76]

Authorities knew the exact amount of their savings. After decades during which gophers and snakes had been busily undermining the earth dam of Lake Zumpango with their burrows, in 1710 guarda mayor Joseph de Valladolid conducted an experiment to find a solution for "the complete lack of permanence of this earthwork, whose expenses are wasted because it must be redone every year." Lake Zumpango's long dam was composed of two segments. One segment ran NS; the other started at the southern tip of the former and ran westward. Valladolid experimented on the latter segment, building "a short section of three varas out of mortar and stone and an equal one of stakes [with earth and grasses], to determine exactly the difference in expenditures that there will be" in rebuilding the entire dam with permanent materials. Multiplying the cost of these samples, he found that making the dam in mortar and stone would cost 29,679 pesos, while the earthwork would amount to 16,625 pesos.

With this information, authorities decided to leave the segment facing the township of Zumpango as an earthwork, rebuilding the segment facing the townships of Teoloyuca and Coyotepec in mortar and stone.[77] Why the difference? Because the Indians of Zumpango were required to maintain the dam of their lake as they had no other maintenance duties, while the Teoloyucans and Coyotepecans were already charged with the maintenance of the Cuautitlan River dam and could not be asked to maintain an additional dam. When they inspected the Desagüe with the viceroy in 1748, maestro mayor Manuel Alvarez and military engineer Felix Prósperi examined the NS earth segment. It passed their scrutiny with flying colors, as "His Excellency, the experts and others who accompanied them applauded the good quality of this work."[78] More than 34 meters long and 6.5 meters wide, this structure was appraised at 12,000 pesos, but because of the arrangement whereby Zumpango Indians "build and maintain this terreplein on condition that fishing, tule, ducks and tequesquite shall be forever free for them," the cost of the structure during its entire history was transmuted into a "savings obtained by means of this provision without detriment to third parties."[79] Paradoxically, then, the more confidence indigenous hydraulic skills enjoyed among Desagüe personnel, the more burdened the Indian townships became. By the same token, in the long run the authorities' reliance on these technologies enabled the indigenous peasantry to keep them alive and through them to retain or struggle for effective control of land and water.

One last example illustrates how far from being a haphazard mixing of elements from different traditions, the act of choosing materials and techniques was fraught with intention. To dig away the hill on top of the tunnel for its conversion into a trench, Indians were lowered into the tunnel's *lumbreras* (shafts) by means of ropes tied from their waists to a beam, itself lowered by means of ropes down the sides of the excavation.[80] There, with iron bars that were also tied to their waists so they would not drop them, they broke the sides, loosening up the terrain with the bars as they were gradually lowered and letting it fall for comportadas of water to sweep the debris away and leave the tunnel floor unobstructed. As the trench widened, they were tied to stakes driven into the tepetate ledge itself. When he worked on the open trench, Serrano wanted the viceroy himself to order "thin tarred rope" (*xarcia delgada y embreada*) made with hemp (*cáñamo*, *Cannabis sativa*) to be sent from Veracruz. He was vexed to have been sent rope manufactured from the wrong plant for these lowering and raising ropes that had to resist both tensile force and friction:

> Given that these cords by which [the Indians] are tied are the ones that work the most, being as they are lowered and raised continually as the work proceeds, and they chafe against the tepetate, they will perforce wear out, and from their splitting or breaking some bad event could fall upon the Indians, given that although the cords that they have now seem strong they cannot be so, for they are manufactured from the main ribs of the leaf of maguey plants, which by nature are weak, unlike those of hemp, which is made stronger and more durable with pitch.[81]

An Old World cultivar, the long, strong, and light fibers of the hemp stem made the plant ideal for the manufacture of sails and naval cordage, for which it was impregnated with tar to prevent water rot. Charles I issued a decree on June 13, 1545, to promote the cultivation of this plant, useful in naval construction, in the Americas. To little avail: attempts to raise hemp commercially in Atlixco, Huexotzingo, Tlaxcala, and Puebla largely failed, and at the beginning of the nineteenth century Alexander von Humboldt was still lamenting the lack of this crop.[82] The laborious and water-consuming process hemp demands to yield its fibers helps explain why Indians preferred to stick with their own plants and were uninterested in hemp other than for medicinal and narcotic purposes.[83] Given this unlikely availability of Mexican hemp, unless he unrealistically expected the fiber to be imported via Veracruz, Serrano seems to have been applying the Hispanic treatment of tarring to an indigenous fiber—sisal (*Agave sisilana*). Sisal is indeed more resistant and was preferable to henequen (*Agave fourcroydes*), which nonetheless had also been tarred and used for traction and load bearing in the Desagüe.[84] For the maestro, the quality of the fiber—and the application of heated and melted tar to an indigenous material comparable to hemp—mattered if he was going to keep the workers hung from it into the gaping open trench alive.

## WORKING WITH THE SEASONS

In addition to appropriating indigenous materials and techniques, maestros also adapted to indigenous seasonal routines. Because the Desagüe's main materials were water and earth, it was imperative that the people who made on-the-ground decisions about timing and materials involved in Desagüe construction and maintenance understand how the biseasonal regime affected the basin of Mexico's terrain, vegetation, labor patterns, and the mechanics of its floods and flood control differently than they did in Europe. Maestros knew best how to handle project demands in view of the seasons and how they affected conditions. In this they were informed by guardas and sobreestantes, who in turn derived much of their knowledge from Indians, who knew which grasses could be extracted when and which needed to be left alone, for instance. In an early modern economy, all work was subject to seasonal variations and was known to be so; the Desagüe was no exception.

As a result, maestros had to know what happened to the basin's different terrains (especially tepetate) under dry or wet conditions, when seasonal materials became available during the year, on what kind of notice they could be delivered, and what might be reasonable to expect in terms of labor according to both the methods used for its acquisition as well as its availability at different times of the competing agricultural calendar. If the maestros who worked on the Desagüe learned to combine the local manifestations of these elements according to the seasons, it was because people in their trade had been working seasonally in Spain already. In the Acequia Imperial de Aragón in the 1580s, for example, desilting and dam repair were done in the summer months before the rains.[85] Masonry structures in the drainage were less subject to the natural cycle in terms of the materials used in their construction, since lime production and stone extraction were not as seasonal, but in terms of setting, inspections, and repairs they did require dry conditions.

Enrico Martínez himself had worked seasonally in building the Desagüe, and in this he was most likely following the practice in the building trades, especially those of masters who specialized in water structures. Afterward, the friars put into place a system of open trench conversion that was excruciatingly seasonal in its dependence on water for debris removal. At all times (and even today in the Sistema de Drenaje Profundo)[86] inspection, maintenance, repair, and construction tasks in the tunnel, the open trench, the causeways, and the riverbeds all had their proper time on the calendar. Inspections during the dry season, for example, allowed maestros, sobreestantes, and wardens to identify repair needs because rivers and streams carried less water and lakes dried up, revealing the foundations of dams and causeways. The tunnel too could only be inspected during the dry season, and even then it was an ordeal. The maestros' appropriation and redeployment of indigenous knowledge reinforced this tradition of seasonally centered technology around the Desagüe. Juan Serrano, for example, explained in his report how working on causeway repair only during the rainy season was better not only for the availability of materials but also to reduce labor needs:

> In the dry season the grasses are torn apart when one tries to pull them up and where one would strike a single blow of an iron bar one now needs six, and it is necessary to have one hundred Indians with pails to water the earth from which said grass is to be extracted, and many other things which may occur and which I will not mention so as not to tire people. To be able to work as quickly as the works demand during the rainy season, six hundred Indians ready and able are needed, and during the dry season which begins in October, one thousand will be necessary in order to leave Lake Zumpango all dried up, where we store the waters from Pachuca and from the Cuautitlan River that go into this [Desagüe] work.[87]

Maestros developed a method of working that was fully subject to nature's biseasonal fluctuations in the basin. This was no small refinement, and a key factor in the acceptability of the maestros' proposals as compared to those of the entirely foreign military engineers. Over the course of the seventeenth century, the friars and maestros' observance of seasonal rhythms in Desagüe activities diffused as a habit of mind among oidor superintendants. In 1691, for example, oidor Pedro de la Bastida was ordering repairs on the causeways of Zumpango and San Cristóbal specifically "at the beginning of the rainy season, which is when they can be performed, and not in the dry season because there are no grasses."[88] The calendar of earthworks maintenance duties replicated those of the Cuautitlan River diversion dam: February for cleanup and widening of river and lake beds; July–August for the repair of dams. February was the middle of the dry season, so water levels would be low, while the early rains presaging the summer sprouted the grasses (zacate) with which many dams were built. Desagüe authorities claimed that these were "the down times in the haciendas, when the workers are lazing about."[89] This was not quite accurate, however, and this fact would have consequences.[90]

The types of tools needed in the Desagüe also varied with seasonal conditions, requiring more metal during the dry season as the earth and the tepetate turned crusty and difficult to excavate. In August 1637, Juan Serrano scoured the Desagüe stores for used tools that he could recycle. Why invest time and labor in the Desagüe's forge to transform four hundred bars of three-fourths varas length into two hundred bars of one-and-a-half varas in length? Because the added length gave more leverage to the tool; the half-pound of steel he used for both tips, strength. He asked that an additional six hundred bars of this type be manufactured "for both seasons." He also needed four hundred spades one-third of a vara in length and one-fourth in width (for the metallic part alone), with edges reinforced in metal sheathing "for the dry season," but he was ordering them now so they could start manufacturing them according to his instructions and have them ready in a timely fashion.[91]

Timing repairs during the rainy season also saved on labor. Despite the use of repartimiento labor in the Desagüe until the beginning of the nineteenth century, labor saving was important for colonial authorities, since even corvée work was not entirely free: a wage of two reales per day was paid, and Indian workers were due food rations (typically meat, maize, and chilies) and shelter.[92] Moreover, any technology that tended toward labor saving was likely to be of critical importance for the Indian towns that were obligated to maintain these dams, in this case Teoloyuca: population decline affected their ability to produce for sustenance, tributes, and other impositions, let alone spare any labor for public works.[93] The trough of the population curve that coincided with the intense first few decades of the Desagüe made labor savings important, if only to smooth tensions with Hispanic entrepreneurs and haciendas angry at the Desagüe for carting off their workers at critical times.

## DYNAMIC CONSERVATIVES

By identifying and synthesizing indigenous techniques, materials, processes, and seasonal rhythms that could be deployed into the Desagüe, master architects served rulership very well indeed. Because of their ties, as architects and as urban dwellers, to the city's administrative, merchant, landowning, and entrepreneurial elites as well as all the religious orders, architects were not in the best position to question the goals of the Desagüe or the technological or agricultural implications of the desired desiccation. It was not their place or inclination to throw doubt on a project that aimed to preserve all that real estate at the seat of power, much of which they had in fact designed and erected.

By appropriating indigenous technologies, the maestros inadvertently propagated them, slowing their extinction in the midst of Indian population decline and the corrosion of their land- and water-based production units in the face of the advancing haciendas, even though they were redeploying them for a project that was as entirely Hispanic as the hacienda. This paradox requires a language that reflects its dynamics. Hardly the product of mestizaje at all, the technology appropriated by colonial technicians, including the maestros, was inserted into a Hispanic technological project whose aims were the radical modification of the environment of the lacustrine region that had engendered the Indian technologies in the first place. In this matter the Indians had apparently little say. The proposal of the Teoloyucans to move the Cuautitlan River diversion dam westward did not challenge the Desagüe as a whole project. They intervened because not only did their fields flood after water breached the mislocated dam, but they also had to waste time and effort in futile repairs every time. Furthermore, despite their evident skill, the Teoloyucans were not allowed to implement their project without the Hispanic maestros' mediation: maestros Montero and Alatorre were ordered to mark out the terrain for them.[94]

By appropriating indigenous technology master architects did more than simply recycle techniques. These practices belonged to an extant pre-Hispanic technology that maximized the usefulness of unprocessed materials and could only be performed during specific seasons. This dovetailed perfectly with, and indeed reinforced, the open trench plan of the friars, in which the seasonal dependence on water became key. The appropriation was in fact far broader: it encompassed a seasonally centered technology that included materials, labor organization, and constructive methods. In the Cuautitlan River area, architects also appear to have been the unwitting conduits for the survival of knowledge regarding the behavior of this river, without which all diversion efforts in its subbasin would have been far more of an act of faith—in Vitruvius, in the Virgin of Remedios of the city's Spaniards, or in the unmuddied lines of a geometrical design.

The guild's maestros were thus the figures who most consistently mediated and translated knowledge to and fro among these distinct groups in the Desagüe, all the while remaining solidly anchored in their urban loyalties and never quite making the transition into lettered expertise. Conversely, the fluid ways in which knowledge moved along a broad social spectrum that ranged from Indians, sobreestantes, guardas, maestros, friars, and superintendants suggests that differences in skill and empirical knowledge between skilled laborer and trained technician, and between both of these groups and Desagüe superintendants, were not perceived in the same way among these people as among later observers and historians.

As time wore on, for reasons that are explained in the next chapters, master architects found themselves playing second fiddle to the Bourbons' technicians, who brought with them a completely different training and set of priorities (imperial ones) and whose opinions about the Desagüe were clearly preferred by viceregal and metropolitan authorities. However, if the maestros' Desagüe role became increasingly that of subcontractors carrying out somebody else's design, it was not because they had failed as technicians. It was because together with the friars earlier on, they had succeeded in designing and consolidating the arrangements that shifted the burden of the protection of the city onto the countryside: they had served their purpose and crown priorities shifted elsewhere.

Throughout, however, maestros—and to a great extent also the guardas and sobreestantes—continued to participate in inspections, repairs, and modifications in the Desagüe and to provide support for essential continuities in the project. Such a sustained permanence, even as their star waned, is precisely how they conveyed their most important form of conservatism—in the literal, non-Whiggish sense of the word—into the Desagüe. The maestros in particular never wavered in their loyalty to the city and its elites, and this led them to accept without ever questioning the desirability, wisdom, or consequences of the means by which these elites protected their wealth from floods. So while these men inadvertently aided the survival of indigenous technology by redeploying it into a drainage that aided the hegemony of the city and the colonization of its hinterland, they had no interest in articulating rural dissent from any group with either the drainage or any other project of urban elites. Against the practices of repartimiento labor and maintenance duties that shifted the costs of flood protection to the countryside, the maestros would never innovate, despite their plasticity and inventiveness in all other regards. As a result, their choices wound up reinforcing the social and spatial dynamics in the basin, including, quite paradoxically, the persistence of the autonomous indigenous peasantry. These choices would help constrain the ability of military engineers to effect change in the Desagüe.

CHAPTER FIVE

# All the King's Men

As Fray Cabrera lay on his deathbed in 1691, the open trench had become "in all truth, nothing more than a ditch, deeper in some parts than in others."[1] Even this derisive description by a royal military engineer was generous: from the air it would have been hard to see it even as a ditch, pocked as it was with unfinished or extant tunnel sections. Because the whole aim of the Desagüe was to protect wealth in the city, and during the friar superintendancy there had been no flooding on the scale of 1629 to reactivate fears, the works carried on driven mainly by inertia. As a result of both of these factors, the Desagüe as a whole was contributing less to the class colonization of water and land in the basin than it might have with a change in conceptualization, renewed flood fears, or both. At the same time, after more than a century and a half of colonization, there was nary a major river in the basin that humans had not tried to alter in the service of the safety of the city, nor a stream, lake, or wetland that they did not use as a source of inputs—such as irrigation water or energy—or things for direct sustenance. Private property relations had penetrated deeper and farther into land, water, and biome. Yet this had not erased indigenous communal forms of resource use predicated upon the assumption that these three elements needed to and would continue to fluidly interact. This was in part because the Desagüe itself was still incomplete, and as a result, what shrinkage of the wetlands had happened was as much the result of other manipulations and of their natural senescence.

Still, there had been attempts to shake things up while Cabrera was still directing the works. Military engineers had started to intervene in the project more consistently, and to inject their specific culture into it. In addition, from 1670 on, audiencia members attended to the Desagüe more intensely, if not necessarily productively. When in 1691, the oidores took over the superintendancy, they did so with broader geographical and jurisdictional powers than any of their friar predecessors.

The new century brought in the Bourbon dynasty, which sought to turn the tide of imperial decline by transforming its American possessions into true colonies. The oidor superintendants would transmit into the Desagüe many of the same cooperative or hindering positions they took regarding the new monarchs' interventionist policy toward New Spain as a whole, and whatever they did in the project had ripple effects throughout the basin.

Which raises a question: If, unlike the friars, the oidores never actually held a tool other than an inky quill or a horse bridle in their hands, how did they shape the drainage? They did so in part with the help of another brand of royal servants—the military engineers. Military engineers had nothing to gain from aligning with local elites and no motivation to cheat the monarchy, on which they depended entirely for their incomes and the welfare of their families. This gave them an autonomy from local interests that nobody else had, and which they used to speak in increasingly uncompromising terms about problems and solutions in the drainage. Although these royal technicians tried hard to leave their mark on the Desagüe beginning in 1614—when Adrian Boot first inspected it—they were able to sink their teeth into the project with vehemence only after 1675. But even then, their proposals, reports, and analyses could push change in the Desagüe only so far.

The more the engineers' proposals challenged the basic structures and practices upon which the social arrangements in the Desagüe depended, the more resistance they faced; the better suited to the priorities embedded in the drainage, the swifter their acceptance. This resistance came from several quarters and did not abate, even with oidores at the Desagüe's helm after 1691. As a result, the Desagüe became a battlefield of sorts over the course of the eighteenth century, as reformers of different kinds who tried to shape it in their image were confronted what it actually was—an extremely complex and vast hydraulic device with long-weathered practices and dynamics encrusting its foundations. Military engineers, savants, academicians, and miners all had many things to say about the project, as did the oidor superintendants, the viceroy, the city elites, and the people who lived in the shadow of the project and were required to work in it, maintain it, or contribute to it.

Whether they accepted a military engineer's proposal, rejected it, or simply ignored it and let it sink in the mud, local technicians, officials, and the broader caste of letrados and savants who had something to say about the drainage nevertheless took note of the way military engineers expressed themselves. The more these groups saw a connection between fluency in the new language of the engineers and royal favor and hence the possibility of advancement and reward in an atmosphere of imperial reform and uncertainty, the more they imitated the language, vocabulary, and representational styles of the military engineers. It was in the discursive realm, therefore, that the military engineers wielded their greatest influence and in which the most notable changes occurred.

This chapter examines the first push for changes in the Desagüe after the departure of the friars, when oidores and military engineers took center stage. The pattern of acceptance and resistance is most clearly seen and best understood when examined by intervention area or topic. For this reason, the practical interventions of the engineers in the Desagüe in the areas of technical drawing and cartography on the one hand and organization of work on the other will be treated with their own chronology, so some retracing of steps is inevitable.

## THE RETURN OF THE OIDORES

When Cabrera died, so too did the delegation of authority: the Desagüe returned fully to the oidor supervision. This was not a burden on the oidores, as it gave them access to paid commissions and superintendancies that maximized their utilities while in office. A return to direct administration by crown officials was also consistent with the initial measures of imperial reforms aimed at improving revenue collection, rule, and productivity. At the same time, it was becoming increasingly clear that the Desagüe demanded a plastic but huge geographical jurisdiction such as only high officials enjoyed: over time, as the interconnected hydrology of the basin and the multiple causes of flooding were understood in greater detail, old and new hydraulic structures in the basin were coordinated to function in tandem with the drainage, when not actually connecting to it.

Starting in the 1740s, there was a mounting awareness that the huge hydraulic machine the basin was becoming could only be run if the Desagüe jurisdiction covered the entire basin. The superintendants would by then be covering more and more ground and in a far more interventionist manner than before. A full tour of inspection would exceed the northwest quadrant by far, and since it was common for superintendants to examine works performed as far away as Lake Chalco, the trek had grown to thirty-six leagues, over which these *togado* superintendants had to "pay attention to everything."[2] With extensive and complicated interactions not just with the hydrological reality but with myriad indigenous townships and Hispanic landed interests that were constantly acting upon it, the drainage project demanded a helmsman sufficiently empowered to make decisions and ensure compliance, particularly since its needs were often urgent and required executive swiftness. Until the implementation of the intendancy reforms in New Spain, only a viceroy or a member of the audiencia enjoyed such powers.

As managers, the oidores came in with perfect understanding of the mechanisms by which the cost of maintenance, particularly on the Cuautitlan River diversion dam, were shifted to the riparian populations in the district. Upon taking over, the first oidor superintendant, Pedro de Labastida, immediately set a budgetary cap of two thousand pesos for the project—thirteen hundred for salaries of on-site personnel, seven hundred for repairs within the Desagüe proper, and zero for friars or priests and for the maintenance of the Cuautitlan River diversion dam or other structures maintained by local populations in the Desagüe district. Any emergency repair and new project within the thirty-six–league inspection circuit—which included the works in Lake Chalco and the Presa del Rey, built in 1627 to regulate the flow of the watershed (*avenidas*) of Pachuca into Lake Zumpango—would require separate authorization.[3] This penny-wise demeanor carried on to Labastida's successors.

During this time of transition, the oidores relied on the guidance of guardas and sobreestantes. Unlike the friars, the new superintendants resided in the capital, not on-site, and as a result the Desagüe risked becoming rudderless and neglected on a day-to-day basis. In recognition of this, in 1696 Labastida created two additional posts for guardas, one assigned to the causeways and another specifically to the trench.[4] The job of all these subordinate guardas was not just to organize and supervise work but also to watch over the hydraulic structures under their specific care to ensure that in the rainy season in particular they functioned as desired, and to quickly report to the guarda mayor any breaches or damage.

The audiencia and the viceroy therefore also specifically equipped guardas mayores with executive tools comparable to those of the new superintendants, which is another reason why unlike the rest of the guardas and sobreestantes they had to be men of rank. A 1704 junta general empowered the guardas mayores to impound the property of hacendados, rancheros, and other non-Indians to reimburse the Desagüe any work its own employees performed that ought to have been done according to the yearly obligations or riparian water users, or to repair damage in dams, causeways, floodgates, water towers, and other installations.[5] Guardas mayores not only exercised these powers but also used them to usurp those of other authorities, particularly on matters of land and water use. In 1730, for instance, the guarda mayor took the liberty of granting the priest of Huehuetoca permission to cultivate lands belonging to the Indians of Teoloyuca next to the Vertideros, without consent of either the Indians or civil authorities.[6] The Teoloyucans' meticulous record of the guardas mayores who either directly permitted subordinate guardas to grow their own wheat on indigenous lands—some of them lakebeds—or turned a blind eye to the practice suggests that the real powers of the former exceeded those they were officially granted.[7]

For fulfilling these responsibilities, guardas mayores earned a yearly salary of six hundred pesos, the same amount as oidor superintendants.[8] They were also allowed to live in the Desagüe building in Huehuetoca, which served as lodging, functional headquarters of the works, and warehouse for all materials and tools as they arrived. This Casa del Desagüe also became a center of civil oversight for all matters relevant to the Desagüe in the partidos of Zumpango, Cuautitlan, and beyond: superintendants and guardas mayores had a jail there that was equipped with stocks to punish the misdemeanors of its employees in the district.[9] Even more than the maestros, the guardas, assisted by the sobreestantes, would ensure continuity after the friars left and the oidores returned. Their powers, continual employment, and physical location on site grounded them as the most effective guardians of practices and local knowledge. This was not always for the better in the technical sense, but it would allow driven superintendants to see their projects through.

## THE BASIN'S HYDROLOGY AT THE TURN OF THE CENTURY

The hydrological sophistication of the officials accumulated over time and through the efforts of others besides the guardas. During the course of the seventeenth century, military engineers had contributed to more refined understanding of the hydrology of the basin as a whole and its challenges. Surveying with more precise instruments and reporting their findings in increasingly systematic and economical reports, they tried to inform audiencia members and viceroys about the intricacies of the Desagüe and its interactions with the hydrological realities as a step to making them understand the reasons for their recommendations. How, then, did these officials understand these hydrological realities and the Desagüe within them?

For a comprehensive view of how oidores might have seen the Desagüe's physical structures and their relationship to the basin as they settled into the superintendancy and into the first years of the eighteenth century, it is necessary to rely on reports from the 1740s. Only then do reports start matching the comprehensiveness of those that had been generated in the first three decades of the Desagüe. The absence of comparably detailed accounts from exactly the time the oidores took over, and more importantly the fact that prior to 1742, oidor superintendants had restricted their activities to routine maintenance, neither adding nor changing structures in the Desagüe proper or its district, allow us to consider accounts from this decade as reliable—if possibly somewhat deteriorated—pictures of how the oidores found things in 1691.

These reports were the result of renewed flood concerns. In 1747, water from the northwest breached the dam of Lake San Cristóbal and caused more than twenty thousand pesos in flood damage in the city. To consider what needed to be done, Viceroy Juan Francisco de Güemes y Horcasitas asked the audiencia lawyer and perpetual regidor of the city Joseph Francisco de Cuevas Aguirre y Espinosa to accompany superintendant Domingo de Trespalacios y Escandón on a basin tour and Desagüe inspection and then give him a synthesis of where things stood with the drainage. This he would read together with Trespalacios's own periodic inspection accounts, the most complete of which dates from 1753. In fact, the viceroy would soon be able to study the basin's hydrology visually, thanks a detailed representation created by Joseph de Páez, an artist better known for his *casta* and devotional paintings, on commission from Trespalacios himself (see Figure 5.1).

According to all these reports and the painting, the hydrological reach of the Desagüe was still what it had been at the time of its creation. In

*Figure 5.1.* Commissioned by Superintendant Domingo de Trespalacios y Escandón, the hydrological and hydraulic detail in painter Joseph de Páez's 1753 depiction of the basin was probably based on Joseph Francisco de Cuevas Aguirre y Espinosa's 1748 report as well as firsthand familiarity.

SOURCE: Instituto Nacional de Antropología e Historia, Mexico. Reprinted by permission.

other words, it served "not just to divert the plentiful Cuautitlan River and the rest of the rivers that flow to it, but also to contain the freshet of Pachuca, and the watersheds of Ozumbilla, and from all those boundaries in Lakes Zumpango, Citlaltepec, Xaltocan, and San Cristóbal," Espinosa explained. This meant that Lake Texcoco could still overflow as a result of receiving excess water from these northern sources. Although it was also still fed water from the southern lakes of Chalco and Xochimilco, which were abundantly fed by freshwater rivers and springs, these sources were not considered "enemies of the city." Keeping this southern part of the lake system alive was important to prevent increases in the price of foodstuffs and materials coming into the city by canoe, so its contribution to the flood threat was balanced out by the benefits it provided.[10] Of the entire basin's rivers, only three of the twelve that earlier generations had condemned to eternal banishment by diversion out of the basin for crimes against the safety of the city had narrowly escaped this fate. These rivers—the Panuaya, Salto, and Amecameca—were in the south of the basin, draining into Lake Chalco. Although they added to the water that flowed to Lake Mexico through the sluices of the causeway-dam of Mexicalzingo, this was a necessary evil. The most feared river, the Cuautitlan, was still diverted into the Desagüe. In addition, to the northeast, water from the San Juan Teotihuacán River was now retained by the 1,245-meter long dam of Oculma. Eight smaller rivers and streams from the northwest quadrant and the north of the basin were now also being channeled into the Desagüe. The target of drainage was still "enemy" water north, west, and east of the city. The Desagüe de Huehuetoca thus bore the protection of the city massively on its back and was about to do so even more as a new phase of growth and therefore complexity would soon open.[11]

Time had changed some of the features of the Desagüe system and not others. The structures the friars had added or modified in the Desagüe proper to improve its functioning or prevent damage survived, if in somewhat ruinous condition. A masonry barrier on the north side of the open trench within the Desagüe proper that Cabrera had built to keep streams from tumbling over the edge of the unprotected trench and carrying eroded debris into the channel still stood.[12] So did a masonry spillway this friar had built on a segment of the Cuautitlan River diversion dam in front of the township of Teoloyuca,[13] presumably to prevent the floodwaters from overtopping the dam and causing a breach.[14]

By contrast, most of the diversion dam that led the Cuautitlan River and its tributaries to the Desagüe was still the ancestral grass-and-earth structure and was therefore maintained by its riparian "beneficiaries" between the Pila Real at Atlamica and a marker known as Cruz del Rey, which as will be recalled designated the point at which the royal domain began, on the albarradón del Rey (see Figure 5.2). Any breach along the three leagues of this earthen structure caused flooding in the immediate fields and opened a direct drainage for the Cuautitlan through the lowland between San Cristóbal Ecatepec and Chiconautla, allowing its water to reach the city. Still, despite recurrent recommendations to rebuild it out of permanent materials, the only part of the diversion dam that was made out of mortar and stone and maintained out of the royal coffers went from Cruz del Rey to Huehuetoca, extending into the Real Desagüe proper.

Flows through the Desagüe system were still handled in much the same way as before. Each of the causeways and the diversion dam were equipped with sluices to regulate the flow of water either into the Desagüe or into lower-elevation lakes. The sluice-operating calendar had become stabilized through experience. The sluices for Lakes San Cristóbal, Zumpango, and Oculma (in the northeast) generally remained closed

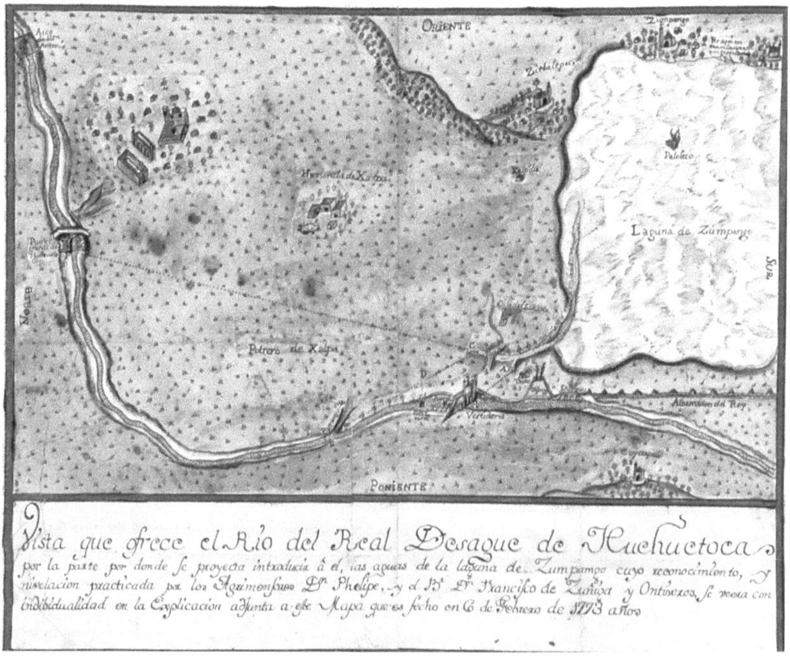


*Figure* 5.2. Francisco de Zúñiga y Ontiveros's 1773 view of the Desagüe proper. The Cuautitlan River appears simply as the "Desagüe River," led by the albarradón del Rey to discharge into the drainage. Note spillways, sluicegates, and the Vertideros warden's house.

SOURCE: AGN, Fomento Desagüe, vol. 1bis, f. 214. Reprinted by permission.

throughout the dry season from the beginning of October to late February or early March. On Ash Wednesday they were opened to vacate water from the higher lakes into Lake Texcoco so as to leave the former empty enough to receive the water of the rainy season. These sluices were opened only one-third to one-half vara to allow for a slow drainage during a forty- or fifty-day period so that the water would be partially absorbed in sandy saline earth on its path to Lake Texcoco, which would by now also have lost some of its water to evaporation. Additional subordinate guardas were now being assigned to each of these points of flow control.

However, the actual functioning of the system had suffered. The silting pool of Coyotepec that Adrian Boot had inspected in 1614 and found even then to be so full of sediment that it was higher than Lake Zumpango, was still often useless. As a result, water from the river at peak season risked breaking straight through to Lakes Zumpango and San Cristóbal in uncontrollable fashion. This risk had existed early in Desagüe history, of course, and this was why a causeway-dam had been built to retain the water in Lake San Cristóbal as the "bulwark of the city."[15] Oidor superintendants would restart construction, destruction, and modification activity in the Desagüe precisely with the repairs of the 1747 breach of this "bulwark." They would soon equip this dam-causeway at San Cristóbal and that containing Lake Zumpango with two spillways at either end to allow excess water to escape and leave the causeways intact, rather than bursting through them. Until then, despite huge outlays of money and Indian calories, the Desagüe had been plodding along with some tweaking here and there, but still was in no shape to guarantee the end of flooding in the City of Mexico, and much less the desired ultimate goal of desiccation. That is what irked the crown and linked the Desagüe's new century with the broader imperial reforms.

## POWERFUL MEN OF THE PEN

As the oidores settled into the superintendancy, the Desagüe would be inevitably drawn into both the efforts of the Bourbons to rein in the administration of the colonies and the colonists' resistance to their reforms. Drawn from the program articulated by enlightened royal advisers such as José del Campillo y Cossío (1741–43), whose extended visit to New Spain years earlier gave him the raw material for his recommendations to rid the empire of inefficiency and make it commercially and productively dynamic, these reforms arrived as the Desagüe was generating increasing tensions among the viceroy, pro- and anti-Bourbon reform oidores, Carlos III's *visitador general* José de Gálvez, and the cabildo. In this context, the jurisdiction and powers of the superintendant became important, as the audiencia itself became the subject of reform in the drive to render it less receptive to creole control and more responsive to metropolitan concerns, and as other corporations with bearing on the Desagüe—the cabildo and the consulado, among them—saw their own jurisdictions and powers come under attack.

Although the superintendants were now drawn from the oidores of the audiencia, and oidores had territorial jurisdiction that exceeded the entire basin and made them well suited for a post that demanded such reach and powers, the fact is that at no time had it been clarified what exactly this meant for the powers of the superintendancy per se. This posed a whole series of problems. Geographically, the superintendants' jurisdictional powers were broad but vague. No one had ever defined exactly how far the jurisdiction of the Desagüe itself reached: Was it to encompass only the *alcaldía mayor* or province of Cuautitlan where the actual works were located, or would it extend to those of Zumpango and Ecatepec, which were part of the hydrological system of the northwest? How about Texcoco, Chalco, Mexicalzingo, and Xochimilco, in the center-east and southeast of the basin, where in the eighteenth century Desagüe superintendants went on tour duty and issued orders to local authorities for works performed in those areas? In 1789, it would be claimed that these works fell under Desagüe jurisdiction because Desagüe monies had been used in the repairs of the causeway of Tlahuac separating Lakes Chalco and Xochimilco.[16]

Time would not clarify things. In 1791, the city council would protest that its own flood prevention work was independent from the Desagüe. According to the city, in 1552 it had created the position of *juez de ríos* "to prevent nearby rivers from flooding roads, common lands and parks . . . leading [floodwaters] through other canals, ditches and aqueducts, and to see that the water does not rise in the internal canals of this city which are cleaned and desilted annually with money from its funds." Since these actions took place inside the city, and the post preceded the Desagüe, its juez de ríos was not beholden to the superintendant of the Desagüe, whose intervention in the city's jurisdiction in the past had been the result of express commissions of the viceroy and were not inherent in the Desagüe's prerogatives.[17] The viceroy and the audiencia, however, had the opposite view: "Since the commission of the Juez de Ríos is directed toward freeing the city from floods, which is the sole aim of the Royal Desagüe," his provisions had to be "in agreement with and subordinate to those of the Judge Superintendant."[18]

Jurisdictional definition was important not only because of the political tension among audiencia, cabildo, and crown, but also because on it depended the financing of the works. Hydraulic works and maintenance such as causeway cobbling and buttressing, as well as canal (acequia) cleaning performed within the boundaries of the city, were paid from its *propios*—its own funds. Those that fell inside the *pertenencias del rey* in the Desagüe de Huehuetoca were paid for by the funds from taxes on wine and butchers, which were called Ramo del Desagüe and were under the administration of the Real Hacienda. Finally, those works that fell in between, and were portrayed as benefiting local interests, such as bridges, water distribution towers, and sometimes (although not undisputedly) roads, were prorated among the neighbors.[19] Thus, political, financial, and hydrological realities were all intertwined, with the fluid and interconnected basin constantly thumbing its nose at the humans who tried to divide and conquer it.

In terms of their powers, the prerogatives of the oidor superintendants were not up for debate. The fait accompli with methods more executive than judicial became the norm at the works starting in December 1742, when Trespalacios became superintendant. This applied not just to matters within the Desagüe itself, or between the project and its neighbors, but even to deeper issues of land and water access and usage within the Desagüe district and any place in the basin where such issues were thought to have an effect on the drainage. So it would remain long after this oidor left the post in April 1764. Unlike his predecessors, Trespalacios vigorously enforced and deepened the Desagüe's control over water and land in its district, also extending the say of the superintendant in these matters throughout the lake system.

Crucially, in 1743, he reinvigorated an order that dated from the time that the friars had inaugurated the water-sweeping method of conveying the open trench debris. This order mandated "that from the Presa Real one third of the water be allowed to flow down the [Cuautitlan] River to the Desagüe for the purpose of keeping the terrain soft for the sweeping and reshaping." While his oidor predecessors might have been lax in punishing infractions of this rule, Trespalacios would tolerate none of it.[20] An Indian or casta who made holes in any dam or inflicted any other damage on Desagüe installations would earn himself two hundred lashes and six years in an obraje; a Spaniard, six years of foreign exile.[21] In this manner, while not at all designed to address agropastoral or any other area of production, the Desagüe from the mid-eighteenth century on would nonetheless still have a deep impact on how Hispanics and indigenes could see and use the basin's water and land.

In addition, from at least 1724 on, superintendants prohibited cultivation on all land less than 41.5 meters (50 varas) from the edge of the water. They correctly argued that cultivation loosened the soil at the margins of the Cuautitlan River and Lakes Zumpango and San Cristóbal, silting them and potentially driving their water over their retention dams. All users of river water resented this, even the hacendados, who in 1748 complained against Trespalacios's enforcement of this rule because "these margins are the richest, and between their length and distance they make up nearly four caballerías of land; and the hacienda will lose its value as its best lands are rendered useless."[22]

This measure was compounded by an even more important one that reached deep into what and where water and land were supposed to be. At some point in the seventeenth century, superintendants had ordered that the parts of Lake Zumpango that the dry season left without water had to be reserved as a "receptacle for the lake" (*vaso de la laguna*). All cultivation or other permanent activity on this lakebed was thenceforth prohibited because this "receptacle" needed to be kept cleared of any impediment to its capacity to store excess torrential water in the rainy season that would otherwise rush into the works and damage them.[23] However, indigenes from townships such as Teoloyuca had used this swath of wetland to cultivate during the yearly recession of water. The rule, therefore, meant the loss of common (ejido) lands because terrain once defined as "lake" was reserved for water alone at all times. Now (in 1747), Trespalacios asked the audiencia to confirm that the seasonally flooded margins in *all* northern lakes were vasos de la laguna.

In sum, under its oidor superintendants from midcentury on, the impact of the Desagüe in human relations with water and land affected even how people could use water and land, and indeed, the very definition of these resources. This is what colonizing the countryside meant. Although the Desagüe was not designed to serve this purpose, it was now nonetheless assisting the process more than ever.

In tandem with this, maintenance obligations became nonnegotiable, with swift punishments for infractors. Trespalacios reinforced the 1704 dispositions that had given the guardas mayores the authority to force any riparian landholder to comply with their allocated maintenance duties and arrest residents for damaging or endangering Desagüe-supervised structures. In 1744, for instance, guarda mayor Fernando Díaz de Ruiloba arrested Juan Muriel Samorano, administrator of the mayorazgo of the Cuautitlan grain mill, for damaging a floodgate of the Pila Real de Atlamica.[24] In April 1748, he took on hacendado Dr. Francisco Jiménez Caro, who refused to comply with orders to maintain his segment of the Cuautitlan River diversion dam, alleging ecclesiastic exemption. Díaz de Ruiloba fined him, impounded an ox team, and finally threatened his administrator to comply with maintenance or spend four years in an African presidio.[25] Asking the audiencia to validate the whole thing, Trespalacios fully backed the guarda mayor's actions when Jiménez Caro complained to the archbishop. Far from ponderous, the *fiscales de lo civil* and *de lo criminal* of the audiencia delivered Trespalacios swift findings in June that legitimized what the guarda mayor had already executed in order to get the work done during the months when it was imperative that it be done, July and August. By August, Jiménez Caro had been subdued into doing his part. Judicial procedure was followed to the hilt in this case, as Trespalacios secured not only a resolution from the viceroy, but even a December 1751 royal cédula retroactively endorsing his actions, publicized by a crier on market day in Cuautitlan.[26] As the eighteenth century wore on, the judicial process these men were following was certainly not slowing down Desagüe affairs, and oidor superintendants would increasingly act as Trespalacios did—with the swift absolutism of a monarch, aided by the trusted guardas mayores.[27]

Oidor José Rodríguez del Toro succeeded Trespalacios in January 1764 when the latter left the Desagüe in 1763 for a promotion to the Council of Indies.[28] The two of them were critical in reviving the completion of the open trench, and in coordinating reports on the hydrology of the basin together with military engineers and master architects that led to a more sophisticated understanding of how human activity had altered the environment and in which direction it should continue to do so. More executive than ever, increasingly enlightened in their culture, the oidores were nonetheless wardens of continuity in the conceptualization and social arrangements of the Desagüe. There is no reason why this should have been otherwise.

## LABOR AND FISCAL CONTINUITIES

To reform the direct and indirect forms of labor appropriation from the hinterland would mean disrupting the arrangements that allowed the city elites to shift to others the cost of their own protection and the crown to make significant savings on the project. In consonance with this, repartimiento laborers also continued to supply the bulk of work needed in the Desagüe proper, while obligated unpaid maintenance continued unabated in the complex as a whole. This had consequences for the works. Working with constantly renewed work crews, or tandas, technicians had little opportunity to try techniques and tools unfamiliar to the workers, who would require training that would be wasted when each crew returned home. Military engineers told oidores in no uncertain terms how coercion and danger produced bad quality and slow work. So far as the laborers themselves went, repartimiento had consequences too, of course. Unsurprisingly, it exposed them to a possibly broader range of exploitative practices than if they had been hired wage workers or even purchased slaves.

Contractors, resident wardens, and overseers were in a good position to take advantage of the captive market of the Indian workers on-site. In the seventeenth century, master architect Juan Serrano had been one of its most enterprising practitioners when in the 1630s he availed himself of Desagüe workers and supplies to cultivate corn and to bake bread that he then turned around and sold, along with chocolate, sugar, candles, and other goods, to the captive market of the Indian workers on-site in true *repartimiento de mercancías* style.[29] Although forbidden, guarda or sobreestante involvement in selling goods to Indians as they came for their Desagüe tandas continued in the eighteenth century and was no trifling issue, especially when sales involved pulque. Not only did inebriation put workers in more danger than they already faced, but, as the gobernador of the Indian township of Coyotepec complained in 1707, pulque consumption also left them without cash to take home for tributes.[30] Such sales of goods on-site only became legal under the consulado contract.

In addition, repartimiento Indians could find themselves working double, for no money at all and not even to the Desagüe's benefit. In 1707, for instance, it was discovered that guarda mayor Pedro Francisco de Moctezuma had been milking Desagüe resources with great creativity. Spending more time overseeing his *hacienda de labor* in La Guiñada, planting in communal lands (*tierras de comunidad*) rented from the Indians of Huehuetoca and working a nearby mine than on his Desagüe duties, Moctezuma brazenly paid his hacienda employees out of Desagüe monies. To divert work crews most efficiently toward his enterprises, Moctezuma only paid attention to the sector of the public work closest to his hacienda.[31] Such practices were hard to control when the guarda mayor enjoyed broad powers that included issuing repartimiento orders, summoning Indian governors, settling disputes among Desagüe workers, and even jailing people for misdemeanors against or merely around the Desagüe. They even had a jail equipped with stocks right in the Casa del Desagüe in Huehuetoca.[32] Under these conditions, there was no reason for these men to behave any more or any less dutifully than their counterparts in the district magistracies. Excesses such as beatings were not unheard of.

The only thing the oidor superintendants from Trespalacios on changed in the manner the city appropriated labor directly and indirectly, then, was its intensity and nonnegotiable character. Direct appropriation though repartimiento was rarely appealable. In February 1743, for example, when the epidemic-stricken townships of Tacuba and Cuautitlan requested reprieve from sending men to the works, Trespalacios turned them down.[33] The indirect manner of appropriating labor also became more extensive, better enforced, and disproportionately more costly to the indigenes. This was the obligation borne by all the Desagüe's rural neighbors to maintain earthen structures such as dams and levees and clean out and sometimes change the channel of rivers and streams that were not in the Desagüe proper but were essential to its functioning, especially along the Cuautitlan River. In 1748, Trespalacios claimed that the diversion of this river had opened up new lands for the expansion of the holdings of haciendas in the region on the old floodplain of the river, and therefore it was only fair that hacendados contribute to its maintenance.[34]

Having established this, he then extended the maintenance obligations of everyone, not just the hacendados, so that they applied not just to this dam but to all earthen structures in the Desagüe jurisdiction. If they were actually real, which seems doubtful in light of eighteenth-century prohibitions on the cultivation of land along the river's edge, the benefits of opening up new lands certainly did not accrue to indigenous townships, for whom the growth of haciendas only brought increasing encirclement and encroachment on their own land and water rights.

Despite the language of proportionality that officials used, moreover, not everyone pitched in equally to the upkeep of the diversion dam. Although duties to maintain works in the Desagüe jurisdiction were shouldered by both Indian and non-Indian producers, there were "notorious and patent differences" in the way that haciendas dawdled and Indian villages complied.[35] While quantitatively the duties listed on Table 2 appear equally distributed, qualitatively they were not: maintenance obligations competed for labor against activities that were necessary to insure

*Table* 2. Maintenance obligations along the Cuautitlan River diversion dam.

| Responsible party | Description | Varas of obligation |
|---|---|---|
| King | out of Desagüe monies | 1,700 [a] |
| Coyotepec | Indian township | 1,258 [b] |
| Teoloyuca | Indian township | 4,590 [b] |
| Visitación | Indian township | 242 [b] |
| Tultitlán | Indian township | 780 [b] |
| Santa Bárbara, San Lorenzo | Indian rural villages | 1,328 [b] |
| Cuautitlan | Indian township | 650 [b c] |
| Azcárate and González | non-Indians of Xalpa | 1,962 [b] |
| **Sabino,** San José, Tecuaque | non-Indian haciendas | 3,160 [b] |
| Cuamantla | non-Indian hacienda | 1,475 [b] |
| **Cartagena** and Cadena | non-Indian haciendas | 462 [b] |
| Córdoba, **Xaltipac, San Mateo** | non-Indian haciendas | 505 [b] |
| Corregidora | non-Indian hacienda | 300 [b] |
| Portales | non-Indian hacienda | 224 [b] |
| Antonio García | operator of 4 ranchos | 168 [b] |
| Rancho de Ribero | non-Indian rancho | 46 [b] |
| Cárdenas | non-Indian rancho | 212 [b] |
| Ozumbilla | non-Indian rancho | 144 [b] |
| Molino de Cuautitlan | hydraulic grain mill | 174 [b] |
| | | Indians, total: 8,848 varas |
| | | Non-Indians (excluding King), total: 8,832 varas |

SOURCES AND NOTES: Hacienda names that appear bold are from Gibson, *Aztecs*, map 9. (a) AGN, Desagüe, vol. 14, exp. 5, ff. 13, 20v; (b) AGN, Desagüe, vol. 30, exp. 4; and (c) Tacuba and Cuautitlan together are elsewhere quoted at 3,053 varas (AGN, Desagüe, vol. 12, exp. 4, ff. 19–19v).

subsistence, tribute quotas, and other exactions; over time, the burden would lessen the attractiveness of remaining in the village for adult heads of household, and finally, labor in Indian townships was not elastic, since hiring extra hands as needed was not normally possible and they had to make do with the able-bodied adults in the community. In addition, maintenance duties and other assignments in exchange for the enjoyment of water and other rights were for some more onerous than Desagüe labor proper, which at least tended to happen in spurts. Hispanic haciendas and ranchos indeed dragged their feet in fulfilling their maintenance obligations. In 1768, for example, in one portion of the Cuautitlan River diversion thirteen Hispanic establishments had failed to do their part.[36]

The principles governing the finances of the Desagüe reflect significant continuity as well. Its income still came from taxes on slaughterhouses in the district of the city, on wine sales, and from occasional levies on property. As early as 1607 and 1630, when the first assessments were levied on property owners to finance the project, authorities had shown an awareness of the defensive character of the Desagüe. This conscious imperative to defend property and rents had remained unchallenged, as expressed in the reasons Viceroy Croix gave when on August 11, 1767, he decreed a new levy on properties and their rents, and mortgages, "for just as the owners are interested in their preservation [from floods], so are said capitals." The list of acknowledged propertied urban beneficiaries of the Desagüe was even more extensive now than it had been in 1607: besides the archbishopric, Inquisition, university, colleges, and religious orders that had been assessed then, now Croix considered the Marquesado del Valle as "particularly interested." The consulado too, "who because of the notable utility that the merchants will receive in having their goods assured and freed from the risk of losses, and the great costs that they would incur moving them to safe places, will be assessed proportionately to their commerce."[37]

In contrast to these three special levies, the taxes were collected continually. Until 1670, both the wine and the slaughter tax were farmed out, the latter by collection district. The revenue was significant. By 1650, the Desagüe had received almost half a million pesos from the Veracruz treasury alone, which was assessed whether the works were halted or not.[38] From 1670 to 1777, administered by royal officials, the wine tax yielded 712,522 pesos and the slaughter tax 606,269.[39] In contrast to the city's wine tax, it was royal officials who administered the *caja del Desagüe,* and from 1670 on they were also in charge of collecting. This money was sent directly to the Real Hacienda account of the Desagüe in the capital, except between 1672 and 1686, when it was kept in the Veracruz treasury and simply deducted from the money Mexico sent to the port city for infantry expenses.[40]

Who paid these taxes? Peninsular wine merchants complained that it was they who bore the brunt of these taxes. Constantly pressing for their elimination or reduction, and thanks also to timely monetary contributions from consulados to the crown, these merchants often gained concessions. In 1649, they won a two-year suspension on the twenty-five peso-per-pipa of imported wine tax, and in 1720 a 50 percent cut in the tax to be charged only on wine disembarking at Veracruz, not that exiting to the interior. Again, in 1733 the Cadiz consulado got another 10 percent reduction on wines warehoused on ships or in Veracruz, a 15 percent reduction whenever this wine was on board a ship delayed for more than six months at sea, and another cut in the tax to eleven pesos and three tomines.[41] As part of the Bourbon attempts to stimulate commerce and production in Spain, a royal cédula of March 24, 1753, sought to reduce taxes on imported wines, aguardiente, and vinegars. To benefit the vintners of Spain while reducing the consumption of allegedly unhealthful aguardiente, this decree aimed to undermine the advantage that wines from Parras and other regions in New Spain had over Andalucian reds: a half liter (*cuartillo*) of domestic was generally two-thirds cheaper than imported (and by the time they reached their markets, probably also spoiled) wine.[42] The decree thus called for reducing levies against Spanish brews and vinegars and suspending taxes charged on Spanish wine in the City of Mexico while continuing to charge taxes on the wine produced in any of the eleven wine districts in New Spain.[43] As a result of this measure, the tax was lowered on liquors (*caldos*) going from Veracruz to the interior, and the applicable tolls (*alcabala*) were reduced from 8 to 6 percent.[44]

Despite the Spanish merchants' protests, it is clear that the tax financially burdened neither them nor Iberian producers because they passed it on to the consumers. Their protestations were aimed at rendering their goods competitive against domestic products and even more so at protecting their own trade against that of Mexico's own consulado. This was clear to critics of the reductions in taxes for the Desagüe, such as the Mexican consulado's ally Francisco Javier de Gamboa, who in 1789 blamed the cuts "on the rejection of the vintners from Andalucía who wanted to avoid paying the five pesos one real per barrel, not being they who pay it but the consumer, who is taxed and deducted the fifth part of every cuartillo." For Gamboa, it was imperative to keep the wine tax precisely because it was "the main income for the Desagüe," whose needs could never be met "with the tax collected from meat purveyors of the district of New Spain."[45]

Above and beyond the competition through taxes between Spanish and Mexican merchants, the Desagüe income was increasingly provided by people who did not even live in the city, own anything in it, or have much to gain from its expensive flood protection. The geographic range in which taxes were to be paid was extended at different times. In response to the imminent open trench works, for example, a royal cédula of April 17, 1637, had confirmed the extension of the jurisdiction of the City of Mexico to a five-league circuit, within which all applicable taxes were to be charged. Similarly, slaughters within twenty-four leagues of the city were subject to the sisa for the Desagüe.[46] By the end of the colonial era, taxes for the Desagüe would be collected not only in Veracruz but also in Guadalajara, Puebla, Valladolid (Morelia), Oaxaca, San Luis Potosí, and Guanajuato.[47] Since these taxes were passed on to the consumer, with both meat and wine taxes the ultimate payer was most likely the population who consumed them. This suggests that like all sales taxes on items of popular consumption today, these were regressive taxes. Although not unusual for the epoch, this was a deliberate strategy that only some officials explicitly resented. From a fiscal point of view, during oidor rule the only shift away from the taxation practices inaugurated in the seventeenth century that made all classes and the hinterlands pay for the welfare of the elites in the capital seems to have been in the amounts and rigor of collection. This was one thing that for all the passionate technological and managerial reformism they brought, the military engineers who joined the oidores in the Desagüe would not try to change.

Clearly, then, as superintendants the oidores made the Desagüe a more formidable creature in terms of the weight of its presence among hinterland populations and the lower classes. This was a result of the expansion of the jurisdiction and powers of the oidores and of the geographic amplitude of their reach, as well as of the greater rigor with which they applied existing arrangements, creating new rules to give them more teeth. But it was one thing to endow the Desagüe with a jurisdiction commensurate to the tasks and appropriately empowered and willing officials, and quite another thing for these officials to be able to shape the Desagüe materially, which demanded an understanding of how the basin worked, some idea of how hydraulic devices functioned, the ability to communicate about and follow up on construction and demolition tasks, and other skills that did not come with a law degree. This is what military engineers could provide, both in reality and in the eyes of the crown. Although during the first years of oidor presence in the Desagüe, the relationship between oidores and military engineers was collaborative, it would not remain so for long. Despite their royal and viceregal backing, most engineers faced resistance to their ideas and proposals at some point or another from several quarters, including the superintendants themselves.

## THE WAYS OF ENGINEERS

Military engineers were different from all other technicians who participated in the Desagüe. For the most part, their allegiances lay with the projects themselves, their distant families, and the king who paid their salaries, not the creoles or colonial officials among whom they worked. As perpetual foreigners they had scant prospects of social integration in their destinations, which accentuated the distinctiveness of the European professional identity and culture based on military discipline and systematic training that they developed over the early modern period. Combined, all of these factors impacted how they intervened in the Desagüe and what kind of reception their ideas were accorded over time.

Typically Italians during the early phase of fortification in the Americas, Flemings or Frenchmen during the seventeenth century, and increasingly also Spaniards in the eighteenth, military engineers were never creoles and always itinerant, moving from post to post. These men hired themselves far from wives and children, who they would often see few times in their lives and who would grow up on their fathers' modest remissions, perhaps to become engineers themselves if they were male or to marry a junior engineer if female. Sometimes, father and son might unite in a Spanish king's service, and many established lineages of roaming engineers. Whatever their nationality, military engineers took both their mother tongue and their own peculiar corpus of technology to their assigned posts.

For the crown, of course, the engineers' distance from colonial elites and their interests was an asset. Not so their professional and national heterogeneity. Although this polyglot band of engineers collectively contributed many strengths to Spanish military and civil construction practice at a critical time of territorial expansion and colonization, from the late sixteenth century on the Habsburgs tried to diminish their predominance in defensive (fortifications) and offensive (artillery) military engineering and cartography. They did this mainly by founding training institutions in Spanish territory such as the Real Academia de Matemáticas in Madrid (1582). Until at least the late seventeenth century, the number, enrollment capacity, and curricular content of these institutions were never up to the military needs of the empire, despite efforts to beef them up by incorporating prestigious fortifications experts such as Captain Cristóbal de Rojas, author of *Teoría y Práctica de la Fortificación*, the first Spanish fortifications treatise, or the captain general of artillery, the Marqués de Leganés.[48]

For a while yet, Spanish kings would continue to meet their field and training military needs by relying on their European governors and viceroys to recruit engineers both in Spanish and foreign domains and on civilian experts to test these recruits in mathematical theory and engineering practice, after which the men would be granted the rank of engineer and shipped off for New World service. For instance, mathematicians such as José de Zaragoza, of the Colegio Imperial, tested engineer candidates like Captain Francisco Pozuelo Espinosa, who passed his exam and was swiftly certified and shipped to New Spain, where he worked with Martín Solís during the oidor interregnum.[49]

By the mid-seventeenth century, this diversity in background and training resulted in a lack of standardization in practice that officials of the increasingly beleaguered empire began to perceive as a strategic vulnerability. Calls for reform in all realms began to proliferate, setting the stage for the systematization of military engineering that would begin to create a more cohesive profession. Reform in military training would have effects near and far, making the military engineers who came to the Desagüe socially and culturally distinct from their professional cousins on the ground—the maestros, guardas, sobreestantes, friars, and oidores. This had its own effects.

This systematization and differentiation began when in 1675, the monarchy founded the Academia Real y Militar del Ejército de los Países Bajos in Brussels, with military engineer and field sergeant Sebastián Fernández de Medrano as director. Although most students there never became military engineers, until the close of the seventeenth century the academy would remain the main center for training of those who did and moved on to serve Spanish kings.[50] Fernández de Medrano was a key military and pedagogical figure in this process. A veteran of the Italian and Flemish campaigns, he would become a prolific author of manuals to first educate potential recruits to the military engineering profession in the mathematics, geography, and fortifications, and then to develop and deepen this training. Thus, his first work, *Rudimentos Geométricos y Militares*, made no mention of modern mathematical tools and went no further than Euclidean geometry because as the author warned, "I write for beginners, not for experts."[51] By his 1708 *El Arquitecto Perfecto en el Arte Militar,* his treatment of mathematical tools included calculus and trigonometry, and he defined the métier as a science, which opened it up even more to contemporary European developments in physics and mathematics.

Fernández de Medrano's work in the academy was influenced by a group of scientific reformers who divulged in the Peninsula the new knowledge and methods in a variety of scientific disciplines that had been developing in royal academies and savant circles beyond the Pyrenees. These "novatores," as they were pejoratively called by their opponents, militated against the prevalence of Aristotelian and Galenic demeanors in Spanish science and medicine. Instead, they propelled experimental methods and an engagement with the new understanding of the universe and its forces with motion, rather than stasis, at its core.[52] Concerning themselves with the relationship between science (particularly the branches of mathematics) and technology (what the "mechanics" did, particularly in civil and military construction), the novatores aimed to reform what Spaniards knew and how they learned it as a way of changing what they produced and how. This meant addressing both a range of theoretical problems in physics regarding the properties and behavior of matter and forces and the basics of measurement of solids, gases, and liquids, and the mathematical tools necessary to their study, such as infinitesimal calculus, trigonometric tables, logarithms, and other elements of analytic geometry.[53]

This new direction was in large part a response to the changing needs of military practice: descriptive Euclidean geometry and arithmetical bases of nonmilitary architectural practice were no longer sufficient for engineers who had to deal with forces and motion as well as statics. Military engineers, particularly artillerists, for whom analyzing and calculating projectile trajectories were essential, needed trigonometry and calculus. The most significant shift to accommodate new needs such as these took place at the time of the novatores and in many ways through them. The culmination of these efforts was the 1727 *Compendio Mathemático en que se Contienen Todas las Materias más Principales de las Ciencias que Tratan de la Cantidad*, by the Jesuit-turned-military-man Tomás Vicente Tosca, who drawing from Castelli, Baraterio, Cabeo, Guglielmini, Schott, and Milliet Dechales synthesized the new science for the practical military engineer.[54]

Novator production did not revolutionize anything, however. Technicians did not all of a sudden and as a cohort suddenly start speaking the language of modern science. At the same time that Tosca and his fellow travelers were asserting the authority of post-Galilean science in architecture and engineering, others affirmed the classical architectural tradition based on Euclidian geometry, arithmetic, and algebra.[55] Still, in the Iberian Peninsula, the weight of state backing turned the tide toward the side of the novatores and their successors in time to greatly influence military engineering, as it was contemporary with the "nationalization" of military engineering that the founding of the Real Cuerpo de Ingenieros Militares in 1710 and the launching of the military academy in Barcelona created.[56] The works of Bernard Forest de Bélidor and Tosca now became the pillars of the training in modern physics taught at the military academy in Barcelona because they were practical in both content and manipulation and because they were regarded as modern.[57] By the second half of the eighteenth century, texts used to train military engineers would be commissioned in Spanish, so that their mathematics, physics, and chemistry content could be tailored to the needs of the nationalized profession.[58] The training now lasted three years, starting with the theoretical and finishing with the practical and draftsmanship. None of this means that the training that military engineers received in the new academy in Barcelona was on the cutting edge of its day in its scientific, technological, or practical content, which it was not.[59] What is important here is that it was distinct, not whether it was the best.

As their performance in the Desagüe shows, these men could only do so much to promote the new dynasty's ability to recast the relationship with the American kingdoms as a colonial one, for the benefit of the metropole or at least in consonance with its stated goals. In the 1720s, Spain still had only eighty-six graduated engineers, of multiple nationalities, with increasing obligations, and only thirteen engineers for all of the Americas. After 1762, the number increased: by 1800, sixty-seven would be assigned to New Spain, while the rest of the dominions (including Florida and the Philippines) shared one hundred fifty-four, although not all complied with transfer orders.[60] Given its importance in the empire, New Spain received special consideration, not so much in the number of engineers assigned to it, which was always too low for all the tasks at hand, as in their quality. Whenever possible, men assigned to this viceroyalty were those "most knowledgeable in fortifications near the water," given the extensive coastline that needed defending, and "dexterous in theory and practice, and reliable," since they would be alone directing projects.[61] Military engineers were therefore stretched so thin that they were unable to supervise their own projects to completion.

In addition, they might eat, sleep, and develop relationships locally, but engineers did all this too impermanently to give them much insight or sympathy for the priorities of colonial elites. Integration by marriage to daughters of well-to-do creole families was extremely difficult. In fact, marrying at all was hard, since it required formal permission from the corps—issued only after various certifications attesting to the *limpieza de sangre* of the prospective bride, certified copies of baptismal records, and the deposit of a dowry with the military pension fund (Monte Pío Militar). Permits for men with imminent embarkation orders might be expedited, but those who met their brides-to-be in their New World posts often had no choice but to marry without a license and hope for the best, since from 1760 on not even viceroys could grant them licenses.[62]

Ironically, the more successful they were, the less these men could integrate locally. Captain Jaime Franck's career illustrates this point well: the more pleased authorities were with his work, the more they moved him around, at the expense of his personal well-being. A German military engineer, Franck's value increased with each posting. He first served in Hungary, Flanders, and Germany itself; in 1682, he was transferred to Barcelona, from where in 1686 he was ordered to New Spain to transform San Juan de Ullúa into a true stronghold.[63] In 1687, Franck was asked to report on the Desagüe, which he did in great detail. As he aged, from 1692 on, Franck began asking to return to Spain, and the king's secretary assured him that he would be "cared for and consoled" when his work was finished. Thus encouraged, but already sick with typhus (*tabardillo*) and malaria (*tercianas*), Franck redoubled his productivity, visiting the Desagüe once more in 1693, drawing up plans for the reconstruction of the viceregal palace that had burned during the riot of 1692 and completing the fortress of San Juan de Ullúa, among other tasks.[64] Fellow military engineer Juan de Císcara, creole savant Carlos de Sigüenza y Góngora, and the viceroy himself praised Franck's fortress.[65] Unfortunately, all this only made Franck more essential to circum-Caribbean defense, where after various other assignments, in 1698—at age sixty—he was finally sent to fortify Pensacola against the French and English, with the "help" of some soldiers and several convict laborers who set fire to his "wretched hut," stealing his provisions, clothes, and savings.[66] His inescapable situation was likely the cause of his "melancholy." In 1702, any possibility of integration or return ended, Jaime Franck committed suicide.[67]

There were exceptions to this general lack of social integration, but they were rare. So far as the crown and the corps were concerned, military engineers were supposed to remain outside of local society. While reducing the potential for collusion and dangerous strategic leaks, this deliberate social insulation of the military engineers inadvertently handicapped them in their pursuance of their tasks, as their stepping on everyone's toes ensured reactive kicks against their more radical proposals. In sum, it was rare for military engineers to integrate enough to gain an insider's view of the causes for the arrangements they so harshly criticized, to identify with the social priorities underpinning the Desagüe that were well known to the people who endured or created them, or to stay long enough to steer their projects as they saw fit against the resistance from other Desagüe technicians and officials. The more foreign and transient, the more strident their criticism and dissonant their proposals.

## DOWN TO GRASS

The proposals of the military engineers were most dissonant when they affected the package of practices and social arrangements that allowed the city elites to deflect the cost of their own protection. In other words, the culture of the engineers posed little threat so long as it was seen as just that—a culture—but when it threatened to take tangible shape, it was infrequently welcome. This section shows how in the Desagüe district, this dynamic was manifested most clearly in the reception accorded to the engineers' way of doing earthworks, which did not fit into the framework used by the indigenous people charged with maintaining these structures. When military engineers advocated the use of materials different from those that had always been used, colonial officials resisted the engineers' proposals—not because they lacked technical merit, but because they would have subverted the system on which city elites relied to transfer the cost of maintenance to the countryside.

Earthworks were what allowed the Desagüe to tap water from the entire northern half of the basin, as most of the dams that controlled each water body and several of the rivers and streams there were made with grass and earth. Although from a point of view of the basic materials and principles involved there was some overlap between the way earthworks were built and maintained in the Desagüe by master architects who appropriated and redeployed indigenous technique—particularly in the crucial Cuautitlan River diversion dam—and in fortifications designed by military engineers, there were also differences. As in the Desagüe, the basic ingredients for a military earthwork were earth and plant materials, which could be either grasses or fascine (bundles of branches), depending on the local conditions.[68] The devil was in the details, however, and the processing of materials either dovetailed or interfered with Desagüe practices.

In fortifications earthworks, the procedure began when, after demarcating the outline of his earthwork, the engineer would untie the fascine and lay it across this demarcation as a first layer. On that, he would pile moist earth and ram it down. As the structure began to rise, he would cover its outer face with grasses "firmly placed and interconnected with binding as if they were bricks." Each "brick" of grass was "a piece of sod from an uncultivated meadow, each usually measuring a foot on each of its four sides and half a foot high," cut from the ground with a spade or adze. Further grassy sods were added on top until this layer was two and a half feet thick, secured with "wooden dowels a palm and half long and a finger wide, which are made from the scraps of wood of the fascine." Another layer of fascine went over this, perpendicular to but not reaching the face of the structure. As these layers were rammed down, the face of the terreplein always had to be higher than the filling to prevent the earth from falling out, and the whole structure had to slope inward one foot over four.[69]

This is precisely how French military engineer Louis Bouchard de Becour wanted to proceed in the Desagüe's earthen dams. Stationed in Veracruz and then Campeche for fortifications work, at the end of the dry season of 1706 Bouchard arrived in Huehuetoca to evaluate the Desagüe upon orders from Viceroy Francisco Fernández de la Cueva, Duke of Alburquerque. Bouchard was to issue recommendations on the best means to liberate the city from floods "forever." He seems to have put great energy into complying with this order, but got nowhere with his fascine ideas, unlike the apparent frictionless way Flemish military engineers Adrian Boot and Xaques de Beste had integrated their own earthworks techniques in hydraulics works in Mexico in an earlier period. What had changed?

Certainly the use of fascine was not new in the Desagüe, in whose record the term "fascine" appears for the first time in 1615, when master Alonso Pérez de Castañeda recommended fixing Lake Zumpango's causeway-dam with "mud, grasses and branches," while master Alonso del Arco called for "some taluses of earth and fascine."[70] It crops up again in 1620, when Boot proposed using "fascine and earth" to reinforce the causeway of San Lázaro.[71] In 1631, viceroy Pacheco instructed master Juan Serrano to repair this causeway-dam again, "for which the manner in which it shall be staked and laid with fascine and terrepleins has been given"—by Boot, most likely.[72] In 1648, Serrano and Xaques de Beste jointly recommended improving the durability of the earth causeways of San Cristóbal, Zumpango, San Jacinto, and Calvario with it. Like these predecessors, Bouchard also used "layers of fresh tree cuttings"; like them and the Indians, moist earth and grasses, even adding "thickets of prickly pear cacti" to the proposal as barriers against livestock. So why was Bouchard ignored?

First, there is no indication that in the earlier context fascine meant anything other than the usage of bundled thin branches, and maestros were referring to structures whose construction or repair they would personally direct. Second, at no time before Bouchard did either the maestros or the foreign engineers suggest using fascine for the Cuautitlan River diversion dam, which was maintained in entirely self-directed fashion by indigenes using their own techniques, except in emergencies. What doomed Bouchard's proposal was not its departure from this precedent, but rather the implications it carried. Given that the Cuautitlan diversion dam had a 1:2 rise-to-run ratio, in asking for it to be raised by two varas, Bouchard was imposing greater labor needs. This was problematic enough, but what made his proposal truly unacceptable were his instructions, which required tools and processes not used in the Desagüe:

> A tier of said sods of grasses which are made of bound earth measuring on the surface one half a foot square, one in length, and triangular on the sides and ending in a point, and they shall be cut very fresh from a site with abundant grass [and] placed in such a manner that the facing formed by the grass shall serve as a talus. Then, this shall be covered with a palm's width or somewhat more of earth spread evenly over the entire width of said causeway and well rammed down. After this, a tier of possum-tree or willow branches very close to each other, which must be freshly cut so they can root and sprout, and this shall be continued in the same manner to the top.[73]

These instructions differ in important ways from local custom, mainly in that they recommended materials and processes drawn straight and unadulterated from fortifications practice and treatises. In his widely circulated texts, Sebastián Fernández de Medrano, head of the Brussels military engineering academy, for instance, explained how the layering with fascine should be done both visually, as in Figure 5.3, and textually, clarifying also that these sods, or "tepes, which in Spain are called grasses" (céspedes),[74] needed to be cut "in moist parts, where the soil is good and grows abundant grass" using "spades designed for that purpose" with which one man could cut four thousand tepes in a single day. He even took pains to show the exact shape the humble tepe should have, as depicted in Figure 5.4.[75] These spades were specially crafted with iron cutting edges, and deserved textual as well as visual detailing, such as that provided in the top row of Figure 5.5, from the widely circulated *Architectvra militaris nova et aucta*, by the Polish engineer Adam Freitag.[76] In fact, Bouchard's instructions suggest he probably had Freitag's method and tools in mind, with which he would have been familiar through either the treatise's 1635 French translation (which circulated widely in Europe as well as in Spanish and Portuguese America) or Fernández de Medrano's works, since this Spaniard drew from Freitag.[77]

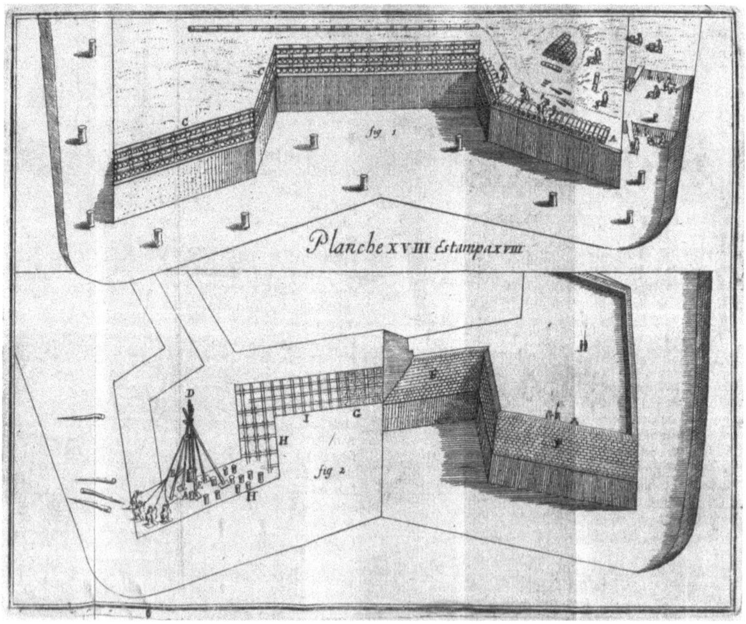


*Figure 5.3.* Sebastián Fernández de Medrano's diagram of how to arrange fascine in fortifications works. On the top right, the fascine bundles are arranged perpendicularly to the edge of the rampart; at the other end, transversal and perpendicular layers have finished the structure's core.

SOURCE: Sebastián Fernández de Medrano, *Rudimentos Geométricos y Militares* (Brussels: Casa de la viuda Vleugart, 1677), 184.

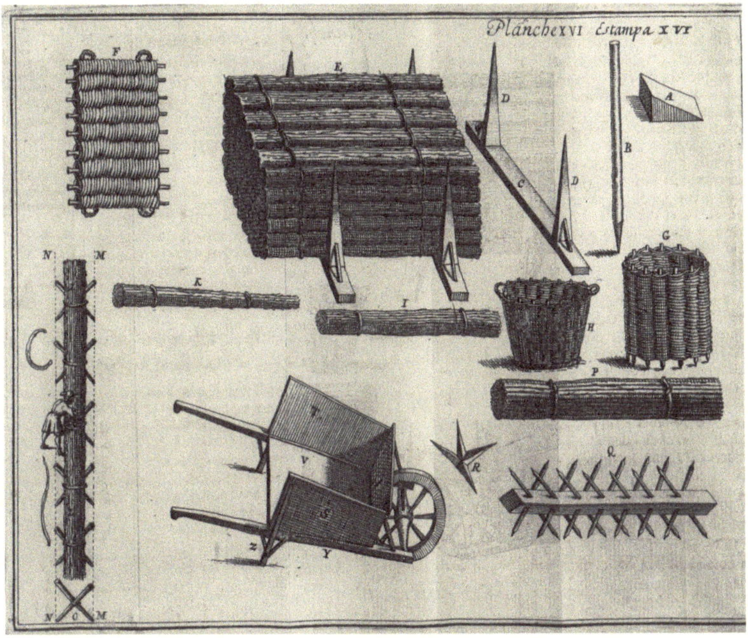


*Figure 5.4.* Materials for building a fortification's earthworks, *tepes*, fascine, and carts.

SOURCE: Sebastián Fernández de Medrano, *El architecto perfecto en el arte militar* (Brussels: Casa de Lamberto Marchant, 1700), 190.

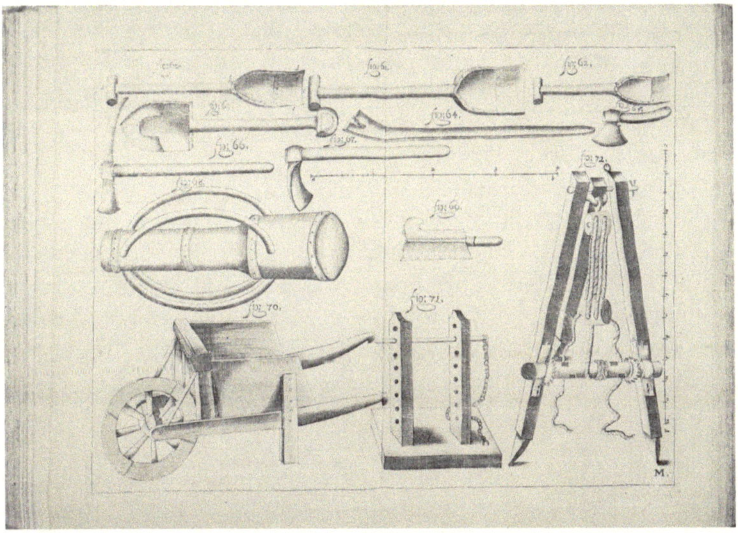

*Figure 5.5.* Spades for sod cutting (top row), and a handheld rammer (center left) used to stamp earth, fascine, sods, and other materials placed in earthworks.

SOURCE: Adam Freitag, *Architectvra militaris nova et aucta: oder Newe vermehrte Fortification, von regular Vestungen, von irregular Vestungen vnd aussen Wercken: von praxi Offensivâ und Defensivâ: auff die neweste niederländische Praxin gerichtet vnd beschrieben* (Leyden: n.p., 1631), 66.

To follow his instructions, riparian water beneficiaries responsible for maintenance of segments of the Cuautitlan River dam would have had to use metal tools. For indigenous landholders, this would have been a difficult proposition, as these tools were expensive.

As importantly, indigenous townships in the jurisdiction of Cuautitlan were subject to what was known as the *repartimiento del zacate*, by which indigenes extracted grasses from their commons to provide officials with fodder for their stables.[78] Like maintenance duties and Desagüe repartimiento labor, this repartimiento also lasted the entire colonial period.[79] While burdensome, the repartimiento del zacate took from the indigenous economy only the labor necessary for its extraction and transportation and the clean plants. It did not take sods. This is important because pulling grasses from moist ground left behind the fertile but fragile topsoil for other uses, while cutting out entire blocks of sod removed the topsoil altogether. So Bouchard's idea to do enough of this to raise the dam by two varas would have increased the loss of soil above and beyond what normal maintenance demanded. This would have been onerous indeed for indigenes, whose communal needs had to be met with ever-shrinking land resources.

In addition, zacate also grew in the shallows of the lakes during the seasonal water recession *if* the Indians had not moved in to cultivate or graze their animals. As guarda mayor Francisco Power would complain in 1781, when the indigenes were not cultivating these shallows they were using them for pasture, which denuded them of vegetation. This forced him and his fellow Desagüe wardens into constant battles with the "uppity Indians" of the township of Teoloyuca, in particular, which had been defying the Desagüe's 1724 definition of these shallows—which lay within their fundo legal—as "receptacles for the lake." The result, Power would despair, was that "we completely lack grasses, which are the only recourse we have for the reinforcement of dams and breaches in said river."[80] When they left grasses in place, it would have been on the sloping banks of rivers and streams, where it would have been crucial then as it is today to promote the infiltration of rainwater into the soil and preventing surface runoff and the loss of soil to erosion.[81] Ultimately, Indians themselves would most likely have resisted Bouchard's requirements. These factors meant that, if forced to abide by them, the Indians of Teoloyuca and other agricultural towns responsible for the maintenance of the causeway might subvert the techniques, and authorities knew this well.

Finally there was the issue of training—it would have taken time and personnel to show Indians how to handle the special metal tools to cut sods of specific shapes and dimensions different from what they already did, and then to place them as Bouchard indicated, and Desagüe officials were reluctant to invest in any of this if they could avoid it. Indeed, upon reviewing Bouchard's proposals, Superintendant Joseph de Luna's felt that the engineer's personal supervision was "indispensable" in the works, not only "to explain the method in which the work is to be done," but to ensure compliance, since it would "only be carried out perfectly under his oversight and knowledge."[82] This personal supervision, of course, could not be guaranteed, given the multiple obligations of the engineers in the colonies.

The costs of all this—in materials, training, and counteracting resistance—obviously contradicted the whole point of leaving the Cuautitlan River diversion dam as much as possible in its traditional construction, which exempted the Desagüe from paying for it. The exquisitely detailed proposal that built upon the fascine technique would remain on paper because, even with Bouchard's unlikely permanent assignment to the Desagüe, investment would be needed on this particular segment of the Desagüe complex, which authorities had long since figured out how to get done for free (to them) and with acceptable results. It was simply uneconomic to change it, but Bouchard would not be in Mexico long enough to realize this. The logic that determined the fate of just this one technique as proposed by three different men applies to everything else they did in the Desagüe.

## "GLOBALIZING" THE DRAINAGE

At the turn of the eighteenth century, as the military academy in Brussels had begun to train and graduate military engineers, and as the institutionalization of the profession advanced toward an actual corps, military engineers started looking at the Desagüe in a different way. They began treating it as a whole, articulating the underlying connection among factors that on the surface seem disparate—technology, matter, energy, people, and finances—and couching their specific proposals in a global appraisal of the project. More closely linked to the absolutist project as whole—rather than to specific works—they were mutating into territorial and project managers who acted as data collectors and propagandists for their science and technology within the physical boundaries of the works under their direction.

Military engineering manuals of the epoch conveyed this perspective, as did the sixty-two different clauses in the corps' Ordinances of 1718, which articulated specifically how engineers were expected to see and act upon material and social realities. After collecting information with field surveys and questionnaires on the territory they were assigned to—including demographic composition (even down to the number of nuns and monks in every single convent), minerals, plant and animal species, cultivated and undeveloped lands, roads and other infrastructure, tax revenue collected, size and location of each urbanized area, and more—military engineers were then supposed to systematize this data and find a practical use for it all. They were to do the former in their cartographic products, through conventional symbols and explanatory cartouches. The latter, in projects that dovetailed with increasing the productivity of the empire, for instance by "trying to put subterranean waters to use, and explaining which rivers might be made navigable or whose terrain and current might conveniently and easily lend themselves to opening navigation channels, canals to feed grain or textile mills and to irrigate fields and lands which for lack of this benefit are unproductive."[83] After all this, whatever resulting project the king approved would be under the overall direction of the military engineer. Clearly royal engineers were now supposed to be much more than simple technicians.

Bouchard is emblematic of how military engineers were as a group consciously articulating on the ground the interests of their sovereigns. They were becoming the designers and project managers of civil and military public works capable of "linking the process of conception-realization, to the much more general, technical imaginary, which sought to organize and quantify flows and movements of every kind, from road traffic to the problems of water flows in rivers or canals."[84] If we include energy in this understanding of flows, the new function of the military engineer meant, concretely, that these men were now responsible for mobilizing science, technology, people, finances, and materials into an overall organization of work (in the mechanical sense of the word—the organization of the transfer of energy onto objects)[85] over a discrete period of time to stamp the priorities of the absolutist state onto urban and rural environments and the relationships between them.

Bouchard thus saw the Desagüe fundamentally as a system whose defects were precisely in the fact that these elements were treated as though they were discrete and with no impact on each other. To him, the finances, the technology, and the management of work, time, terrain, and hydrology were all bound up and needed to be reformed as a package. "The manner in which the work ought to be performed henceforth to remove the earth and rocks and get it all turned into an open trench with greater diligence, ease, less work and without exposing the Indians to the risk of losing their Lives," Bouchard felt, involved several articulated measures. The exclusive reliance on the water-sweeping method had to end because it was inefficient, since while aiming "to avoid major expenses . . . this manner of working . . . has cost much more than if the earth that was in the bottom had been carried up to the very top." This financial short-sighted technology should be replaced by "scarped bridges, which is a road made for the purpose of allowing one to go up and down without risk, as is practiced in all of Europe." In turn, this was related to the treatment and yield of the workforce

because to expect twenty or thirty Indians to labor hanging from ropes, each having in his hand an iron bar or another instrument with which to excavate the earth of the slopes is nothing but to expose them to death because of the ease with which the stakes [up on the ledge, to which the ropes are tied] fail due to the weakness of the terrain, which is reduced to dust after the sun hits it. . . . Fear and apprehension among the workers has been the reason why they have not worked all they could, as they have not considered themselves to be safe. And also because of the great amount of dust that rose from the terrain in which they worked, so that they could barely see each other, so that all day long they did nothing or very little . . . which has been the reason for the great expenditures incurred and the death of an infinite number of Indians without it having been possible to finish the works in all perfection.

To him, the Desagüe had been ill surveyed, designed, and managed from the beginning to the moment he saw it.[86]

The more inextricably the engineers articulated the social, technological, scientific, constructive, and strategic into a tightly glued-together package, and the more comprehensive their view of projects, the harder it became for officials in the Desagüe to selectively comply with their proposals. This starts to become visible at the beginning of the eighteenth century, with Bouchard's intervention, and is clearest in the Desagüe proper—that is, from the Vertideros, through the open trench and extant tunnel segments and on to the outlet at Boca de San Gregorio. For instance, on the matter of workers' safety, Superintendant Luna did feel that Bouchard's proposal was "the worthiest of attention." In fact, everyone was aware of this problem, from the viceroy who had seen how the Indians worked during one inspection to the master architects who were regularly on-site.[87] In 1712, master architect Pedro de Arrieta, for example, certified that the terrain in which the stakes were to be driven to be "so fragile that at the thrust of the wind it breaks apart."[88] Yet, it is clear that either Luna did nothing about Bouchard's proposal to end this, or the guardas undermined it, because during a 1718 inspection of the works in the open trench, Arrieta and other masters of architecture, guardas, and the new superintendant discussed how "it shall be necessary for [the Indians] to work hanging from some ropes fastened to stakes that will be well secured on the surface," all the while poring over Bouchard's own map![89]

Bouchard's assessment and proposals were repeated by subsequent engineers. In 1743, Gaspar de Courseulle, a Fleming who having worked on Havana fortifications had now been transferred to the mainland, arrived in the Desagüe.[90] Courseulle too called for the expeditious, economic, and safe management of the open trench excavation and provided both a drawing and a bill of quantities for the materials needed for the ramp and scaffolding system and the treatment and transport of the debris. This bill included four wagons of a cubic vara capacity, "with their wheels reinforced in the manner of naval artillery gun carriages and four windlasses"; "cordage of ixtle fiber to make four whim-ropes"; three hundred leather bags; thirty-six rammers; one hundred wooden spades; twelve wheelbarrows; twenty-four wooden maces covered in iron and other equipment for five hundred working men. In total, this would cost 861 pesos.

The drawing has not survived, but the bill suggests that during the dry seasons Courseulle hoped to haul debris up ramps with the help of capstans, in an adaptation of the method depicted by Fernández de Medrano in Figure 5.6, which was standard in fortifications manuals of the epoch. Also in the dry season, explosives could be used "in the manner we employ them with the mines and countermines in the attack and defense of fortifications to blast the terrain," Courseulle explained, "so that the paraboloid or funnel-shaped figure to be blasted do so along the charge and line of least resistance."[91] Implementing all this would mean substituting the work

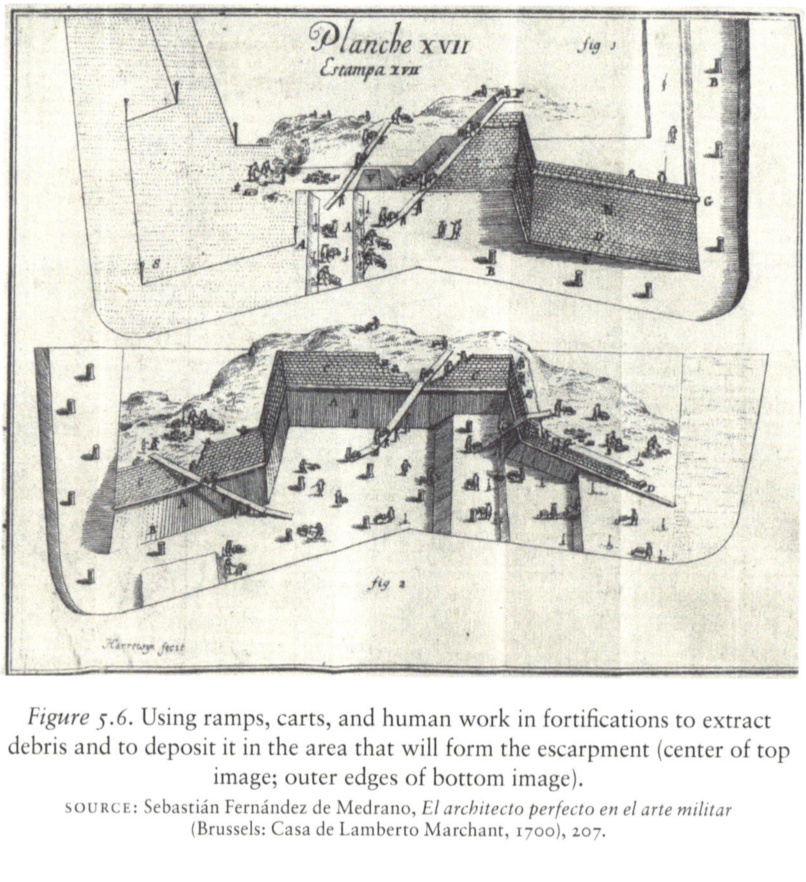


*Figure 5.6.* Using ramps, carts, and human work in fortifications to extract debris and to deposit it in the area that will form the escarpment (center of top image; outer edges of bottom image).

SOURCE: Sebastián Fernández de Medrano, *El architecto perfecto en el arte militar* (Brussels: Casa de Lamberto Marchant, 1700), 207.

done by the mechanical energy of water with that performed by humans, machines, and explosives, enabling the excavation to proceed in both dry and rainy seasons. This "has never been done before," Courseulle naïvely emphasized in his report, constraining progress on the trench to the wet season "and for this reason the Desagüe is in the referred state."[92]

He too failed to execute his plans. Upon reviewing it, retired guarda mayor Joseph Alfonso de Valladolid told Trespalacios, "I have never seen nor practiced myself, in the fourteen years that I was interim Guarda Mayor, any employment but that of iron bars, spades of steel and of wood, sacks of leather, thick rope, and some iron maces (*almadanetas*), to whose handling the workers, who are Indians, had been accustomed."[93] Trusting this, the superintendant promptly rejected the engineer's ideas. Clearly, with their occasional visits, the king's men could do little against the practices and arrangements that the guardas and sobreestantes held watch over every single day.

As a result, by 1755 nothing had changed in the work methods, as can clearly be seen in Figure 5.7, which shows the vigorous survival of the Indian-dangling method.[94] Nor had work methods changed by 1764, when military engineer Manuel Santiesteban again insisted on ending

*Figure 5.7.* The "Indian-dangling" method of work in 1755. Unperturbed, Superintendant Trespalacios (facing viewer), guarda mayor José Antonio Palacios (to his right, pointing), a subordinate guarda, and a notary inspect the Bóveda Real, the mouth of the extant Desagüe tunnel. Note also remaining arches and stone-and-mortar revetment inside this vaulted section.

SOURCE: AGN, Desagüe, vol. 15, exp. 7, f. 306. Reprinted by permission.

the Indian-dangling method, demanding "all convenient precautions for the safety of the workers . . . with escarpments or flats proportionate to the height and width of the lower plane . . . and robust scaffolding with planks and ramps" to ensure the happy marriage of safety with efficiency.[95] The essence of the new global approach had not been implemented, but neither had it been emphatically rejected: it had simply been ignored and allowed to sink into the thick mud of a package of practices that was molecularly bound with the priorities of the elites in the city.

## COMMUNICATING ACROSS TIME, SPACE, AND INTERESTS

Arguably the biggest impact the military engineers had in the Desagüe was where they communicated more pedagogically and less threateningly: the realm of language, which lent itself to appropriation for local purposes in ways that their material and organizational proposals did not. There were also "pull" factors making officials on the ground, creole technicians, letrados, and savants more amenable to these particular aspects of the engineers' innovations. During the course of the eighteenth century, the more royal policy and preference weighed on the side of the "moderns," the more local Desagüe technicians and officials began to emulate the culture of the engineers. Still, while the engineers could transmit some aspects of their culture with deliberate pedagogy, their deeper allegiances and understanding of their roles were incommunicable because they were hostile to local arrangements.

The sheer size and complexity of the Desagüe, as well as the fact that several layers of authority were involved in it, made visual and verbal communication essential to decision making, particularly when large outlays of money were involved. In addition, because the Desagüe proper was a royal project, the manner in which technicians and officials communicated about it had to be amenable to being transported and read across long distances and through the passage of time. In the visual realm, military engineers gave the oidores the first glimpses of the bowels of the drainage and introduced methods of representation that amounted to a new language. The concrete effect on Desagüe structures of these novelties was rather muffled, even as oidores, letrados, and city elites eagerly emulated both the language and elements of the culture associated with it.

Once it became an expectation that all military engineers assigned to New Spain would inspect the Desagüe and leave recommendations for it, Desagüe graphics accrued a new functional layer. The drawings and plans of these royal technicians were now supposed to provide enough detailed guidance over the long term and in the absence of the engineer for all contractors—typically master architects—to be able to successfully complete the project.[96] In addition, when projects involved royal expenses, these plans were crucial to the communication with the Council of Indies and its approval, which followed a close scrutiny by other engineers, not just bureaucrats. This expectation overlapped with changes in the increasing codification of the technical drawing and cartography taught in the military engineering academies in the eighteenth century. As more royal engineers intervened in it more often and as their communications practices were emulated by others, the diversity of representational practices in the Desagüe would tend to diminish over time while the number of visual representations of the project or parts of it would increase.

The earliest extant technical drawings of the Desagüe were by the aforementioned Captain Jaime Franck and are shown in Figures 5.8 to 5.11. Asked to inspect the Desagüe and pay particular attention to the problems of strength and circulation in the extant parts of the tunnel, Franck toured the Desagüe district and inspected the tunnel, trench, and other structures with the guardas during early March, at the end of the dry season. He then presented his findings to the viceroy in two reports that intertwined text and images that together constituted a different language in Desagüe communications.

*Figure 5.8.* Jaime Franck's depiction of the Bóveda Descubierta.
SOURCE: AGN, Desagüe, vol. 7, exp. 3, f. 112. Reprinted by permission.

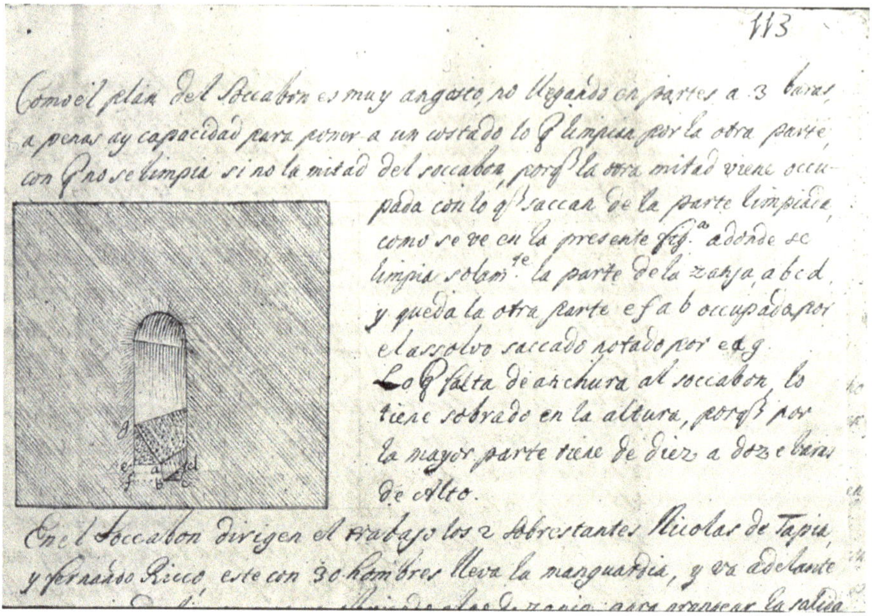
113

Como el plan del Socabon es muy angosto, no llegando en partes a 3 baras,
apenas ay capacidad para poner a un costado lo q limpian por la otra parte;
con q no se limpia sino la mitad del soccabon, porq la otra mitad viene ocu-
pada con lo q sacan de la parte limpiada,
como se ve en la presente fig.a adonde se
limpia solam.te la parte de la zanja a b c d,
y queda la otra parte e f a b occupada por
el assolvo saccado notado por e d g.
Lo q falta de anchura al soccabon, lo
tiene sobrado en la altura, porq por
la mayor parte tiene de diez a doze baras
de Alto.
En el Soccabon dirigen el trabajo los 2 sobrestantes Nicolas de Tapia,
y Fernando Ricco, este con 30 hombres lleva la vanguardia, y va adelante

*Figure 5.9.* "What the tunnel lacks in width it has in excess height," Franck explains, using an image inset.

SOURCE: AGN, Desagüe, vol. 7, exp. 3, f. 113. Reprinted by permission.

The first report was an overall assessment of the Desagüe proper and the water bodies and structures feeding into it where the engineer used images to inform the viceroy of the salient points. Although overall the tunnel was still quite solid, there were two areas of concern, to which Franck drew attention by inserting two drawings within the text. The first illustrated problem was the Bóveda Descubierta (Figure 5.8), a vaulted section some thirty-three meters in length which to Franck should have collapsed after the open trench excavation had completely denuded it of the earth cover above it and yet had held up for forty years already, as he learned from guarda Alonso de Quesada. The second was the need to restore the bed of the tunnel to the depth that Enrico Martínez had given it, the minimum required for the Desagüe to drain water from Lake Zumpango and the Cuautitlan River, albeit not Lake Texcoco. The narrowness of the tunnel made this extremely difficult, as the text and the second image showed (Figure 5.9), placed side by side on the page to enable a nonexpert reader to recreate a reality in his mind out of the interaction of the two:

Because the ground plan of the tunnel is very narrow, at times not even reaching three varas, there is barely any space to place to one side what they clean from the other, so that only half the tunnel is cleaned, because the other half is occupied with what they remove from the cleaned side, as can be seen in this figure where only the *abcd* part of the ditch is cleared while the other part *efab* remains occupied by the removed silting denoted by *eag*. What the tunnel lacks in width it has in excess height, since for the most part it is from ten to twelve varas high.[97] (Emphasis added.)

A few days later, after the sobreestantes leading the cleaning of the tunnel floor came to the 235-meter-long Techo Bajo section and found that a segment had caved in, leaving a dangerous obstruction on the tunnel floor, the viceroy asked Franck to inspect the site and report back. In this second report, Franck used the same exposition techniques as in the first report, employing interrelated text and images. The first image is at the bottom right of the first page. This placement is deliberate, as Franck needed the space above it to provide the viceroy with the basic information he would need to understand the problem the image and the text to its left focused on.

The Techo Bajo, given that name because of its low ceiling, was the weakest point in the tunnel due to the composition of the rock that it traversed. Unlike the sections he had inspected before, which had resisted the devastations of neglect and time in part thanks to the fact that they were dug into hard tepetate, in this long section soft tepetate prevailed. This material, Franck explained, had developed a thin, hardened crust on the outside due to the humid conditions in the tunnel, but since the humidity did not penetrate into the rock, beneath this apparently solid palm's width of crust, the rock was so friable that it crumbled to the touch. To compensate, the original builders of the tunnel had begun to revet this section with stonemasonry arches starting at both ends, but had left this reinforcement unfinished for "lack of money or some other incident," but clearly not because of ignorance.

Having thus written the background information his reader required, Franck now drew a sectional profile of the Techo Bajo, lettering its contours, adding a vertical scale in varas, and placing a textual explanation on its left (Figure 5.10), which informed his reader that "the effect of said collapse can be seen in the present figure, in which *abcde* is what fell from above, leaving on each side a wall-rock (*respaldo*) four and more varas thick as can be seen by *ba* and *ed*" and that his proposal "thick timbers could be placed at intervals along the entire gallery, and upon them some manner of prop timbering (*puntalería*) to support the ceiling for some time to prevent further collapses" had been rejected by the maestros, guardas, and sobreestantes.[98]

The last two figures in this report (both shown in Figure 5.11) indicated the other dangerous segments in the Techo Bajo sector, as before guiding the gaze of the reader, now to a huge crack running along the wall-rock. Franck's "Figure 2" showed how "the dotted line along *abc*

*Figure 5.10.* "The effect of said collapse can be seen in the present figure, in which *abcde* is what fell from above." A scale in varas is attached to the right margin to aid the viewer in seeing the magnitude of the problem in the Techo Bajo section of the tunnel.

SOURCE: AGN, Desagüe, vol. 7, exp. 3, f. 116. Reprinted by permission.

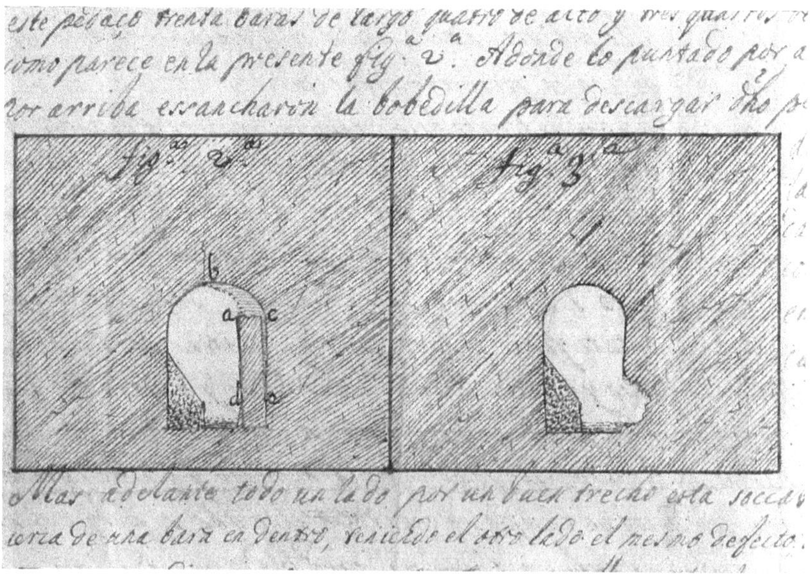

*Figure 5.11.* The other afflictions of the Techo Bajo—debilitated tepetate side supports.

SOURCE: AGN, Desagüe, vol. 7, exp. 3, f. 116v. Reprinted by permission.

shows [where] they widened the vaulting from above to unburden said false slab, removing what is contained within said letters and leaving the slab somewhat inclined" so that *abed* might easily collapse. To the right of "Figure 3", the text explained that the base of the wall-rock was eroded to a depth of one vara. Given all these problems, Franck saw no alternative but for the open trench conversion to continue, but giving the sides of the trench three-quarters of their height in, not as had hitherto been done, "without observing any rule or proportion."[99]

There is a logic to the engineer's choices among the range of visual and textual techniques he would have been familiar with. Audience and purpose played a determinant factor in Franck's choices. His audience was the viceroy—a bureaucratic authority, rather than a builder or senior engineer. What this official needed to know, therefore, were not the technical details of execution but what the fundamental problem in the tunnel was and what his options were. To guide the viceroy by pointing his gaze to contours by means of letters, Franck resorted to the familiar language of Euclidean descriptive geometry in both reports, much as Andrés de San Miguel's open trench depiction or, with much less sophistication, Rodrigo Díaz de Aguilera's sketch for the Coyoptepec silting pool dam had done before. All three technicians used visual language familiar to their contemporaries that sought to evoke various Desagüe realities, not reproduce them as with a mirror.[100] Unlike the local technicians, however, Franck's purpose was not to compete for a contract or for patronage, since his royal employ neither allowed this nor made it necessary. He could therefore afford to muddy Euclidean space with dirt, shadows, and confinement where Andrés de San Miguel had chosen to leave it pristinely blank. Franck did this with lines, filling in the space to evoke both a third dimension and the materials that mentioned in the text (two kinds of tepetate, silt, wall-rock). The lines are different because they served different functions. The dotted lines in Figures 5.8 and 5.11, which in other maps of the epoch would indicate work that was slated for demolition, were here used to denote demolition that that already happened. Continual heavy lines denoted contours; light and parallel or stippled, wall-rock; tapered, depth. Rubble, silt, and stonemasonry arches were depicted realistically.[101]

These representational techniques evolved. Starting in 1718, the successive ordinances of the newly founded corps reflected the progressive refinements of visual communication, becoming increasingly specific about the purpose and content of plans and drawings. By now, to approve any works paid out of the royal treasury, the king wanted to see, in duplicate, a "project, plan, sectional profiles, elevations and lists explaining in detail its constitution and the reasons for its execution, as well as the location of the terrain." This plan had to include or be accompanied by a report with a vast amount of detail, including "an estimate of the cost [of the project] according to the normal price of wages and materials of the country, detailing also the quantity and quality of all the materials that will go into the work, where they might be found and prepared," all using the French *toise* or foot, the Castilian *fanega*, the Castilian pound for weight, and specific units of Castilian money. Plans thus needed to be as detailed as possible, through sectional profiles and different views of the more complicated objects in the structure. This was not only because these authorities had to decide whether to approve the defenses in question solely on the basis of what they saw and read on paper, since site inspections were impossible, but also because the engineer's own career depended on his merits as a planner and builder of defenses and cartographer. Once approved, the engineer would draft instructions about the "quality, mixing, and placement of the materials, and all other circumstances by which the works shall be executed, in order to both ensure their good quality and durability, and to prevent doubts and controversies when they are executed by contractors."[102]

This language spread to nonengineers. Key media for this transmission were the contracts (*asientos*) that master architects, masons, carpenters, and others signed to execute an engineer's plan after it had been approved. Just as in fortifications work, in the Desagüe the engineer had "to very precisely instruct the maestro mayor who will carry out the works . . . on the methods, precautions, and other circumstances."[103] In fortifications, contracting was a means of speeding up construction, by laying recourse to several local artisans and entrepreneurs and their networks of supply for labor and materials to complete the massive amount of work. On strategic projects an engineer would be on the job too, personally supervising materials, management, and workmanship; not so in the Desagüe. Despite the risks, contracting was an avenue for technology and professional cultures to migrate across ocean and society, creating the opportunity for the diffusion of the metropolitan technologies borne by engineers, but also for increasing tension between the professions.

Franck's exposition brought this new visual lingua franca that was beginning to differentiate the practices of military engineers from the craft traditions of other technicians into Desagüe communications.[104] Until then, its usage had been as sporadic as the presence of royal engineers in the works—a scale (*pitipié*) in varas and color to differentiate planned construction from existing structures appeared in a 1628 contract with instructions by Adrian Boot for maestro Bartolomé Bernal for work to be done on the causeways of Chapultepec and Sanctorum (see Figure 5.12). While common in fortifications, the language and conventional signs engineers used were as new in the Desagüe as the relationship of military engineers to it. But if local contractors were to follow the engineers' stipulations in

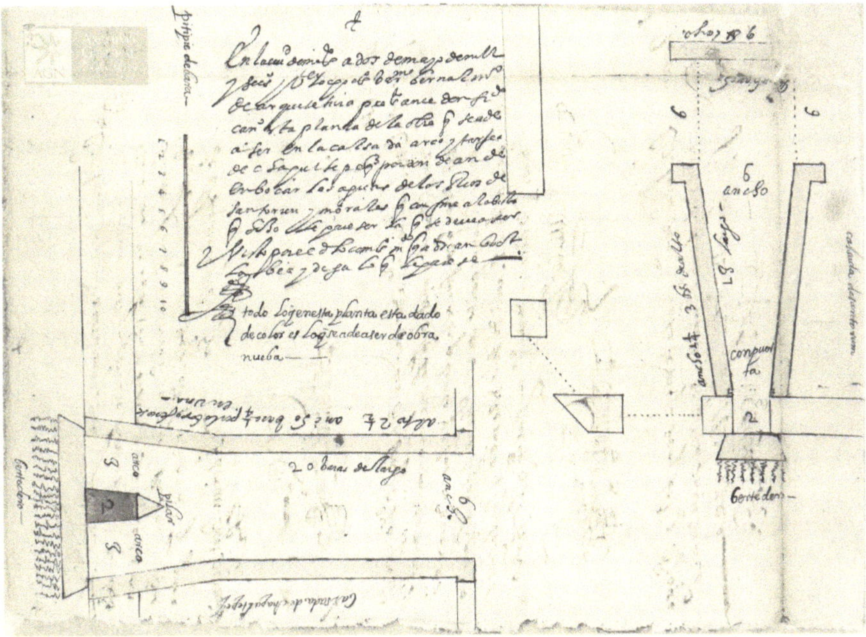

*Figure 5.12.* Adrian Boot instructs master architect Bartolomé Bernal how to read his plans for spillways on the causeways of Chapultepec and Sanctorum: "All that in this plan appears in color is what is to be done as new construction," providing also a scale in varas (center and left of image).

SOURCE: AGN, Desagüe, vol. 3, exp. 4, f. 244. Reprinted by permission.

these contracts and work "according to the plan, which is to serve as instructions for the optimal success of the works,"[105] and if authorities were to verify this, they all needed to be able to read the new lingua franca.

This was not easy at all. The new language was constantly evolving, becoming increasingly economical, precise, and difficult to follow for the nonexpert as it invented sectional profiles, more and more conventional symbols and color codes whose meanings were not often explained on either the plan or the text of the reports. Without something to make them legible, over time, plans and proposals could have become more and more impenetrable as their language grew increasingly compact and symbolic, after the corps was founded. Instructions from the Council of War circulated to the chief engineers in the corps in every province, viceroyalty, and territory set very precise standards for admission to the corps. As of 1737, candidates would only be admitted by exam and after passing all aspects of mathematics relevant to warfare, including analytical and practical geometry and calculus, besides "surveying, machine-making and operation, hydraulics, theory and practice of regular and irregular fortifications" and of offensive and defensive warfare in general. Besides proving they were single, healthy, and of good behavior and honest upbringing, candidates would be tested in draftsmanship.[106] The engineers carried these changes in culture and language—in plans with sectional profiles, color codes, sequencing of depiction and other conventions, and in economically worded reports and proposals—with them into the Desagüe.

So how did contractors in the Desagüe learn how to read, interpret, and apply the military engineers' language and even inject it in structures and plans of their own, and how did officials check the correspondence between a built object and the plan that was supposed to have guided it? This is puzzling because in New Spain there were no academies comparable to the ones in Brussels or Barcelona, where military engineers learned their language, until the late eighteenth century. Instruction could not be gained from books, since military manuals did not provide enough detail on the subject, or from ordinances, which being only textual had obvious limitations. Military engineers did not take apprentices locally, nor did creoles go to Spain for training and then return. Although the exact answer to this question may be inaccessible, there are some hints about how the military engineers' language spread to their nonmilitary audience.

When the aforementioned military engineer Luis Bouchard de Becours went to the Desagüe in 1706, he drew a "map with all its measurements from the entrance that is called House of the Vertideros to the outlet named Mouth of San Gregorio *with the profiles taken at intervals* so as to demonstrate the width and depth according to the measurements as is shown on said map, examining the drop of the waters through its entire length" (emphasis added).[107] This map showed both existing and proposed structures, the latter including two additional diversion dams to prevent two streams from dropping into the open trench and eroding its sides.[108] Unfortunately, this piece has not survived, but his plans for fortifications work in Campeche have. Given his reference to a comprehensive map of the Desagüe that included sectional profiles, the Frenchman probably treated the Desagüe in the manner he dealt with the city of Campeche and its fort in the two images in Figure 5.13, which display elements of the lingua franca the nonmilitary viewer might have seen.

As in the top image in Figure 5.13, Bouchard may have included topographical detail, showing neighboring townships, fields, rivers, and so on, and he probably used color. Like Franck, Bouchard looked and threw light from the left, drafted in graphite first, used perspective (soldierly, not cavalier) selectively, including sectional profiles to focus the gaze of his audience. In both of the Campeche images Bouchard relied on color conventions: brown for earthworks, red for existing masonry work, yellow

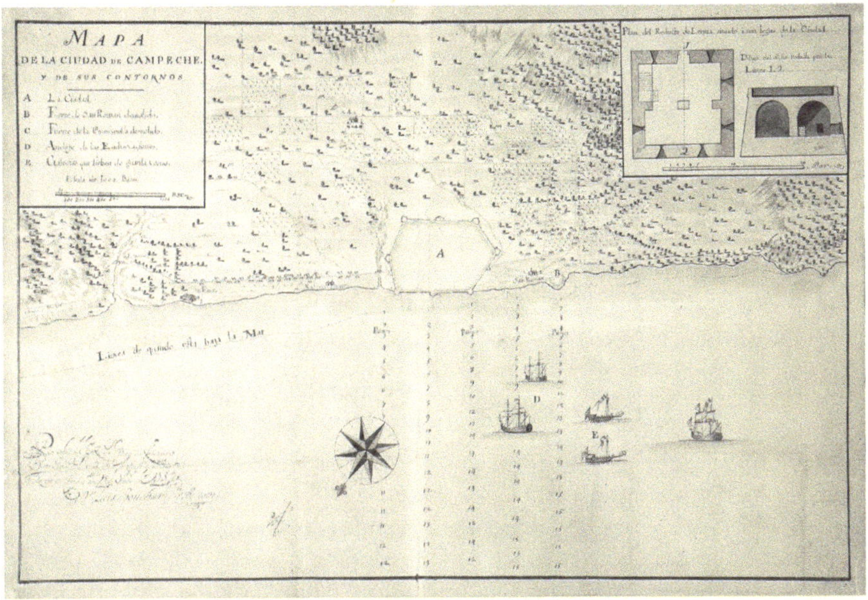


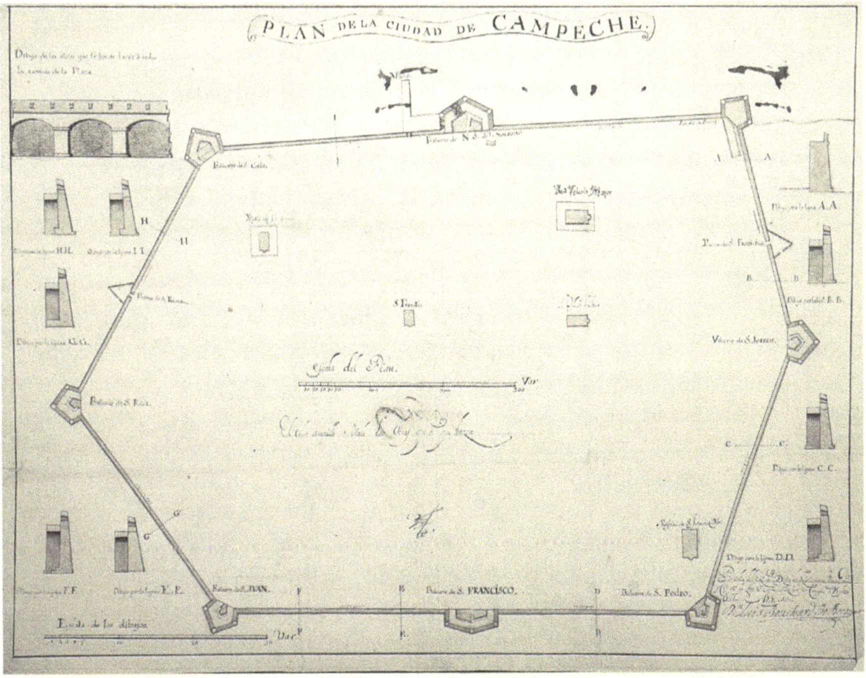


*Figure 5.13*. Luis Bouchard de Becours's 1705 *Plano de las fortificaciones de la Ciudad de San Francisco de Campeche con indicación (en color amarillo) de las obras que hay que hacer* and *Mapa de la Ciudad de Campeche (San Francisco de) y de sus contornos*. His Desagüe map likely used the same conventions.

SOURCE: España, Ministerio de Educación, Cultura y Deporte. AGI, MP-México, 98 and 99. Reprinted by permission.

for new work to be done, and each color was applied as a wash. All of this, which he might have deployed on the Desagüe map, was well on its way to codification in both French and Spanish military engineering plan making, in works such as Hubert Gautier's *L´art de laver, ou nouvelle manière de peindre sur le papier suivant le coloris des desseins qu´on envoie à la Cour* (Lyons, 1687).[109] These works both registered and propagated a practice prescribing black India or China ink for outlines except in masonry structures, "a light, flat tint of burnt umber, burnt sienna or tobacco juice" for earthworks, green for turf, verdigris or light blues for "wet ditches, rivers and the sea," gamboge yellow for works that were planned or incomplete (see also Boot's Figure 5.12), and shading that assumed that light fell from the "top left-hand corner of the plan," taking "left" and "right" to mean those directions as they were seen from the fortress or city, not the country.[110]

Both Franck's and Bouchard's plans and reports were for the viceroy primarily. In Bouchard's, which called for specific work to be done, after the viceroy approved it the engineer would have drafted a separate instruction for the contractor, probably with a cheaper version of the same plans. Both Bouchard and Franck supported these unfamiliar figures and codes with textual prompts teaching how to look—Franck: "as can be seen by ba and ed"; Bouchard: "The yellow color indicates the work to be executed" in the center of the fort and "Drawing along line [G.G., H.H., and so on]" as captions for each profile (bottom of Figure 5.13). Bouchard's large plans are physically separate from the written report that accompanied them, so the image as a whole required more incorporated text to facilitate its interpretation. Thus part of the answer to the question of how the language spread seems simple enough: the engineers were particularly careful in how they worded and drew their plans and proposals, as well as how they verbally explained to contactors and officials while they were still present on-site, deliberately teaching this new "literacy."

But this language could not spread without reception. Even at the highest levels of the imperial bureaucracy reading the plans was not easy, in part because although ordinances were passed about the conventions, scales, and other signs on the plans, not all engineers were versed in the same ones, some having trained abroad. In addition, if only two copies of the map or plan had been produced—one for the Council of Indies and the other for the viceroy—then any consultation about the details of the plan required that all actors involved in a project meet where the plan was physically located, or be sent a copy. Neither alternative worked very well, and since in 1757 even the War Department (Secretaría de Guerra) was "deprived of the plans and drawings that it needs to function," engineers were henceforth required to produce three exact copies of each plan—one for the Council of Indies, another for the War Department, and the last for the viceroy or governor general—so that everyone involved in a decision could refer to the same map, "which being the same for everyone will not mislead anyone's understanding."[111] The authorities themselves thus propelled the diffusion of this new language, multiplying the reach of each plan or map along four different lines of communication across time and space, hence pushing the need for each official to learn how to read them or reveal their ineptitude.

In the Desagüe, all indications are that while the language itself was appealing and its incorporation unproblematic, if uneven, the practices linked to it were a different matter altogether. In accepting the form of the military engineers' communications, Desagüe authorities and technological actors often emptied it of substance. Because language is only a tool whose function is determined by those who use it, officials, letrados, and technicians in New Spain who absorbed it then redeployed it to support the social structures and dynamics in place in the Desagüe and the basin.

## SPEAKING VOLUMES WITH VELOCITY

These eagerly imitated novelties in visual and textual communications were accompanied by less acceptable ones. Hydrometry was one of the innovations where emulation did not come until nearly a full century after it was introduced. This delay both contributed to and resulted in the enduring allure that Vitruvian axioms and Aristotelian categories continued to enjoy among Desagüe technicians seeking to explain the misbehavior of water and materials in the project's structures. Conversely, it also helped maintain hydromechanics at the margins of the Desagüe and of scientific hypothesizing with experimentation.

Friars did not use cubic units to measure water even though they had full knowledge of this means of measurement. For the military engineers, in contrast, cubic units became as essential a tool as their compasses when it came to designing objects for water management. Unlike all other technicians in the Desagüe, these men saw a connection between the design and dimensioning of water-conveying or -retaining structures and both the volume and the flow of the liquid. As a result, when they evaluated existing canals, dams, spillways, trenches, and other structures in the Desagüe, military engineers applied to water the same simple tools others used to measure solids. These tools were simple arithmetic and cubic varas. If the friars' abstinence from cubic measurement of water had both causes and consequences in their understanding of how water behaved and why, so too did the engineers embracing them have significance. Both demeanors were of great consequence for the Desagüe and for the path and content of a Mexican science of water.

During the oidor interregnum, military engineer Francisco Pozuelo Espinosa measured the amount of water that was desirable to release from Lake Zumpango into the Desagüe in cubic varas, using this datum to then determine whether the spillway and canal that was supposed to convey it was capacious enough and how much it needed to be desilted.[112] Fiscal Martín Solís, who was working with Pozuelo Espinosa, did not mention the measured water, but still entered into the record that 123,301 "cubic varas" of earth over the tunnel had been scheduled for demolition. Cabrera, their nemesis, reacted with extreme vehemence to this deviance from the tried-and-true method of using area units to describe volume, deriding Solís as "the inventor of all this confusion about measurements of cubic varas." Cabrera proudly affirmed the friars' adherence to linear measurements in the Desagüe, stating that cubic units

> are not the style in these works, because since the year of 1637, when the open trench was started, until these latest measurements that were taken for the recent excavation in the year of 1675, no text or record of the inspections of these works whatsoever will be found in which this term, *cubic varas*, is utilized, because the measurements which have been carried out since long ago were only of varas of length, width and depth, which is what is truthful and sure.[113] (Emphasis original.)

This was patently not true and only makes the tenacious defense of linear units more mysterious, particularly since military engineers and officials were using these units in their correspondence, reports, and even joint site inspections with master architects and friars. Moreover, even the king's ministers were starting to send unmistakable signals about the desirability of embracing all the "confusion" of cubic units. Seeing the problems that lack of measurement appropriate to engineering tasks was causing for design, practical functioning, and reporting, in 1690 they ordered that Desagüe technicians were to

> carry out a reasoned calculation (if cubic measurement is not possible) of the amount of water that fills up the lakes during the period of their maximum rise, and that calculating these with the discharge of the sluicegates they commensurate the bed of the open trench so that the discharge is larger than the received quantity, so that running through a greater receptacle the waters flow more gently, which will result in fewer collapses in the lips of the trench.[114]

Royal bureaucrats learned about the importance of hydrometry for hydraulic engineering design from the droves of military engineers that had been in the crown's service, building and theorizing about fortifications, water-supply systems and canals, as well as measuring and surveying land and water in all three dimensions. By the time this instruction was issued and reached Mexico, these men had "trained" authorities to understand the language of their proposals and reports, including units of volume—the cubic toise, pied, pouce, or vara. These units would permit the greater compenetration between advanced mathematics and analytical geometry and hydraulics, which could provide insights into water and matter that Vitruvian axioms and Aristotelian categories could not. Yet, cubic units and higher mathematics, not to mention new understandings in hydromechanics, remained on the margins of Desagüe practice and theorizing even as military engineers continued to be imitated in all other aspects of their culture. Old and new scientific and empirical approaches to water in motion would continue to coexist for a long time in the Desagüe. In addition, explicit appeals to the science of motion did not generate new hydromechanical questions and therefore knowledge around the Desagüe.

During this period, this is evident in a consultation over whether to harness the phenomenon known since the time of Andrés de San Miguel as "summons" (*llamamiento*), which described how gravity "beckoned" water in an inclined canal. In 1755, maestro mayor Manuel Alvarez drafted a proposal that called for using this summons for the purposes of ensuring the water's swift transit through the tunnel, where it tended to stall. His idea was to extend and deepen the Desagüe de Huehuetoca from its terminus at the Boca de San Gregorio northward to the Salto de Tula over a distance of four leagues with the help of mining specialists from the Real del Monte and Pachuca districts. Viceroy Revillagigedo had this proposal forwarded to the son of deceased guarda mayor Joseph de Valladolid, who approved it, and then on to the Jesuit Colegio de Tepotzotlán, in the Desagüe district, where Fray Martín Ignacio de Iriarte evaluated the wisdom of pursuing this "summons."[115] Iriarte issued a report that for the first time in the Desagüe enunciated a hydraulic problem and its solution in explicitly post-Galilean terms, drawing from hydrodynamics and hydrometry.

This Jesuit reasoned that the lowering of the level of the Salto de Tula would result in a canal four times steeper, carrying more water at a greater velocity. Desagüe records since the seventeenth century consistently argued for the desirability of a brisk flow for water in the Desagüe because it was argued that this prevented silting and because the scouring effects of rushing water with debris was often regarded as an aid to the excavation in the open trench. Based on the figures given to him, Iriarte concluded that if the gradient were increased by a factor of four, twice the amount of water would flow in the same amount of time. "The truth of this rests," he continued, on

> an incontestable theorem of hydrometry, which states that the velocities that water has in different segments of an inclined canal have between them a proportion that is the subduplicate of the perpendiculars that go from the sections to the horizontal. . . . [A]n equal quantity of Water . . . cannot pass through all sections of a River if the differences between the sections is not compensated with the difference of the velocities.[116]

The report not only reached for scientific law over anecdote but also addressed fluid dynamics, specifically with regard to velocity as a factor of discharge. This was qualitatively different from all prior discussions of water with regard to the Desagüe, in that it argued by reference to universally applicable postulates of hydrodynamics, as then understood. It was the first time that laws of motion were applied to water flowing in the Desagüe. What it was not was anything new, in scientific terms, since Iriarte was quoting verbatim from Proposition 40 in the third book of Tosca's 1727 *Compendio Mathemático*, the *Tratado de la Hydrometría e Idrología*.[117]

Fray Iriarte's assertions were neither disputed nor built upon, despite their evident relevance to the open trench and subsequent canals. He himself seems to have been content with *applying* existing knowledge, in a gesture that was becoming habitual among the letrado and learned circles in New Spain. Neither was there a recorded attempt by others at the time to dig deeper into the conduct of water particles in a flow or the forces that govern the motion of water in the constraints of a canal. The economic reasons for this lack of interest will be discussed later, but in cognitive terms, this is likely to do with the fact that to study water along these lines would have required quantifying it in all its dimensions, not just the wetted areas, as was customary. Military engineers continued tying the measurement of water to hydrodynamics every time they had to assess how to improve the movement of water through the basin and Desagüe structures, but unlike what they did with the language of their reports and plans, they rarely explained the principles behind their practical recommendations. In contrast to other areas of their intervention, they never made much noise about the scientific laws or mathematical methods behind their proposals, so why scientific hydromechanics and hydrometry might be connected and relevant to the drainage was not terribly clear outside a reduced circle of savants and well-read oidores. Nonetheless, this knowledge continued to have an impact, and what more became of the linking of hydrometry and hydromechanics is discussed in the following chapters.

Throughout the eighteenth century, the technological language of the military engineers found far less resistance than the implementation of its substantive contents. Master architects and officials would often accompany the royal engineer, observing and assisting him as he used his instruments and drafted his notes and drawings. Technical drawings were the maximum expression of this drive for efficiency, and because they are such a compact medium, the significance of their component parts can easily be overlooked. How the drawing was oriented, whether it had a scale, what colors or line qualities each element showed—all conveyed a signal to the viewer, who in turn could only understand it by becoming privy to this exclusive language. Technical images thus acted as an avenue for the diffusion of a specific scientific and technological culture, as well as for professional differentiation, and enjoyed an eager and receptive audience, even among the most recalcitrant members of the bureaucracy, who would soon become fond of flaunting their newfound wisdom.

The increasing intervention of the engineers in the Desagüe and other works, such as the cathedral, roads, sewage, the tobacco factory, and street paving from 1762 onward would create and multiply the possibilities for conflict between them and the administrators and artisans involved in such projects. The skirmishes that ensued help us understand the source of the resistance to these metropolitan technicians. Of all the professions in the Desagüe, engineers came into closest contact with masters of architecture, but the conflicts also involved superintendants and guardas, and in each case there were struggles over professional specialization and prerogative, as well as social and economic interest. Despite the clashes, and through the clashes, however, engineers provided alluring examples of the European Enlightenment culture.

CHAPTER SIX

# A Show of Patriotism at the End of the Trench

Coinciding with a new wave of flood threats in the City of Mexico, the British capture of La Havana and Manila in 1762 during the Seven Years' War (1756–63) made that year something of a turning point in the Desagüe. The Bourbons and their advisors reacted to this shockingly concrete proof of the empire's loss of hegemony by accelerating the implementation of long-considered reforms to stem the tide of decline. More freshly minted and career military engineers than before immediately received orders to embark for the colonies to bolster defenses, infrastructure, and territorial control. This increased both their presence in the drainage project and the drive to complete it for once and for all. It was not an isolated measure, as the Bourbons' broader program aimed to reverse the decline by promoting economic growth in agriculture and manufactures in the Peninsula itself. This required bolstering the power of the monarchy vis-à-vis all other social sectors and turning the crown's New World kingdoms into proper colonies through a variety of measures designed to rein in the autonomy of their elites, recapture American wealth, and ensure markets for Spanish goods. Although their outcomes did not always match the intentions, the measures of Charles III and IV and their ministers touched every aspect of the colonial order and the link with Spain, from production to trade, administration to taxes, religion to science, urbanism to art—and of course, engineering, public works, and nature itself.

In the basin of Mexico, these efforts took shape amid urban denizens' renewed worries about the city's vulnerability to flooding. There had been above-normal rainfall in 1707, 1732, and 1747, but the rainy seasons of 1762–64 brought so much water that floods seemed imminent and inescapable. As feared, the city endured flooding these years, and although the floodwaters never reached the catastrophic heights they had in 1629, they did rekindle the desire to rush the open trench conversion to completion and reconsider the long-postponed aim of a general desiccation. In 1767 royal officials decided to put the entire open trench completion out for public bidding. After a tense discussion over the benefits and costs of privatizing this critical component of the Desagüe proper, the contract was finally given to the consulado. This corporation of the wealthiest merchants of the viceroyalty became not just the contractor but also the authority in the Desagüe proper between 1767 and 1788, when it finally declared it had "completed" the conversion of the tunnel into an open trench. But what happened in the Desagüe during these years was not just the consulado's doing. Rather, it was the result of how the actions of all actors on the ground and in the halls of power directed or refracted the penetrating beams of Bourbon Enlightenment, weakening, focusing, or deflecting them altogether to serve their own interests.

For a long time, different sectors of the elite within the viceroyalty had shared fundamental agreement that wealth and rents in the capital must be preserved and that the burden for this should be borne by others. Within this consensus, however, there were conflicts over matters of authority and prerogative, finances, social arrangements, and technological and scientific culture. The crown increasingly favored some actors as better agents of its new priorities, while relegating others to secondary roles, and since the Desagüe was after all a royal project it could not but get caught up in the struggles over who benefited and who lost out as the changes trickled down to the project's most material, human, and cultural dimensions.

These conflicts involved not just the Desagüe's officials, technicians, contractors, workers, and neighbors, but also the letrados and savants of the city, who were either called upon to assess possibilities in the project or intervened in the debates on their own accord. Invited or not, the opinions these men gave were colored by their status as privileged creole creatures of the city, and they had both cultural and social implications. Both the presence of the consulado in this broader context and the interventions of not just local savants but also the ever-more-numerous royal military engineers had effects on the Desagüe and its relationship to society and culture. With the reactivation of the open trench, serious reconsideration of the possibility of extending it for a general drainage, and a push for several complementary projects of embankment, dam, and causeway construction and repair throughout the basin, as well as much theorizing about water and matter, the initial years of the consulado contract mark the culmination of the first wave of potential change begun in the previous period.

## *POR ADMINISTRACIÓN*, OR *POR ASIENTO*?

Above all the human actors in the Desagüe and their struggles, there was the nonnegotiable natural fact of flooding. Quite unproblematic in the countryside, flooding could ravage the Hispanic city. In the absence of major climate changes, the city had escaped major floods since the oidores took over mainly because of the river diversions accomplished up to that point and the natural senescence of the enclosed lake system—not because of the smooth functioning of open trench, which was still in deplorable shape. It also helped that at least from 1742 on, Superindendant Domingo de Trespalacios y Escandón had enforced, with a draconian hand, the haciendas' and townships' maintenance duties on earthen dams and had created measures that ensured the proper functioning of the storage lakes of the northwest quadrant. Thanks to this, Lakes Texcoco and Mexico had not been inopportunely fed water from the northern half of the basin.[1] Although these aspects of Trespalacios's zeal assuaged fears, they also threw light on just how much neglect water-management works and the trench conversion had endured before his term, renewing awareness among city residents about the precariousness of their well-being and the fact that the Desagüe had yet to be completed.

When the three back-to-back seasons of unusually heavy rainfall in the early 1760s reawakened dormant fears about catastrophic flooding, the incomplete state of the Desagüe became worrisome. Inspections of the basin were making it clear that while the Cuautitlan River diversion had been enough to preempt catastrophic flooding over the previous century and a half, it might soon be insufficient "because the lakes no longer have the same capacity . . . due to the abundant silt that they have received in this long time," as Superindendant José Rodríguez del Toro explained to the viceroy in March 1764.[2] As if emerging from a long slumber of complacency, crown officials revived the old goal of a "general and perpetual desiccation." This demanded that they first insufflate the stalled open trench works with new breaths of money, people, and expertise to ensure its completion even if they had to ride roughshod over the prerogatives of its resident elites and drag them into contributing to their own welfare. The problem was: How to do this, and at what and whose expense?

There were a number of choices that needed to be made to deal with these questions. In terms of who should pay, there were other decisions to make besides those based on class biases. The viceroy might decide that cost would be borne by the crown (through the Ramo del Desagüe, a division of the royal treasury in Mexico), the city (through its taxes and property rental income), the "beneficiaries" (though a prorated contribution based on their real estate), or some combination of all three. Making this choice became considerably more complicated as the eighteenth-century oidores spread their purview apace with the increased understanding of the basin as a system and of the interdependence of the structures that regulated the flow of water throughout it.

Jurisdictional conflicts muddied things further, beginning in the winter of 1763, when Trespalacios was bracing for the possibility of flooding the following summer coming from the southern lakes. After inspecting the dams and causeways in those lakes, maestro Ildefonso Iniesta Vejarano recommended rebuilding the dam-causeway of Tlahuac that prevented water from Lake Chalco from freely entering Lake Xochimilco, and reinforcing the causeway-dams of Iztapalapa and Mexicalzingo, to the tune of forty-three thousand pesos. But some of these works lay outside the boundary markers (*mojoneras*) of the city, so the vecinos balked, arguing they were not their responsibility, and while they were asked to contribute, they scored a victory when the costs were prorated among the haciendas, ranchos, and Indian townships there.[3] As this triggered protests from the latter, Trespalacios's successor, Rodríguez del Toro, reviewed the decision in March 1764. "The purpose of these works is none other than to imprison and dam the waters of the lakes so that they do not descend upon [Lake] Texcoco and flood this City," so unlike "the building of bridges, habilitation of roads, and the like," Rodríguez del Toro reasoned, "the hacienda owners and the townships derive no individual benefit from them, but instead evident pain and damage, as their lands will be flooded from the retreat [onto land] that the waters must necessarily make once dammed, impeding their cultivation and the growth of grasses for livestock." He thus reversed Trespalacios's order because "it would be harsh under these circumstances to press and compel" the haciendas and townships to pay for such expensive works.[4]

So who would pay—urban denizens or riparian landholders? Neither, as it turned out. Discovering that the Ramo del Desagüe was flush, with more than 171,000 pesos in the black (a result of frugal spending over the previous decades of stringent enforcement of maintenance duties), Rodríguez del Toro conceded that the Ramo del Desagüe could well afford to shoulder works in the southern part of the basin.[5] This was justified because while the individual gains of the townships and haciendas around Lake Chalco from the Tlahuac causeway project were null, "the universal gain that the whole kingdom accrues as a result of the preservation of its capital is undeniable." If this principle had been invoked to try to prorate the cost of the protection of the capital among populations just outside its limits, then "so too does it lend itself for the prorating to be extended to all of the jurisdictions, citizens and townships of the entire kingdom, as all of it has a stake in keeping this city unharmed."[6] Since the Ramo del Desagüe was fed by the regressive taxes on wine and slaughterhouses, this meant that not just the whole kingdom but most particularly all of its lower strata would pay for the work. The vecinos won out.

Lastly, the viceroy would need to decide whether to order the works executed directly by crown or city governments (*por administración*) or through a contract with private parties or corporations (*por asiento*).[7] On mayor royal projects such as the Desagüe proper, a military engineer (if available, and if not, the maestro mayor) would plot out the design, conditions, and materials. After the public bidding process, the winning individual or corporation would name a guarantor, whereupon payments could be issued to him out of the Ramo del Real Desagüe, the drainage treasury account.

Contracting at all scales was a "common and current practice with which many expenses are avoided," as Trespalacios put it in 1747.[8] The benefits for the exchequer and the contractor are clear. The exchequer saved itself the risk of overhead resulting from Desagüe technicians underestimating costs.[9] The contractor made profits commensurate to his ability to cheat the granting entity, the workers, or both. Although contracting was occasionally done even in the sensitive open trench, in structures within the Desagüe complex that did not use repartimiento labor or were assigned to riparian populations to maintain, contracting by public auction was even more frequent.

With these advantages, however, came significant problems. Because an open trench contractor would automatically be allotted repartimiento labor, the temptation to divert this workforce for private enterprises or overexploit it was great and apparently irresistible. In addition, Desagüe superindendants knew that "the contractors defraud the workers, be it in the manner or the amount of the payment of their wages or overworking them excessively to gain time and *reales*."[10] Moreover, superindendants believed that "the quality of job-work is not convenient, because [the contractors] do not usually perform it with the necessary solidity and permanence because the master [architect] or other private party who is given the contract avoids consolidating it as is required for the sake of his own gain."[11]

This practice of contracting is one of the reasons why although the oidores reformed certain things about the Desagüe, they scarcely advanced in the open trench until the 1760s and were unable change the overall shape of the project. Any decision about what to do about the open trench and any complementary works in the Desagüe district was thus subject to the same complex considerations as the Tlahuac project involved. Once acquiesced to by all propertied and letrado sectors of the elite, these decisions expressed the consensus as well as the balance of power among them; once implemented, they defined the lay of land so far as the elites were concerned.

## MULTILATERAL TENSIONS

Everything the monarchy and its local officials decided about the Desagüe directly or about things affecting it in diverse realms fell upon this lay of the land, so to speak. One of these realms was the choice the Bourbons made about who to trust to design and implement their policies—bureaucrats to change the administration of rule and military engineers to change the physical structures of rule. Both types were Europeans, and the latter the single most important group capable of transmitting new knowledge and also applying it with the single aim of supporting the crown (and their own careers with it, naturally). Much was a stake in who shaped these physical structures—who, concretely, directed public works aimed at rearranging, capturing, or organizing nature. Before the founding of the Real Academia de San Carlos and the Real Seminario de Minería, the only technicians the crown trusted to be solely responsive to its interests were military engineers. Guild masters of architecture were loyal to local, urban elite interests, and these coincided with those of the monarchy sometimes but not always, and certainly not when the Bourbons began making clear their intent to displace the creoles from positions of power and minimize the role of the bodies that they did run, like the cabildos.

It is therefore hardly surprising that in 1768 a new set of ordinances for the Corps of Military Engineers were issued that, while not excluding creole technicians from projects "not belonging to the department of war, and which are paid for with moneys from my Real Hacienda, from the public, from communities or private parties," clearly set aside all fortifications and other strategic works for military engineers alone.[12] Desagüe superindendants became aware of these preferences and tended to abide by them wherever possible, despite the fact that they were not required to. It would become increasingly the practice for superindendants to meet alone with engineers called in from Veracruz and other locations to assess what work needed to be done and to draft instructions to be passed on to guardas mayores and master architects alike.[13] "It appears the Royal Will is that an engineer be named with preference over an architect," Superindendant Trespalacios would remark to Viceroy Revillagigedo.[14] It is not difficult to understand this in the context of the Bourbon reforms. While alienating, these measures clearly sent the signal that engineers had something that Madrid liked and expected, and this in itself was an incentive for emulation, not least as a means of advancement. Masters, lawyers, and other bureaucrats were increasingly speaking in the language of engineers even while they occasionally disparaged the substance, and frequently undermined the implementation, of the engineers' recommendations.

During the seventeenth century, it had been customary for engineers to participate in inspection teams together with learned men and masters of architecture. Although at times each might compose a separate report, engineers were not asked to opine on the quality of the work of maestros in the Desagüe, only on the state of the project in general. But now even reputable letrados and maestros mayores might find their expertise subject to the often harsh judgment of engineers, who enjoyed stronger royal backing than ever before. This could in no way be flattering, or good for business or patronage and paid positions, as the maestros' work came under greater scrutiny for technological adequacy and budgetary honesty while the competency of the creole learned elites was questioned. The more disparaging the engineers grew, however, the more resentment they generated, the more their proposals were mired, and the more the fundamental character of the Desagüe remained the same.

These tensions of course involved Desagüe officials, the vecinos, and viceroys as well, since what was truly at stake was not the relative prestige of individual engineers or architects. The tensions were brought to the surface when as part of the disquiet that put the works in the southern part of the basin in motion, in February 1767, Viceroy Croix ordered Captain Ricardo Aylmer to evaluate the Desagüe proper. A few days later, this military engineer reported back, explaining why he thought the project had been carried out "with a surplus of ignorance." With the help of a plan, which is shown here in Figure 6.1, he described the interspersed segments of trench and tunnel that carried the excess water from the Cuautitlan River and other sources as a tortuous obstacle course. From Vertideros water flowed through a trench segment with slopes so steep they neared perpendicularity, which, given the type of terrain they had been dug in, stunned Aylmer. The water traversed this trench segment to enter the Bóveda Hermosa, a 477-meter-long tunnel segment, then it went into another misshapen trench segment of 229 meters, where again

*Figure 6.1.* Ricardo Aylmer, 1767, view of the Desagüe on its terrain.

SOURCE: España, Ministerio de Educación, Cultura y Deporte. AGI, MP-México, 240. Reprinted by permission.

it squeezed into a tunnel stretch of 571 meters, and so on—trench for 23 meters; tunnel for 173; trench, 7; tunnel, 23; trench, 37; tunnel, 10—until it was finally liberated through the Boca de San Gregorio to flow down the exit canal and away. Neither was Aylmer impressed with the tunnel segments, where in addition to problems already cited by his predecessors, Desagüe artificers had "not taken the sober precaution of lining the floor." In sum, he was

> dismayed to see the complacency with which the city of Mexico rests upon this Desagüe, deluding itself that this is a universal preventive against the rigors of flooding that could damage it from those parts and apparently resigned to the large annual sums spent on maintaining its present state, which in fact it must have not known about, since in over a century it has not progressed at all toward chasing away with the open trench and free flow of the waters the damages that with great probability could ensue, resting always on the zeal and diligence of those who were charged with caring for and preserving this important work as they were paid to do.[15]

Moreover, the superindendants were responsible for this state of affairs, Aylmer intimated, as "those commissioners have either not acknowledged how imminent the danger is, or for one or another reason have defrauded the trust placed upon them by hiding the constant risks." Besides the danger of collapses within the tunnel sections, these "very probable" risks emanated from the problem that in any moderately rainy season the water that could not squeeze itself into the narrow tunnel sections piled up at their mouths to a height of up to twenty-five meters, as had happened the previous year. Then, "with their violent rotational movement they could easily break away a clump of earth too large to pass through the mouth of the tunnel, and with this blocked, the waters would retreat to Lake Zumpango," and from there on to the city. "No amount of human diligence could stop such a fatal event," concluded Aylmer.

For Aylmer, then, the trench had to be completed forthwith. This meant uncovering all remaining tunnel segments, widening the trench floor to eight meters uniformly, and giving the sides 45-degree slopes. In total, this would entail the removal of "four million cubic varas," and cost some 1.2 million pesos, including salaries, materials, tools, housing barracks for the workers, and of course, the engineers' favorite "ramps for the extraction of earth." Convinced by the engineer's forceful exposition, by May 4, Viceroy Croix had decreed the execution of Aylmer's recommendations. In light of the apparent urgency, and losing patience with the cabildo's hemming and hawing over paying its share, in September 1767 Croix decided to put the 1.2 million-peso trench completion project to public auction—pointlessly, since the only entity that could possibly post the necessary bonds and shoulder such a budget was the consulado.

Although a firm believer in the need to protect the city from floods, one of the biggest critics of this fast-paced completion of the open trench was Rodríguez del Toro.[16] For him, Aylmer's proposal would have enormous costs. Its continual labor demands would require repartimiento Indians from distant provinces. The Indians would have "to abandon their families, huts, and meager belongings," while the works would suffer too because being from distant lands, these workers were "not used to and not trained in such works" and so would "necessarily die in great numbers, and also endure the continual labors which the interested party will force on them." Finally, it would all require "the delicate decision to impose levies on the estates of ecclesiastics and seculars."[17]

Moreover, Rodríguez del Toro continued, it was technically unnecessary, since despite the opinion of "every other engineer who has come to this capital in the last fifty years, and yet three more in '64," the vaults were still strong after 160 years and would withstand many more. Aylmer's 1.2 million estimate only served "to enrich the same people who before now had estimated the costs at less than a sixth part of the current estimate, in considerable detriment to the public."[18] Instead of abiding by the engineer's opinion, del Toro urged, the open trench should return to the practice of demolishing each year a small portion according to the financial resources of the Desagüe and the seasonal methods of work, where "the works were carried out without haste by the Indians of the region themselves who are accustomed and have practice in the works of the Desagüe already."[19] All this for well under three hundred thousand pesos.

A pitched battle ensued. Aylmer dismissed Rodríguez del Toro's argument by referring to the clauses of the bidding conditions that prohibited ill treatment, adding that any Indian peasant could do the job of digging and moving earth from birth. Moreover, Rodríguez del Toro, he argued, was "resentful and downhearted" because his opinion came second to the engineer's, "when it is not his profession, and it is mine, and as such His Majesty entitles me with a salary and promotions commensurate to the diligence I evince, stationing me wherever I may carry out the functions that are exclusive to this job." Aylmer thought it was "very funny" that del Toro should think that his naming as Desagüe superintendant entitled him "to perform the most intricate functions in this commission as if with the nomination he had been infused with the science."[20] Aylmer had a point: Rodríguez del Toro, as a graduate of the University of Salamanca, had purchased his position as oidor in 1741, while Aylmer had secured his own career through merit.[21] Rodríguez del Toro, Aylmer concluded, had better "go practice his talents in his own tribunal."

This heated exchange could not articulate more clearly the perception that engineers had of themselves as royal servants, members of a professional corps trained in academies that instilled science and technology as

*Figure* 6.2. Ricardo Aylmer's longitudinal and cross-sections of the Desagüe formed the basis for the technical instructions in the consulado contract.

SOURCE: España, Ministerio de Educación, Cultura y Deporte. AGI, MP-México, 241bis. Reprinted by permission.

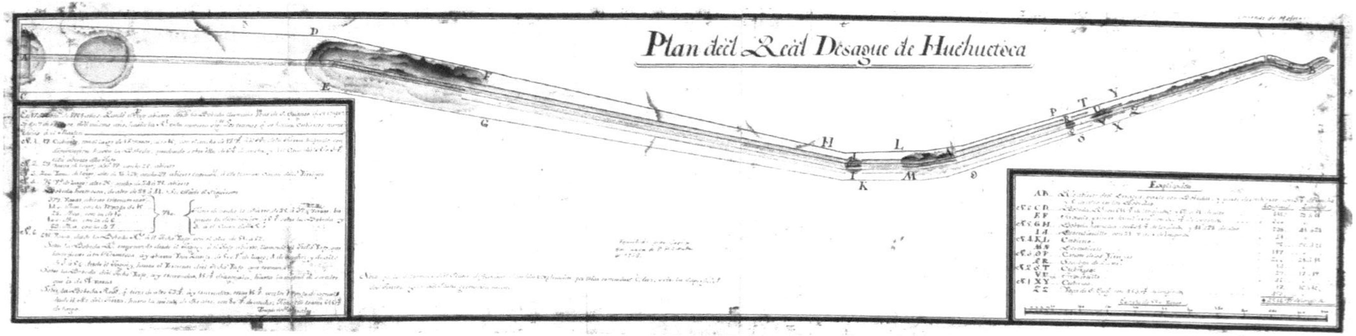


*Figure* 6.3. Joseph de Urrutia's plan of the Desagüe in 1768, showing two sets of measurements for the longitudinal plan in the center. Left, traditional methods yielded distances in varas on the terrain; right, triangulation with instruments provided latitudes and longitudes. In the center, the engineer explains the usage to the two methods is the reason why the two sets of distances differ.

SOURCE: España, Ministerio de Defensa. ACEG. Reprinted by permission.

bases for authority, on a mission to apply their methods and "science" against all odds. And these "odds" included not only the material aspects of their projects but also the interference of any who they felt refractory to their expertise. A dismissive demeanor toward officials and technicians they felt lacking in this knowledge in the Desagüe and even beyond was quite widespread among the engineers.[22]

Despite this opposition to the royal engineers by rather influential people, Viceroy Croix would not be swayed: if what Aylmer warned about came to pass, he would be responsible. Besides, the consulado seemed willing to do its patriotic duty, perhaps nudged on by the generous terms of the contract and by the fact that in bidding to do the work themselves, the merchants avoided a levy "proportionate to their commerce"—which would have come on top of that assessed on their urban real estate, as stipulated in the viceregal decree approving the open trench completion.[23] Before the end of the year, the contract was signed. For eight hundred thousand pesos payable over five years, the merchant guild committed to finish the open trench as indicated and to do so in five years.

For the contract, Aylmer documented the state of the works at that moment in the longitudinal and sectional profiles, shown in Figure 6.2, which visually supported the technical stipulations in the text of the contract. For good measure, as work began in 1768, another punctilious military engineer, Joseph de Urrutia, provided a second longitudinal plan with two sets of measurements (linear on the terrain and triangulated with instruments) as well as the latitude and longitude for the segments of the Desagüe under contract (see Figure 6.3).

Predictably, Aylmer saw the same defects in the seasonal water-reliant and Indian-dangling method of work as his predecessors: it was not just a waste of human resources because of the high risk of death but also detrimental to the works, as it took so long that erosion took its toll before the shapes could be consolidated. The trench conversion, Aylmer prescribed, should therefore be completed swiftly by excavating in all seasons, using human energy (provided by repartimiento labor as an option, but without precluding wage labor) or mechanical means to extract the debris when water was not available to sweep debris away. This, of course, meant ramps and devices such as winches and carts similar to what Courseulle had proposed decades before. This condition would be seconded in February 1768 by the Walloon colonel engineer Charles de Witte, who reviewed Aylmer's plans and all past decrees, judicial orders, reports, and proposals of the Desagüe. Witte added that all excavated materials, "shall be taken out and placed away from the edges, to prevent them from falling in the River," instead of "trusting that the current or force of the water can carry them away." Apparently feeling authorized by his successes in the canalization of the Jarama River in Spain the

1740s, where he had faced similar problems as those of the open trench, Witte expressed contempt for the authors of water-sweeping method, which was a "huge mistake," just like

> the ill-formed ideas of the learned and Practical men of this Country who imagine that the current of the water carries the earth that is thrown into it; on the contrary, although they appear swept away, they only litter more in other parts further down where they cause more damage than before and are four times as costly to remove, since following their thoughts they create gullies to give the waters unheard-of and extravagant velocities, and from this kind of ideas and thinking nothing good can come about . . . uselessly spending millions.[24]

Equally predictably, the corporation had particular problems with the methods of work the engineers stipulated—that is, the overall project management. Yet the provisions about work (in the physics sense of the word) were really the only onerous thing about the contract. The other clauses were actually rather generous. The consulado was given the right to the sisa and wine tax owed the Desagüe during its contract.[25] In addition, the funds for all its public works came from its rights to the *derecho de avería* collected in the customhouses of Veracruz and Mexico, which gave it a yield of seventy thousand pesos annually.[26] But what exact profit its merchants actually made from their Desagüe involvement is unclear since the contract allowed them to document their accounts "in the manner of commerce and without the scrupulous detail that is required for the accounts of the Royal Exchequer," the word of the prior, consuls, deputies, and councilors of the corporation sufficing. Exempt from treasury oversight, the corporation was allowed to police its own accounting process without any interference whatsoever from crown officials, such that "under no condition, motive or pretext will the entries of the already approved account be inculpated."[27]

More significantly, the contract gave its administrators jurisdiction and powers akin to those of the oidor superindendants. In addition to all Desagüe installations, stores, and supplies at hand when the contract was signed, the repartimiento labor, maintenance, and supply obligations that indigenous townships provided the Desagüe were also transferred to the corporation, whose administrator on-site would "enjoy the privileges enjoyed by the Desagüe."[28] This wholesale transfer of powers to the corporation did not have immediate precedents, but it did have an architect—Francisco Javier de Gamboa.

A lawyer with a keen interest in the sciences, Gamboa was a relentless ally and consultant for the merchant guild beginning in 1755. An oidor from 1774 to 1783, Gamboa had in 1761 presented a vast program that would have allowed Mexican merchant capital to control mining. This was the *Comentarios a las ordenanzas de minas*, a critique of the outdated sixteenth-century mining code.[29] Apparently drawing from his late colleague José Sáenz de Escobar's 1706 *Breve tratado de las más principales Ordenanzas de Minas*, which circulated in subsequent manuscript copies,[30] Gamboa discussed the difficulties confronted by Mexican mining. Although there were technical problems, which he addressed by explaining and advocating the patio method of refining that was then in practice, Gamboa argued that mining's main problem was investment, and only the consulado had the capacity to solve this by creating a bank with stockholders. This bank would be a Compañía General Refaccionaria de Minas, funded and run by the consulado, who would be in charge of not just the finances but also the administrative aspects of the recapitalization of mines.[31] The banking proposal was not implemented, but many of Gamboa's ideas went into the new royal mining ordinances of 1783.[32] It is unlikely that the consulado would have proceeded in such an important contract negotiation as the Desagüe without Gamboa's advice.

The logic of this advice can be found in Chapter 25 on "The Jurisdiction in Mines Cases" of the *Comentarios*. For Gamboa, the way in which the old ordinances granted mine administrators and managers broad rights to govern the mines and "all things relevant to them," including the right to hear civil and criminal cases and to mete out punishments, nullified the authority of the district magistrates and other government officials.[33] To him, this was a problem for the industry, as in taking on such responsibilities, miners would be encumbered with a costly apparatus that replicated what the crown already provided. Fortunately, this had not been implemented in the Indies. Gamboa's insight that full authority and power in the mines would have had an avoidable cost to the miners themselves is the likely source of the clauses in the consulado contract that granted the corporation all the privileges of jurisdiction without any of its costs: unencumbered by bureaucrats and detailed record-keeping, its powers were not only cheap but also impervious to the oversight and performance reviews that could restrain the power of government magistrates.

Interestingly, upon reviewing the terms of the consulado contract as minister in the Council of Indies, where he had been serving since 1764, Trespalacios objected to the technical provisions, not the administrative ones. He accused Aylmer of "reporting with no knowledge whatsoever" and reprimanded Viceroy Croix for not calling in the chief engineer Manuel Santiesteban, who was in New Spain, resorting instead "to one who was new in the Country, and a foreigner who could not even understand the language and the provincial terms used there."[34] So it would appear that for all their training, the military engineers continued to be either ignorant of or unwilling to accommodate to the social and political arrangements in the colonies, and Aylmer himself may have served as a pawn to the very corporation his king would try to rein in.

For their part, the consulado merchants were all too willing to accommodate new discourses of patriotism, utility, and royal service without necessarily matching their words with deeds. Individual members of the consulado who had grain haciendas and other agricultural enterprises in the Desagüe district may have had individual interests in swaying the drainage toward personal gain in their manner of conducting the works and using resources.[35] More importantly, since as a corporation it had almost absolute freedom to act as it pleased in the drainage works, the consulado could do many things—from changing the social arrangements it had inherited, to ending the ponderous pace of the water-sweeping method for debris removal. If it wished, it could even present plans to change the very character of the project, turning it into a multipurpose artifact. The contract opened up many possibilities for the consulado, its members, the Desagüe, and the viceroyalty as a whole. In the Desagüe, these possibilities remained alive for decades after the consulado left the works. Again, what the consulado did with them was determined by the relative power and tenacity of the different groups with a stake in the Desagüe, including the merchants themselves, of course.

## A CORPORATE PATERFAMILIAS

One possibility was to alter the prevailing social arrangements in the public work for the sake of profit, if not for a less wasteful use of Indian labor. Did the arrival of these versatile entrepreneurs, the consulado administrators, change the manner of recruiting and using labor in the Desagüe? Did they realign Desagüe labor for the purpose of preserving the lives of its indigenous laborers and making them more freely available for other uses? It would appear that more attention was given to management of public perception than to actual improvement of working conditions.

The corporation claimed that it used its contractual Desagüe privileges wisely, improving the lives of workers "as a diligent Father of a Family does with those in his charge" with kindness and largesse:

> If they sicken they have a Doctor, Apothecary and assistance at no cost whatsoever; if they get hurt while working they enjoy their whole wage in addition to their food and treatment as if they had worked the entire time of their disability; . . . if they die, the Priest gives them a charitable Ecclesiastic burial without charging them at all; if the deceased leaves a widow, in order to somehow dry her tears and compensate the harm done, the widow of an overseer receives fifty pesos, and a laborer's a team of oxen with all its trappings. . . . Special care is taken that the workers are given rest periods and that they are not fatigued too much.[36]

Many in the elite willingly believed this. But not all contemporaries were taken in by the merchant guild's tooting of its own horn. The contract had barely been signed when, without directly naming the consulado, Superindendant Rodríguez del Toro expressed his fears that contractors trying to profit were known for their relentless pressure on the workers, which forced the crew captains (*mandones*) to routinely beat the Indians to urge them on.[37] Such fears about superexploitation appear to be well founded at many levels.

For example, although the consulado committed "to contribute to the workers their [repartimiento] wage free of any withholdings, without mistreating them, or forcing them," it also ceased providing Indians the food rations that had been customary in the Desagüe labor draft, "leaving them free to provision themselves of their food where it is most convenient for them"—that is to say, purchased on-site from agents of the merchants themselves.[38] Given the precedents of guardas illegally selling goods to Desagüe Indian workers in what amounted to the traditional coerced sales of goods that district magistrates were known for (*repartimientos de mercancías*), there is little reason to believe that merchants, who after all were the purveyors for this form of trade, would abstain from the splendid opportunity to engage in it without the intermediation of district magistrates (corregidor, alcalde mayor, and later *subdelegados*).[39] All the more so when these workforces were not negligible—871 men a day on average between December 1770 and July 1772, and 257 men on average thereafter.[40]

The consulado's contractual right to summon Indian governors and laborers and to order and supervise maintenance duties even on structures that were not part of the contract, such as the Cuautitlan River diversion, thus had consequences for the Indians. But it also caused problems with the guardas of the Desagüe and district magistrates. This in turn affected the project itself. The right to summon Indians obviously treaded on the toes of magistrates, whom superindendants had traditionally relied on to summon the parties to maintenance duties and supervise their work. The Cuautitlan alcalde mayor complained that not only was this a usurpation of the prerogatives of his post, but it was actually detrimental to the Desagüe itself, since under the new administration seven haciendas, six ranchos, and two townships assigned to maintenance duties had done sloppy jobs, or none at all, on their assigned Cuautitlan River diversion dam portions.[41] Unfortunately, there was nothing the alcalde mayor could do, since Article 5 of the contract had given the consulado all civil jurisdiction over all its employees and workers, and excluded district magistrates. Meanwhile, because all of the tools of the Desagüe had been turned over to the consulado, the guardas of the Desagüe who still had to maintain its structures outside the open trench could not do their jobs.

Superindendant Rodríguez del Toro got the viceroy to order the consulado to take over maintenance of structures such as the dam retaining Lake Zumpango, which were not within the terms of its contract, and although the corporation would be paid for this separately, it still dragged its feet.[42]

As it turns out, this jurisdiction extended even to criminal matters, and this also affected the way the consulado saw and treated workers. As of 1770, the consulado's Desagüe administrators could also hear and judge misdemeanor cases among minor employees and jail alleged felons in the Desagüe headquarters in Huehuetoca, which were equipped with a jail and stocks. The administrator was then supposed to give notice to the viceroy, who would "order him what to do," district magistrates were supposed to send any employees of the works they apprehended for law breaking to the consulado administrator.[43] All this had repercussions for the already hazardous working conditions. In 1770, an accident in the consulado's open trench works killed three Indians and injured five, but the consulado refused to allow civil authorities to investigate the case, citing precisely the contract clause that gave the corporation exclusive jurisdiction over all matters pertaining to the works, the workers, and the resources within its purview.[44] Thus, while the audiencia remained the ultimate authority in all Desagüe matters above and beyond any contracts, and its oidor superindendants continued to have jurisdiction in all basin-wide matters relevant to the drainage that were not within the Desagüe proper, in the high-stakes area of the Desagüe district as a whole, this jurisdiction was in practice curtailed until the consulado left the works.

## WATER OR MEN?

In taking on the Desagüe in large part to please a restless monarch, the consulado was motivated to minimize the outlay its Desagüe contract required and maximize what profit potential it had. This did not augur change: since the consulado enjoyed the relatively cheap repartimiento workforce, which largely fed itself at its own cost, it was advantageous to continue to use human energy as opposed to purchased fodder-fed animal energy to extract the excavated debris from the trench. Only rarely were winches (malacates) used.[45] Despite the corporation's contact with the mining sector, its Desagüe administrators did not consider adapting mining machinery for hoisting, such as that shown in Figure 6.4, which took no more than two to four men per adit or shaft to operate, and that at depths far greater than the Desagüe's. Naturally, with no more elaborate conveyor technology than ladders, the deeper the trench got, the slower debris extraction went, since each Indian had to climb up to fifty meters with his load and then supposedly carry it the stipulated

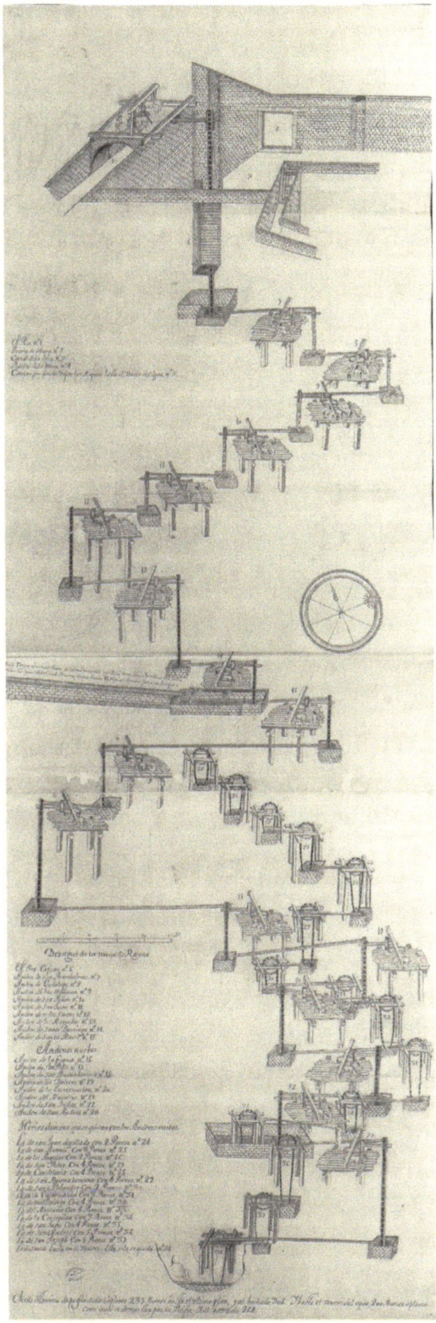

*Figure 6.4.* Hoists used in the Rayas mines of Guanajuato since 1704. Such machines allowed extraction of debris and water from depths far greater than the Desagüe's using minimal manpower (in shafts here numbered 24–28, for instance).

SOURCE: España, Ministerio de Educación, Cultura y Deporte. AGI, MP-México, 96. Reprinted by permission.

twenty-five meters away from the edge to prevent it from being swept back in. The administrators never considered using alternative methods, such as the scaffolds and mules to facilitate and expedite the removal of excavated material that had been proposed by military engineers Gaspar de Courseulle and Manuel Santiesteban in 1743 and 1764 respectively.

Barring investment in draft animals, winches, scaffolds, and like equipment, the merchants could not possibly complete the project on time at the work rate of one-quarter man-hour per arroba (11.5 kg) of debris extracted from the trench by hand. This fact did not deter the consulado from relying on human labor, however, as the merchants either deluded themselves into thinking the works would progress as planned or simply counted on receiving an extension of the contract and its terms. Importantly, the continued reliance on human labor went hand in hand with the water-sweeping method of debris removed established by the friars. Removing debris entirely by hand was so painfully slow that the consulado could not resist the temptation to use the energy of the water flowing through the Desagüe for debris removal, when enough water was available to do the job. The continued use of water for debris removal had unfortunate consequences, but not for the consulado.

In reverting to the water-sweeping method, the consulado imposed renewed restrictions on the riparian users along the Cuautitlan River. In approving the transformation of the Desagüe into a single open trench back in the 1630s and choosing to rely on water to sweep the debris from the channel, authorities had turned water into a critical element in the relations between their city and the hinterland for the rest of the colonial era. Since the most reliable source of water was the Cuautitlan River, the Desagüe was given one-third of the water of the diverted Cuautitlan River, measured upriver from the Pila Real at Atlamica and at the expense of the shares of the river allocated there to indigenous townships and Hispanic haciendas and ranchos in the area. By the time the consulado stepped in, this colonization of the river—its appropriation for the Desagüe and the city—had gone so far that it was often simply called Río del Real Desagüe.[46]

Consulado administrators inherited the benefit of these rules. In 1770, the consulado's project administrator Antonio Barroso y Torrubia installed two sluicegates to control the entry of the Cuautitlan's water into the Pila Real of Atlamica (extant in the twentieth century, as shown earlier in Figures 1.5 and 1.6), and made Pila beneficiaries pay for them.[47] To ensure complete availability of water for the open trench, he padlocked every Pila outlet and sent the keys seven leagues away to the City of Mexico for safekeeping.[48] This galvanized all the beneficiaries of the Cuautitlan River's water normally at odds with each other into a united front against the consulado's man. Barroso y Torrubia's actions, they claimed, had dried up the Pila, depriving crops, herds, and humans of water. Disease and death were spreading. Superindendant Rodríguez del Toro accused the consulado of dispossessing the water users illegally, since it had "no authority other than for the reconnaissance of the levees and the issuance of instructions pertaining to them and to the misuse by the landowners." Viceroy Croix ordered the sluices opened and new installations built to restore the usurped water.[49] The distribution of water at the Pila Real of Atlamica had been firmly grounded in customs that became engraved in the Desagüe record before the consulado contract (and recently, in 1743, 1761, and 1762).[50] Although deprived of complete control of the Cuautitlan's waters, the consulado nevertheless continued to receive its allotment. Its ability to use the water-sweeping method of debris removal was constrained but not eliminated; it merely meant that the work would proceed more slowly.

On the eve of the expiration of the contract and with the trench still far from completed, as the consulado itself expressed in writing and in the plan, shown in Figure 6.5, the merchants appealed for an extension and,

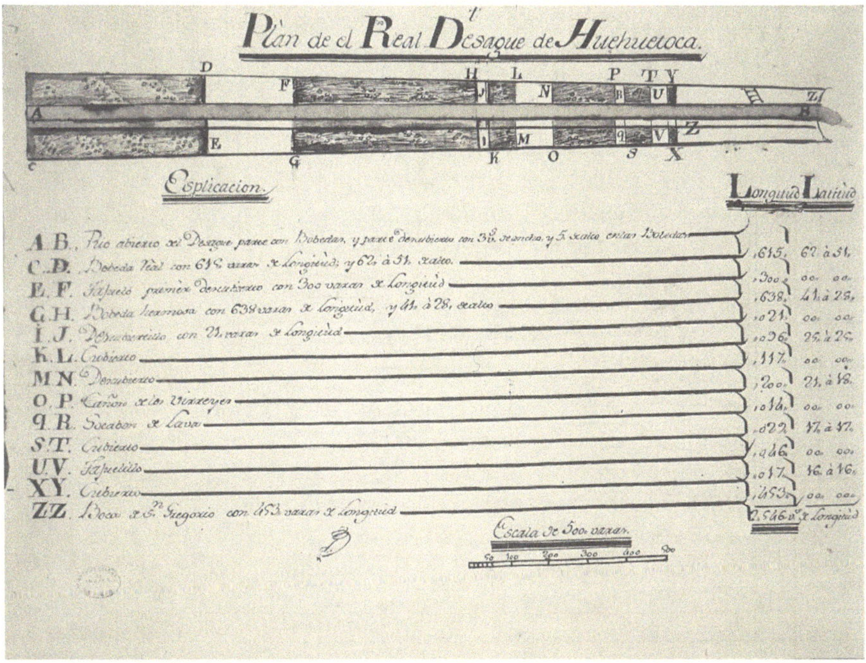


*Figure 6.5.* The state of the consulado's open trench conversion in 1773, when it requested a contract extension to continue working with the water-sweeping method. The extant tunnel intake at Bóveda Real appears on the left, and the outlet at Boca de San Gregorio on the right; white segments show "finished" sections.

SOURCE: España, Ministerio de Educación, Cultura y Deporte. AGI, MP-México, 689. Reprinted by permission.

more importantly, for an amendment to the method of work. In February 1773, Superindendant Rodríguez del Toro seconded the request, so in July Viceroy Buccareli approved it and a royal cédula of February 12, 1774, confirmed an extension to 1777, when it was again also rolled to 1789, giving the consulado an "indefinite extension of the contract to all the time it needs." All this so that the merchants could work exclusively by the water-sweeping method established by the friars.[51]

In sum, the officers of the merchant guild won close to absolute control over their use of resources but, unlike the oidor superintendants, with little accountability. It took exceptional collective action and the intervention of the viceroy himself to shake this control. With technological, financial, administrative, civil, and criminal jurisdiction added to the very tangible control over much land and people in the Desagüe district, the consulado's privileges there were unprecedented. Under these conditions, little existed to stop it from acting as it wished in the Desagüe district, and that it certainly did. But to what effect? Did the consulado's intervention inaugurate a new epoch for the project, for its function, for the human relationships to and within it, or for its impact on scientific and technological culture in New Spain? Again, it seems that the chasm between the patriotic, enlightened discourse embraced by consulado members and advisors and reality was only widening. The remainder of the chapter addresses these questions, beginning in the realm of culture, where huge possibilities opened up: with a large letrado caste with scientific and technological inclinations hungry for patronage, the consulado had a great many men at its disposal whose skills it could tap for the Desagüe's benefit.

## A SAVANT'S ALTERNATIVE: PYROPHYLACIA

Control of the Desagüe gave the consulado an opportunity to patronize technological and scientific activities among savants so inclined. The European Enlightenment had penetrated the colony through reformers, engineers, texts, objects, and measures; it spread by emulation and conflict, and it was reworked and built upon. As abundant scholarship has made clear, the letrados were both receptive to and critical of the new culture, interacting with each other and these curious imports in their salons, publications, and government offices as well as in or apropos of public works such as the Desagüe. A number of periodical publications and pamphlets appeared. In turn, these members of the letrado caste might show themselves useful and thus secure incomes outside the state by turning to the drainage as a site for the production of knowledge. What follows explains the type of knowledge that emerged from this conjuncture, among whom and why.

Some in the intellectual elite of the city responded to the 1762 fears about floods by presenting solutions to the problems as they saw them. Their proposals and commentaries were characterized by an attempt to deploy their own scientific culture toward the Desagüe and its problems. The floods that peaked in 1604–7 had ushered in the technological leap that Martínez's tunnel represented. That of 1629 had precipitated another major shift spearheaded by the friar technicians, towards a design of less technological complexity, greater human and ecological input, and different scientific challenges. As the tunnel receded out of sight and out of mind, its specific hydrodynamic challenges had receded too, leaving intact the Vitruvian truths and the tendency to address the Desagüe's hydrodynamic problems with statics: Would the new flood threats change this?

The creole savant José Antonio de Alzate y Ramírez was one of the letrados responding to the challenges of the 1760s.[52] In July 1767, at the beginning of his publishing career with polemical tracts on water in the basin and during the tensions between the city and royal authorities over who would pay for the completion of the Desagüe open trench, Alzate presented to the cabildo his own drainage proposal. This *Proyecto para el desagüe de la laguna de Tescuco* dismissed the Desagüe and any other project to extract the water from the basin as excessively expensive and impractical. Instead, Alzate proposed, with the help of a map shown here in Figure 6.6, that excess water could be led by a simple trench (E) with sluices (F) from Lake Texcoco to the foot of either of two volcanoes in the peninsula of Iztapalapa (A and C), where it would pour into the "large concavities" that he was certain would be found therein since "there have been and there are underground fires." Although some would think this project was "a fantasy or a dream," he warned, "others, owing to its novelty, will see it as feasible. . . . More unfeasible projects than this were deemed sufficiently deserving as to be published by the chronicler Zepeda."[53]

"Extravagant" (Alzate's own words) or not, this proposal is significant for what it says both about the kind of expertise contemporaries believed to be relevant to the Desagüe and about the many ways the letrado caste intervened in the Desagüe works or in shaping how it was conceived. Alzate's idea was influenced by the explanation of the circulation of air, fire, and water offered in *Mundus Subterraneus* (1665),[54] a work by the Jesuit polymath and teacher at the Collegio Romano, Athanasius Kircher. Understanding the Earth to be a huge organism reproducing the same mechanics and forms as the "microcosm" of man, and drawing from both Plato's caverns and Aristotle's four elements, Kircher thought that air, fire, and water circulated in the interior of the Earth, which had an immense fire at its core, as was only proper in Catholic orthodoxy. All three

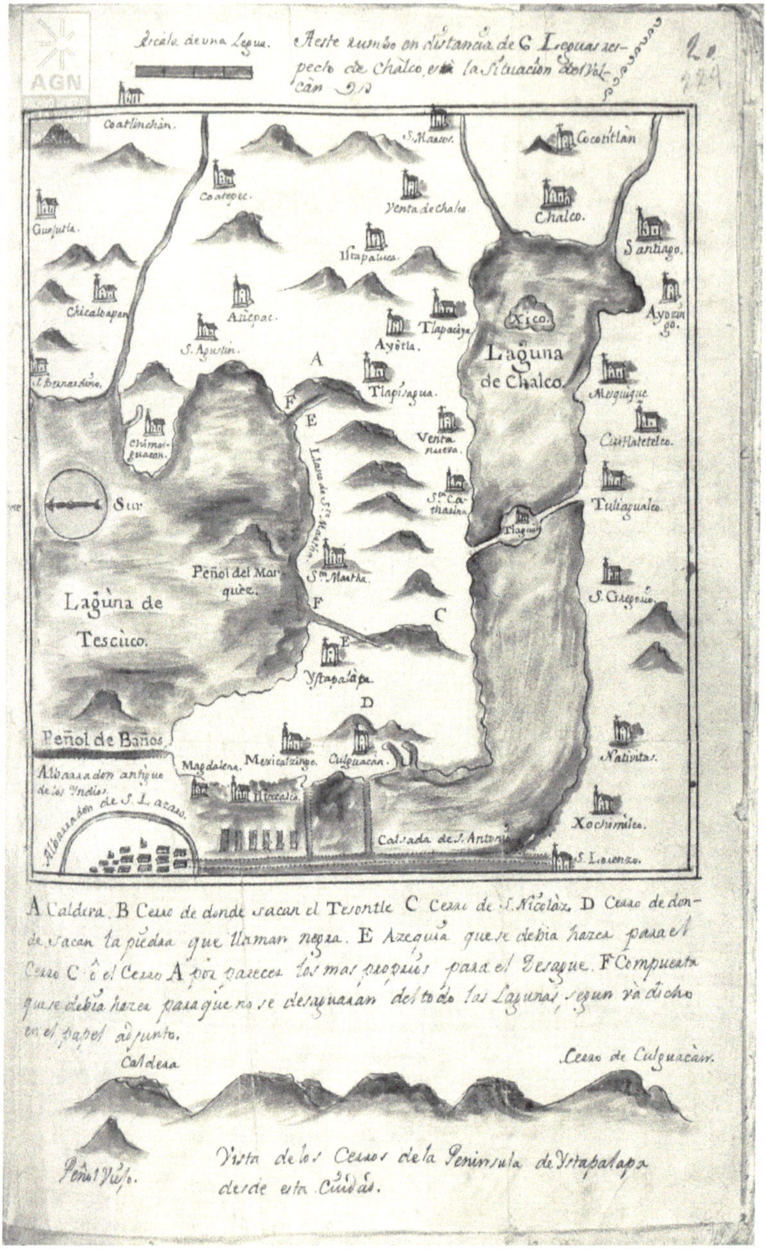


*Figure 6.6.* José Antonio de Alzate y Ramírez's 1767 proposal to abandon the Desagüe de Huehuetoca and instead drain Lake Texcoco into the "large concavities" found at under volcanoes in the peninsula of Ixtapalapa (A and C).

SOURCE: AGN, Desagüe, vol. 17, exp. 10, f. 224. Reprinted by permission.

elements interacted, forming earth and rocks, and flowed through a series of chambers and vessels, respectively named *aerophylacia*, *pyrophylacia*, and *hydrophylacia*.[55]

Kircher's writings while at the Collegio Romano had been quite influential in Spain, where they had impacted the *novatores* of the end of the seventeenth century, particularly the astronomers.[56] Copies of the *Mundus Subterraneus* had arrived in New Spain the same year it was published, sent by Kircher himself as a result of his correspondence with the creole priest Alejandro Favián, to whom the Jesuit dedicated his 1667 *Magneticum naturae regnum*.[57] In addition to Favián, sor Juana Inés de la Cruz, Carlos de Sigüenza y Góngora, and others in their circle were avid readers and diffusers of Kircher.[58]

Some of these readers sought practical applications for the ideas in *Mundus Subterraneus* and subsequent works. When in 1727, Juan Antonio de Mendoza y González, an accountant and astronomer resident in Puebla, claimed to have designed a machine capable of draining the deepest of mine shafts, he explained the functioning of his invention with a combination of the Cartesian theory of celestial vortices with Kircherian aerophylacia and hydrophylacia. When the tips of the "speedy spirals" of vortices formed by opposing winds touched the hydrophylacia, "they function like spongy clouds." As the "copious drops" falling from these spongy clouds met the "water of the hydrophylacia, rarified by the adjacent heat [and] ascending in a vaporous mist, it forms a Typhoon, with all the characteristics of a Maritime one." It was the power of the of these "spirals" that de Mendoza y González sought to harness in his "mixed suction and traction pump," which he claimed was capable of raising water from as much as the "300 varas the deepest of the straight mine shafts possesses."[59] Similarly, the consternation caused by the catastrophic 1755 Lisbon earthquake sent many to seek explanations in the *Mundus Subterraneus*, which thus enjoyed a revival in the Iberian realm and particularly in the tremor-prone Mexican and Andean highlands.[60] Among those drawn to these theories was Alzate, although his interest in pyrophylacia lay a bit closer to home, as he speculated on the utility of the theories for the drainage.

Since there were underground fires beneath the entire city, the whole basin must be riddled with cavities. A series of facts, which Alzate connected into an organic whole, proved that these fires existed: the volcanic eruptions that periodically afflicted the area; the heat of the waters around the Peñol del Marqués, which, not being chemical in origin, had to be physical; the conical shape and mineral composition of the mountains, in addition to the concavity at the top, since "this part, having been left without a foundation [after an eruption], sunk on the part that it found hollow"; the presence of tezontle ("burnt and calcinated mud") in places such as mountain "B" alluded to in his map's legend; the emanations of dense smoke from a hole dug by a landowner and a surveyor in the dome of La Caldera (mountain "A" where "constant subterranean noises" were heard), and finally, the fact that rainfall did not accumulate in this dome, but either evaporated due to the heat from the fire beneath or filtered through the ground. Alzate deduced that "in their inferior part there must be large concavities, as it is necessary that where there is fire there must be air, and it is obvious that air only circulates where there is room or extension for it to do so." It only remained for such cavities to be detected by simply digging a hole into one of the mountains: the terrain above it would collapse, revealing the presence of the desired hollow.

Unlike most Desagüe project proposals, but like Mendoza y González's, Alzate's plan included neither a budget nor an estimate of the number of workers and time needed for its completion. Far from it. Alzate's perspective in this regard ran counter to the attitude evinced by the military engineers who produced reports during this period. The engineers' reports exhibited at their core a drive for rationalization that sought long-term durability and efficiency in the utilization of materials, time, and labor. Alzate's attention to these matters was polemical, not substantive. Cost, he admitted, was a potential drawback of a project whose success he estimated as probable but not certain. There was also the potential for fatalities in the construction work. For the military engineers, unsafe work was inhumane as well as irrational and inefficient. Not for Alzate: if in the mines "many people die, and they do so without much more than looking for some treasure," then it was legitimate to ask, "is there any mine that is worth as much as this City, being as it is one of the foremost in the Globe? In the defenses of this City many people die, because the public good demands it."[61]

The cabildo thanked Alzate for this proposal and passed it on its procurador general, who promptly shelved it.[62] The city councilmen evidently regarded it as an "extravagance," as Alzate later bitterly complained, and modern historians have also characterized it as rationalist in a "merely speculative" way, a product of "ill-assimilated readings."[63] However, just three years before Alzate's proposal, the viceroy had named a technical commission made up of military engineer Manuel Santiesteban, surveyor Juan de Alba, and maestro mayor Ildefonso de Iniesta Vejarano to inspect large cavities in the ground off the northern shores of Lake Texcoco and west of the Oculma reservoir into which water from the San Juan Teotihuacán River seemed to disappear, depicted by Iniesta Vejarano as point L in Figure 6.7. Asked whether the desiccation of Lake Texcoco was possible by this means, the commission made a timed experiment by leading the river's water into the cavities. They concluded that it was not, and that was the end of all discussion of sumps among Desagüe technicians and officials.[64] But clearly not among all sectors of the reading public.

*Figure 6.7.* Three years before Alzate, maestro mayor Ildefonso Iniesta Vejarano used this image to explain why he believed that a “sump” drainage into cavities, this time into one north of Lake Texcoco, would not work.

SOURCE: España, Ministerio de Defensa. ACEG. Reprinted by permission.

From the speculative, Alzate swung to the practical in his Desagüe concerns. In 1785, he proposed the creation of a navigable canal connecting Lake Texcoco with Totolzingo, seven leagues to the north. Referred to the Desagüe superindendant Miguel Calixto de Acedo for comment, the proposal explained how like the canals connecting the southern lakes to the city and in imitation of the Canal du Midi in France, Alzate's would substitute sixty to seventy canoes for the fifteen-hundred-plus mules that made their way daily to the Provincias Internas to the north through the town of Totolzingo. This would drastically reduce freight costs and make transport easier and faster. Alzate had a point: mule trains were fast becoming an objective limit to the expansion of commerce radiating to and from the capital.[65] As the volume of internal trade increased from 1760 on, so too did the competition for pasture on already crowded land and this was more noticeable between Guadalupe and Totolzingo because the saline terrain produced little edible vegetation. Moreover, like most other roads north in the rainy season, the road between these two points became a muddy quagmire of stuck mule trains.[66]

Characteristically, Alzate provided neither empirical data about the terrain the canal was supposed to traverse nor its length, width, depth, slope, shape, and exact path. Nor did he explain how the canal would actually work, given the seasonally variable level of the lake. Equally characteristically, he stated that although "the people who are destined to the works of the Canal must be free," the freedom of the investors would be superior: they could dismiss useless workers and prohibit them from quitting. Claiming to already have backers, he explained how a company of stockholders would own the canoes, charging fees per load, not per canoe; private parties could own canoes on the canal and would pay a fee when using them to transport freight; royal cargoes would be free of charge: all other freight, regardless of whether it only used the bridges over the canal, had to pay. Landowners and towns who built canals connecting with Alzate's would have to pay too, in proportion to the "utility" obtained from it. Beasts of burden were strictly forbidden along the banks, "unless the investors decided to use the European method of animal-drawn barges." People on foot could enjoy full and free access to the tree-lined embankments.[67]

Unfortunately for Alzate, Acedo was not impressed. The savant had neglected the facts of terrain, hydrology, and impact on the Desagüe. First, the dry season Lake Texcoco had little or no water in it. In addition, the lake was "much less extensive than in former times," when even with more terrain to occupy the water had still flooded the city. The bed of the lake had also risen due to "the unceasing silt and rubbish that are introduced in it by the Rivers with their powerful freshet, and the filth that is continually coming into it from this City." As a result, the bed of Lake Texcoco was barely lower than the city. Furthermore, "for Mexico to flood it is not necessary for the water from the Lake to come into it, it is enough that the former rise in level so as to impede the exit of the water that Mexico has within."[68] Rejected by Superintendant Acedo, the project did not prosper as a royal venture. The cabildo, in turn, felt that much as canals were a wonderful idea, it was not in favor of this one. As a lure for entrepreneurs seeking investment opportunities—such as the merchants had become now that their trade monopoly was threatened by the free-trade Reglamento de Comercio of 1778—it did not work either: the consulado did not bite.

This outcome should not obscure the significance of the canal proposal in the current context: while with the pyrophylacia volcanic drainage he dipped into past scientific paradigms, with the canal he seemed to stride along with the Bourbon's usage of new technology and science to further economic productivity and commerce. What moved Alzate to intervene in Desagüe matters in this apparently Janus-faced way? From a philosophical perspective, the volcanic drainage seems consonant with the reputed hostility of this savant toward the universal explanatory power of European scientific systems promoted by Bourbon institutions, inasmuch as he saw them applied to represent American realities unfavorably.[69] This might explain in part why it was so difficult for Alzate to link his projects to the new science of water, which participated in the creation of such scientific systems, and why he instead reached for elemental explanations from a previous scientific paradigm. At the same time, there are social reasons why Alzate chose Kircher and not, say, Guglielmini, Torricelli, or even Castelli, to authorize his volcano drainage proposal and why devoid of any scientific discussion at all the canal is presented simply as a project that would generate "utility" at this specific time.

Writing of the volcanic drainage in July 1767, and then defending his stance in 1787,[70] Alzate was part of "criollo attempts to oppose their culture to that of the peninsulars" and to Europe's conception of America more generally.[71] A bitter time it must have been for the creole intelligentsia, as peninsulars with possession of the new metropolitan scientific and technological culture were the only ones advancing and gaining state patronage, and as the only institutions where creoles had been able to access and shape this culture at all—the Jesuit colleges—were now coming to an abrupt end. Bitter and desperate indeed for the likes of Alzate, who were now on their own to confront the *visitador* (royal inspector-auditor) José de Gálvez and the military engineers, the shock troops of the Bourbons' campaign to demote the New World kingdoms of Spain's own "universal" or "composite monarchy" to proper "British-style 'colonies,'" to borrow J. H. Elliott's terminology.[72] Alzate's Desagüe intervention coincided with the reactivation of the open trench according to plans by foreigners who were once again applying European techno-scientific systems to the Desagüe and proposing to end the seasonal regime of labors and the wasteful treatment of workers defended by creole elites and, to add injury to insult, being rewarded with the king's favor and salary while so doing. More threatening still, Gálvez was implementing reforms that aimed to upend creole dominance, and with such swift apparent success that he would propagate them to other dominions during his tenure as minister of Indies (1776–87).

It would appear that over the span of his engagement with the Desagüe, Alzate was waging a covert and overt culture war on several fronts at the same time, with different weapons depending on the theater, but with a single clear goal. As an opponent of the universal value of philosophical systems, he could not countenance viceroys and reformers who adhered to Campillo y Cossío's *Nuevo sistema de gobierno económico para la América* while continually shutting down his critical scientific gazettes; or military engineers, for whom the Desagüe was subject to the same physical laws (scientific systems) as any other hydraulic project anywhere and who loudly blamed creole ignorance and indolence for its deplorable state. At the same time, he struggled to profile himself as no less a provider of "useful knowledge" than his foils.[73] Swinging from the scholastic features of the Catholic enlightenment that the Jesuits had imparted to the sons of Mexican elites to fervent utilitarianism, Alzate's ultimate goal was to preserve a creole-dominated social order with the city and its elites at the center, and of course his own position within it as a patronized letrado "Patrician who must look out for the good of the Patria."[74]

With his pyrophyliacic drainage solution shelved when "far more inaccessible projects than this one have deserved publication,"[75] Alzate's stakes in the Desagüe (and by extension those of many in the letrado caste as well) seem to have been less whether the drainage might be interrogated to reveal something about matter and forces than whether by intervening on its fate, more immediate and preferably pecuniary rewards might be reaped. Coming at a time of heightened awareness and apparent official receptiveness to proposals about the project, Alzate's interventions in Desagüe affairs were thus not so much efforts to change realities as they were attempts to legitimate criollo knowledge and find and secure patronage in a shifting sociocultural environment. Equally worried about their own privileges relative to those of the European invaders, the vecinos and merchants were uninterested in his canal, which was of no clear benefit to them, or his pyrophylacia, which they correctly identified as hopelessly out of synch with the intellectual mood of the source of all favor—the state.

## A DIFFERENT ANSWER: FLUID DYNAMICS

Unlike Alzate, other savants did manage to use the Desagüe to position themselves correctly in relation to the new mood of neomercantilist utilitarianism, applying the new laws of motion and matter to the project to produce perhaps not original but at least useful knowledge about the basin. In 1773, after several years of public discussion of a possible general desiccation and five years of work on the open trench conversion, the consulado commissioned Joaquín Velázquez de León to assess the possibility of extending the Desagüe de Huehuetoca to Lake Texcoco, adapting the existing work as needed. Velázquez de León was at the time in possession of the records of the Desagüe because Viceroy Juan Francisco de Güemes y Horcasitas had asked him to write a history of the project. A former professor of mathematics at the Real Universidad in the City of Mexico, he was now a legal agent (*apoderado*) of the mining guild. Before responding to the consulado, and in contrast to Alzate, Velázquez de León set out to gather precise data on the basin with Joseph de Burgaleta, a trusted surveyor, and two young assistants. Velázquez de León's resulting report is extensive, providing descriptions of the instruments and methods used, as well as a discussion of common pitfalls to be avoided in survey operations.[76]

Velázquez de León began his operations by preparing his instruments, which included a vara rule crafted out of solid dry wood with square angles, encased in tin and measured with the original seventeenth-century vara kept in the Casas de Cabildo; a fifty-vara cord that was prepared with oil and wax to minimize expansion and contraction; and two air-bubble levels, an English one and a French level made in 1768 by the renowned precision instrument-maker "Munciur" Canivet, the "Instrument Engineer" of the Royal Science Academy in Paris since 1756 and reputed to be one of the best available anywhere.[77] Velázquez de León took great pains to document the fine pedigree of his instruments as part of his attack on the accuracy of maestro mayor Iniesta Vejarano's 1764 survey of the same terrain.

After completing his survey, Velázquez de León reached the conclusion that a canal from Lake Texcoco by the city to Bóveda Real would only need one-fifth of a vara of slope for every one thousand varas of its fifty-thousand-vara length, so long as its bed was clean. Since the bed of the trench at Bóveda Real was already ten varas beneath the level of the estimated beginning of the canal at Lake Texcoco, it needed no further deepening. Contrary to the massive dynamiting and excavation project dreamed up in 1755, for Velázquez de León the only obstacle remaining between the status quo and a general desiccation was an adaptation of the existing trench from Bóveda Real to Vertideros and the excavation of a slightly sloped canal to Lake Texcoco.[78]

He imagined the possible Desagüe extension as a canal from Lake Texcoco to the Salto de Tula, where the Desagüe waters met with the Tula River, on the other side of the northwestern sierras. The entire span from Vertideros (where the Cuautitlan discharged into the southern portion of Lake Zumpango, sometimes referred to as Citlaltepec Lagoon, and from there entered the trench and extant tunnel portions) onward, including the still-in-progress open trench excavation, would have to be deepened to enable the water to be driven by gravity alone all the way from Lake Texcoco. The length of this canal was 51,761 meters, close to the survey conducted by Alonso Arias and approved by Enrico Martínez in 1611, but 2,396 meters shorter than the 1764 measurements done by maestro mayor Ildefonso de Iniesta Vejarano. The measurements were double-checked through the geometric method, by triangulation, which yielded a distance of 52,040 meters to be covered by the canal project. This difference in his own two measurements was acceptable to Velázquez de León, who attributed it to the potential for error from the distension of the cords and from small variation in the cardinal direction of the measurements.

As significant as the empirical results were, something else was afoot. As a result of his survey, Velázquez de León concluded that "for water from the Lake of Mexico to flow through the Canal de Huehuetoca it is not necessary to reduce the level of Bed of the Bóveda Real by anything at all and that the latter and the small mass of terrain it still has above it are incapable of causing any retreat or detention of the water of the River in Vertideros or any other location."[79] Not only was the general desiccation possible, in other words, it was far more feasible than the project estimated in 1764, when an architect had conducted the survey.[80] Slope was needed not to make water flow, but rather for water to overcome the obstacles in the terrain more easily, to ensure that it remained entrapped in the canal, and to obtain velocity, if that was desired for some other reason. This conclusion was revolutionary in its time and place because it was a complete and scientific refutation of the paradigm that the seventeenth-century technicians had found in the architectural writer Vitruvius and that had been repeated reverentially ever since: for water to flow it needed a minimum slope of half a vara in every hundred. In contrast, Velázquez de León was saying "that for a river to run to its terminus it is not necessary that its bed have any incline and this is the same that Archimedes demonstrates in his book *de incidentibus aquae*." Archimedes, whose work on hydrostatics had been the basis for the reinvigoration of the Renaissance science of waters, helped Velázquez de León make a distinction between the practice of science and that of art.

Archimedes had been "great man who flourished many centuries before Vitruvius" and had been "as good a hydraulician as the other may be an architect."[81] Moreover, by then introducing the late-seventeenth-century Italian mathematician Domenico Guglielmini, who studied the laws governing the velocity of water in canals, what Velázquez de León did was to assert the authority of science over that of art in Desagüe matters, specifically the science of both hydrostatics and hydrodynamics. This was completely new in the Desagüe.

Also new—and unique at this moment—was the manner in which this creole emulated the visual language of the military engineers. By the time Velázquez de León began to peruse the Desagüe record for the history he was supposed to write and to provide an answer to the question about the plausibility and ease of a general drainage, a solid corpus of Desagüe plans and reports by military engineers had developed, with the more recent additions of Manuel Santiesteban, who taught engineering draughtsmanship to younger members of the corps, and Ricardo Aylmer, who had written and drawn the stipulations for the consulado contract. A closer look at Aylmer's communications explains what aspects of the engineers' culture Velázquez de León and others in his savant cohort found most amenable and more generally why the substantive content of the engineers' textual and visual communications would not be obeyed, while their formal aspects continued to be imitated.

Aylmer's and his colleagues' textually and visually transmitted stipulations were still hammering at the problems that irked the engineers the most (design, labor usage, efficiency, and so on). They additionally rendered increasingly clear the widening chasm between the foreign and local technicians and their respective career prospects. These characteristics, which contrasted to the plan he had washed earlier to show the general state of the works, were the result of the need to provide detailed instructions to the contractor and eliminate any potential for ambiguity originating from the interpretation of words alone. The consulado administrators and any technical personnel they employed were supposed to tack close to the text of the contract and the "Profile taken over the length of the plan," shown in Figure 6.2, specially the problematic sections demarcated by dotted cut lines 1-2, 3-4, 5-6, and 7-8 in the four profiles. Aylmer adhered closely to conventions by using unbroken black lines for actual contours and dot-dash for imaginary, analytical lines; yellow for what needed to be done (the 45-degree sloping on the sides), red for existing vaulting and masonry structures, smooth and dappled browns for alternating earth and open trench sections, verdigris for water. The lighting was also conventional, from left to right.

All this indicated precisely where each element was supposed to go and what it should look like. But "reading" this information required a literacy with the language of engineering draftsmanship far more refined than that which Franck and previous engineers had used because in the meantime communications standards in the corps had accrued a larger "lexicon" and a more complex "grammar." Unlike their brethren from Franck and Pozuelo's time, engineers were now being handpicked for their ability to show "plans, sectional profiles and terrains neatly and with the proper colors," besides other mathematical, technical, and personal qualifications. And they had to successfully complete the fourth year of their training, which was all about descriptive geometry and draftsmanship, in order to be able to graduate and enter the corps.[82] This only made their communications language more sophisticated and specialized, with ever more uses for colors, signs, trigonometry, perspective, and analytical geometry.[83] Unlike engineers of earlier generations, Aylmer did not pedagogically explain his usage of lines and color. This did not help narrow the growing chasm between royal engineers and local technicians, which was made worse by the fact that until 1783 the latter had no institutions where such a language could be learned. Still, creole technicians and savants like Velázquez de León and Desagüe employees were motivated to do their best to abide by the terms and plans of the military engineers and to emulate them at least in style, if not in substance. The consulado, in contrast, was not.

Following the practices demonstrated by Aylmer and other engineers, Velázquez de León also used a sequence of sectional profiles (Figure 6.8) and a longitudinal cut (Figure 6.9) to accompany his report. But the emulation is not complete. Figure 6.8 has a scale in varas, and sectional cuts to show the path of the water from Vertideros to just beyond the Boca de San Gregorio are numbered 1-10 to guide the reader, but there is no overall plan with dotted lines to show the exact place where the cut was to be made. Some color codes are used—brown for earth, verdigris for water, red for existing masonry works—but others were not. There was no outlining in black ink. And, although the light comes from the left and the view is from the city outward, the perspective is unclear. Most significant line, "A," serves to project the surface of Lake Texcoco and show the depths of Desagüe structures relative to it, thus communicating visually the main conclusion of the report— that a general desiccation was possible as things stood because of the universal applicability of European scientific systems—which his observations and measurements corroborated. This borrowing of a representational language for analytical as well as communication purposes was new among the creole participants in the Desagüe, as a comparison with maestro Iniesta Vejarano's 1763 depiction of hydraulic works in part of the basin (Figure 6.7) shows. This guild master adopted none of the language of the military engineers he often interacted with.

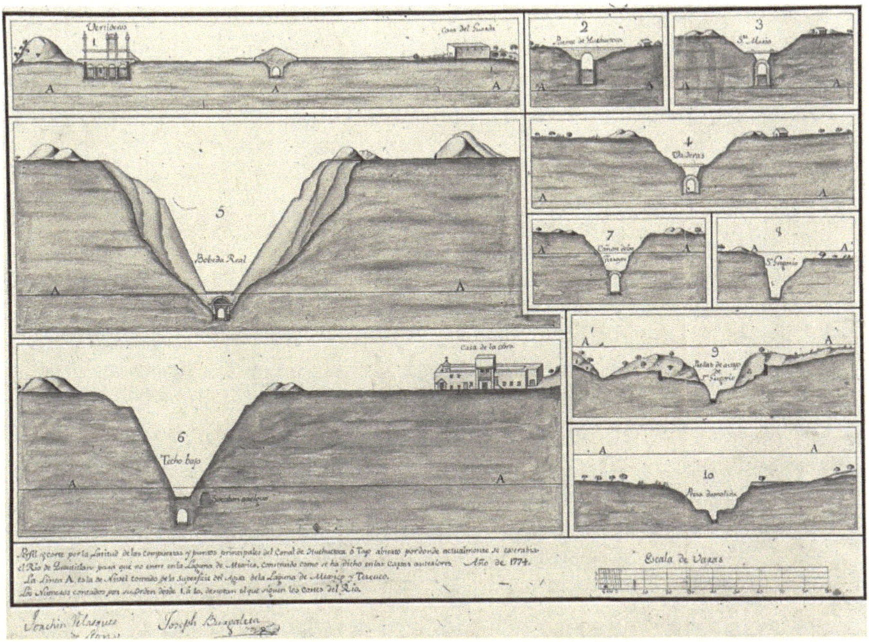


*Figure 6.8.* Joaquín Velázquez de León's projection of the level of Lake Texcoco on the open trench.

SOURCE: AGN, Fomento Desagüe, vol. 3, f. 19. Reprinted by permission.

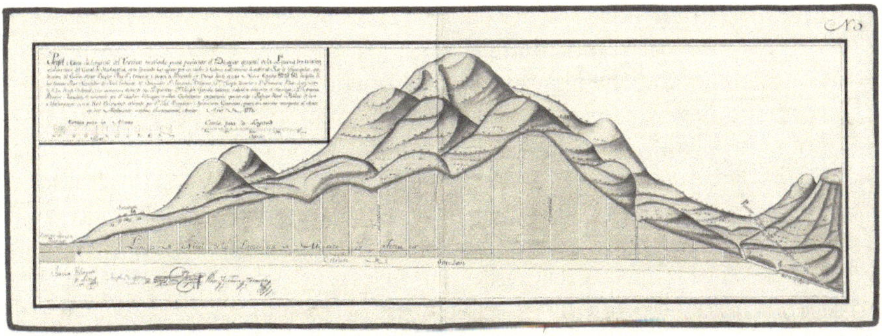

*Figure 6.9.* Joaquín Velázquez de León's 1774 *Perfil longitudinal del desagüe general de la laguna de México.*

SOURCE: España, Ministerio de Defensa. ACEG. Reprinted by permission.

New as it might be in the Desagüe, all of this was emulation nonetheless. Indeed, Velázquez de León's demeanor toward the scientific foundations on which his work was based was one of deference. Besides imported instruments, to conduct the survey he also claimed to use the methods described in Jean Picard's 1684 *Traité du Nivellement* as published by Philippe de la Hire in 1728, and Jérôme Lalande's 1764 *Traité d'astronomie*. The result was the first complete, and to that date most accurate, survey of the entire basin, an application of European science to more deeply know the American reality. What it was not was an addition to, a critique of, or an engagement with the science itself, and in this attitude as a learned man Velázquez de León did not differ from his predecessors and contemporaries in New Spain. So whether they accepted or rejected new European sciences, systematic thinking, and their practical implications, neither Alzate nor Velázquez de León used the Desagüe or any other hydraulic project to their own autonomous science of water.

The pluralism of cultures and forms of knowledge in the project continued in all its realms, from hydrometry to communications technology, however. By now, measuring land and water with discrete units for surfaces and volumes had made water both increasingly permeable to mathematical analysis and open to the tools of descriptive geometry. This was something the military engineers continued to infuse into the Desagüe. In coming to the conclusion in 1770 that Lake Chalco in the south of the basin, not the Cuautitlan River in the northwest, "should be regarded as the main enemy" of the city, for example, royal military engineer Nicolás de Lafora calculated their respective threats in terms of cubic varas of water discharged per year into Lake Texcoco.[84] Lafora would also use his calculations of volume to determine the dimensions the causeway of Mexicalzingo and the positioning and operation of its two sluicegates in order to best regulate the movement of water and canoe traffic north from Lake Chalco, with consideration of the effect of this volume on the velocity and smoothness of the flow through the sluicegates.[85] Despite these various demonstrations of their usefulness, only very late do we find creole technicians using cubic varas to describe water, as when in 1801 master architect Castera flattered Desagüe superindendant Cosme de Mier y Trespalacios for stopping a rush of 182,722,400 "cubic varas of water" from the surrounding lakes from flooding the city by raising the height of the causeway-dams by a vara.[86]

This illustrates the parameters within which master architects in the Desagüe adjusted to the impositions of the new reformist regime. Dependent on commissions to shape the built environment, these men were affected differently than the savants and letrados, who depended more heavily on patronage and state incomes. Nevertheless, creole architects did realign themselves by emulating constructive and communications styles. Consider the visual language employed by maestro mayor Ignacio Castera in Figure 6.10 and his predecessor, maestro mayor Iniesta Vejarano (Figure 6.7), relative to that of military engineer Miguel de Costanzó in the three Figures 6.11. Vejarano's plan has virtually none of the attributes of the plans of military engineers; Castera's has sectional profiles whose locations are designated with dotted lines on the longitudinal cut, a scale in varas, an orientation, explanation and color codes approximately consistent with those used by military engineers. Since both men had issued from the guild and presumably had very similar training, what explains the differences in their respective visual languages? Intervening between their two epochs was a decisive Bourbon institution.

In 1777, Charles III had begun to signal his intentions to reshape the built environment toward neoclassical tastes and functions by reinforcing the role of the Real Academia de Bellas Artes de San Fernando in Madrid (and later those of San Carlos in Valencia and Mexico) in publicly funded civil construction and architecture. "To prevent the waste of moneys on public works which, being required to serve as adornments and models, stand only as monuments to deformity, ignorance and bad taste," the king ordered magistrates and town councils to submit "elevation plans and sectional profiles of the planned constructions" with their written explanations to the Academy of San Fernando, whose architects would examine them for free, "indicating the merits or errors contained in the designs or the best means for their successful realization." Meanwhile, royal decrees in 1782–85 struck at the guilds by ordering that the architecture was to remain "entirely free for any national or foreigner to practice without any impediment or contribution whatsoever."[87]

*Figure 6.10.* Maestros of the late eighteenth century tried to abide by cartographical and draftsmanship conventions diffused by engineers and the Real Academia de San Carlos. Here maestro mayor Ignacio Castera's 1789 "Plan of the ground and profiles of the works in the Real Desagüe de Huehuetoca" partakes of some of them.

SOURCE: España, Ministerio de Educación, Cultura y Deporte. AGI, MP-México, 415. Reprinted by permission.

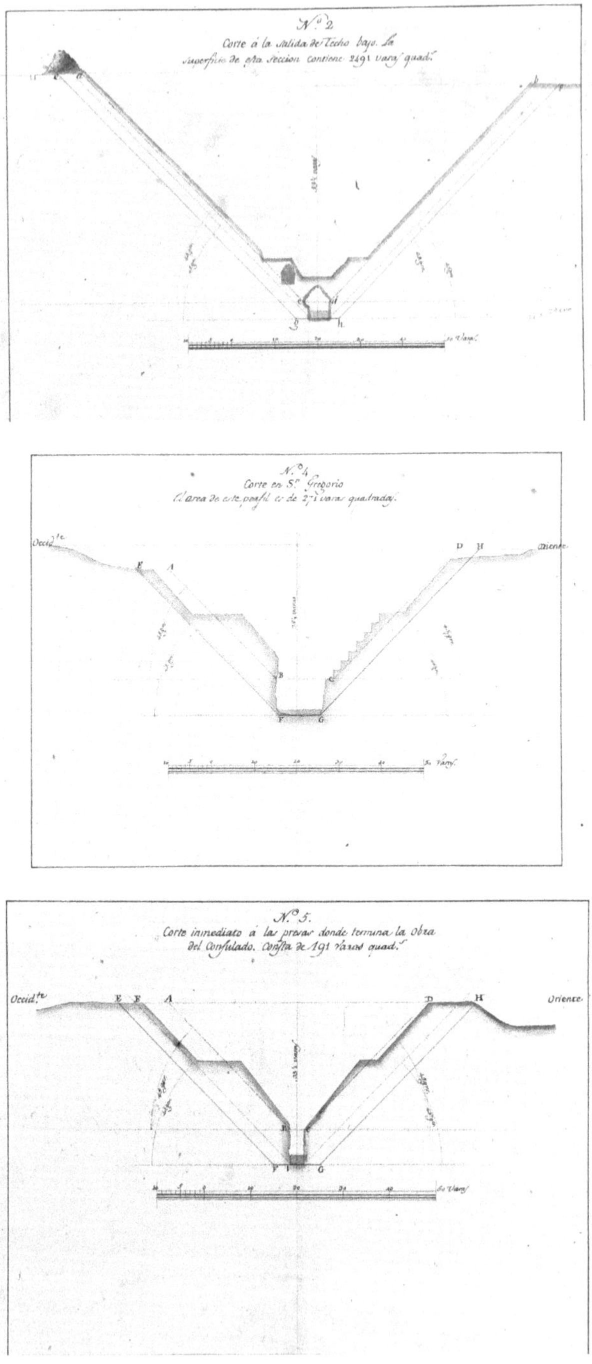


*Figure 6.11.* Three of Miguel de Costanzó's 1788 five cross-sections of the Desagüe, using all conventions current in the corps of engineers at the time.

SOURCE: AGN, Desagüe, vol. 24, exp. 3, ff. 103–8. Reprinted by permission.

The same spirit guided the foundation of the Real Academia de San Carlos in Mexico in 1783, while further royal decrees gave the antiguild regulations some teeth. In the Mexican academy by-laws, guild architects were required to present a plan with elevations and sectional profiles for a "considerable building" in order to be recertified as an "academic by merit" (as opposed to by training), after which they were still presumed to be of inferior skill since they were required to attend classes in the academy as often as possible to "increase their proficiency." Without this, they were not allowed to bid for public projects, or private value properties.[88] Vejarano had been free of these requirements, but Castera no longer was. Having just endured a thorough criticism of his draftsmanship from military engineer Miguel de Costanzó, who was serving as the Academia's mathematics professor, Castera was under great pressure to adjust his visual language in order to secure his certification as an academic by merit.[89] With the outcome of this process still uncertain, as is clear by the fact that he did not include this title among the others in the heading of the plan, Castera had strong reasons to imitate as best he could the visual language of the engineers. Many of the principles for plans and techniques of civil construction taught in the academy originated in military engineering in the first place, reinforced by the presence of five high-ranking military engineers in the rosters of the Madrid academy.[90]

Still, however much creole savants and technicians tried to learn the new aesthetic and communications language and function within this atmosphere of more rarified opportunities for advancement, in the end their efforts to embrace the new standards were often dismissed. While valuing Velázquez de León's legacy to the Desagüe and knowledge of the basin, as late as 1802 military engineer José Cortés had little positive to say about others, particularly the master architects who were even then responsible for much of the work in the project. Evaluating the plans and sectional profiles for the project, he found them lacking clarity and detail, which could lead to "great errors in judgment" on the part of officials. With an air of resignation, he added, "If the plans and profiles cannot not be drawn and washed with greater delicacy, then at least an effort should be made that they are presented untorn and neatly and cleanly executed."[91]

## NAVIGATING THE DESAGÜE

The consulado years in the Desagüe were pregnant with possibilities to infuse the project with economically generative uses, such as navigation. Three elements contributed to this conjuncture: the consulado was, after all, quintessentially concerned with commerce; military engineers were now under royal mandate to proactively search for ways to make the colonies productive; and, finally, broad lettered sectors were keenly interested in emulating at least part of the military engineers' culture and in engaging with Enlightenment notions of utility. Urban elites had heretofore never considered the possibility of giving the Desagüe an economically generative purpose. But now the conceptual tools and wherewithal existed to change this: the Desagüe could conceivably be made navigable. This was, after all, the time when the Bourbons were pushing canalization projects to articulate the urban and rural in the Peninsula through commerce and agriculture. For this purpose, in the 1750s, they had sent the naval officer and scientist Antonio de Ulloa to study these canals abroad and to import specialized hydraulic engineers to create comparable grids in Spain. Along with the hegemonic drive that propelled Bourbon Madrid to step up canal construction, the designs of the Canal de Castilla and the Canal Imperial de Aragón reflected multiple goals, balancing at least in theory the needs for irrigation water in fields adjacent to the canal with the minimum level of water required for adequate navigation conditions.[92]

Making the Desagüe navigable was a reasonable expectation because in New Spain, the consulado had a record of lavishing its "patriotic zeal"—as many a contemporary and more than one historian called this apparent largesse—on various other royal and municipal public works, among them roads and infrastructure important for transportation, including causeways and dams. In 1741, for example, the consulado had repaired the causeway of Guadalupe, connecting the City of Mexico to the pulque-producing region to its northeast, and prior to that it had fixed other roads into the capital.[93] Within the Desagüe complex, the corporation also had to its credit the fulfillment of a 1747 contract to extend and reinforce the dam of Lake San Cristóbal for six thousand pesos.[94] As in any other enterprise invoking patriotic discourse then and now, none of these works were performed at the personal or corporate sacrifice of the proponents or contractors, of course, and there were always gains to be made by taking them on. In the words of Revillagigedo, "It is the consumer who actually contributes for all these services alleged to be the consulado's, coming as they do from a general tax on goods, whose cost is increased by the value of the new tax."[95] In its contract to complete the Desagüe open trench too, there was something slightly more prosaic than patriotism motivating the consulado's actions and the contract provisions it secured. During the term of its extended contract, the consulado could hope for two gains in pursuing navigability for the Desagüe—the kings' favor (through the flattery of imitating his canalization drives in the Peninsula) and money.

A consideration of the possibility of reconditioning the Desagüe's open trench as a navigable canal would point to a desire to give the drainage an economically generative purpose through a commercial function. The corporation's Mexico City merchants controlled vast networks of trade that distributed Mexican and imported goods from Asia and Europe, and of all the routes these goods traveled, one of the most profitable connected the City of Mexico to the mining districts to the north through the bishoprics of Michoacán, Durango, and Guadalajara. Even though it was nominally a royal road, the valuable cargo that traveled this route was not always safe and during the rainy season it was liable to get stuck in the impassable mud churned up by hundreds of hooves on overworked causeways and denuded roadside pastures. Despite the amount of commercial traffic it supported, this *camino real* hardly deserved its name.[96]

Because extant transport options were problematic, connecting to Querétaro, Guanajuato, and other districts with a waterborne segment through Cuautitlan was in fact present in the imagination of the consulado's men. In a 1781 proposal to open the oft-discussed drainage canal from the northwestern extremity of Lake Mexico to the Desagüe, the prominent merchant Antonio Barroso y Torrubia argued that it would also provide a navigable alternative to the expensive mule-train freighting of goods coming through Tula.[97] The timing of this proposal is telling. Barroso y Torrubia was then prior of the consulado, but in 1770 he had been administrator of the Desagüe contract. As such, Barroso y Torrubia and his successor, Joseph González Calderón, had worked with Velázquez de León as the latter drafted his report on the prospects for a general drainage canal tapping Lake Texcoco and Lake Mexico directly, even participating in the surveying of Lake Zumpango.[98] At the time Barroso y Torrubia hinted at a navigational use for the projected general drainage canal, the consulado had won an extension of its trench contract until 1789, which gave it time to reconceptualize it if it wished.

Meanwhile, however, changes and reforms disagreeable to the merchant and creole elites were unfolding. On the one hand, the 1762 temporary loss of Manila and Havana had prompted Charles III to siphon resources out of New Spain for various urgent needs, including fortifications and defense in the Caribbean. Then came the *Reglamento y aranceles reales para el comercio libre de España a Indias* of 1778, whose Article 53 announced the intention to destroy the monopoly over trade of the existing consulados and reinvigorate commerce with new, competing ones in Spain charged with promoting productive activities of interest to the crown. New Spain was not yet encompassed by this *Reglamento*, but the writing was on the wall. From 1778 on, merchants were increasingly motivated to turn to agriculture and mining because their ability to reap large profits through the shipping and trading of goods appeared threatened.[99] However, these circumstances do not seem to have prompted members of the Mexico consulado to start "sinking their capital" into lands in the basin of Mexico: the merchant hacienda-buying spree both in the districts the Desagüe affected (Cuautitlan and Ecatepec) and the rest of the hinterland of the capital had taken place before 1779.[100] Possession of these estates in prime grain-producing areas did not prompt their consulado proprietors into imagining an irrigation component for the Desagüe, which their corporation was in the process of reshaping. Instead, perhaps the time had come to take on some other activity, one that could also be a major gesture, such as giving the viceroyalty a navigable canal for the greater glory of the king?

In fairness, at the time the corporation was involved with other projects that appear to be about transportation. In 1786, the corporation was busy refurbishing three of the causeways into the city—the Piedad and San Antón causeways that led to the road to Cuernavaca and Acapulco, the port out of which City of Mexico merchants controlled the profitable Manila trade, and the Vallejo causeway that led to the *tierra adentro* road to Querétaro, Guanajuato, and beyond via Cuautitlan and Tula. Repairing these causeways would compensate for the effect of the *Reglamento del Comercio Libre* of 1778, the postponed implementation of which hung on the horizon, threatening earlier monopoly gains on other routes.[101] Even though a mercantile desire to improve transportation may have infused these causeway works, they were at least partially brought about by the 1785 harvest failure, which sent droves of hungry rural poor into the city and pushed officials to beg the consulado to put the troubling mass to work.[102]

The consulado was not in principle opposed to linking navigation to the Desagüe. Indeed, in 1789, the year after completing the open trench conversion, its official position was

> that the project for a general drainage would not only prevent the danger of flooding in this Capital, but it would also bring about other utilities to the public. Because if a navigable canal were built, it would increase and facilitate the traffic and transport of goods and products from the Provinces of Tula, Cuautitlan, and others in the North and Northwest that come into Mexico and are twice as dear because of the cost of freight on mule trains.[103]

The tangible hydraulic shape in which the consulado left the Desagüe, however, contradicted this discursive allegiance to water-borne transportation to the interior. In October 1788, having broken through to the bed of the tunnel in every section shown in Velázquez de León's ten profiles, the consulado declared its job completed and asked for certification of fulfillment of the contract and the nullification of the security bonds.[104] Accordingly, Viceroy Manuel Antonio Flores, a close friend of the consulado, named a commission to inspect the works in light of Aylmer's conditions in the contract. This commission included military engineer Miguel de Costanzó and maestro mayor Ignacio Castera. After inspecting the open trench, Costanzó reported that the consulado had failed to widen the trench channel from the original three varas to ten, as stipulated by his colleague Ricardo Aylmer. Using two longitudinal views of the Desagüe as well as five sectional cuts to convey this point (Figure 6.11), he also explained that while a channel eight and one-third meters wide might accommodate small vessels, nothing but the tiniest barge would ever hope to be safely towed over a channel barely two and a half meters wide. This barrier to navigability was compounded by "the great slope of the floor or bed where the waters flow."[105]

In other words, the condition in which the consulado had left the open trench precluded any possible adaptation for navigation, unless locks were installed. Reshaping these features—that is, combining Aylmer and Velázquez de León's ideas on shape and slope—could conceivably open up the Desagüe for navigation during the wet season, turning the overabundance of water that urbanites saw as a liability into an asset that not only would relieve the reliance on muddy, overstressed roads but also join with then existing and possible lacustrine communications to create a route covering eighty kilometers from Chalco to Huehuetoca, and from there communicating with the road to the north.[106] So why the rush to have its Desagüe open trench conversion declared complete, when the bed of the channel was still narrower and the sides steeper than stipulated in the technical specifications of the contract, making any adaptation to navigation impossible?[107]

The answer would seem to be that for all its patriotic zeal, the consulado's interventions in transport and urbanism were aimed less at creating new opportunities than at seeking the best possible strategic position in an environment of threatening changes to their monopolistic privileges. This is apparent from the overall goals of the corporation during this time, which were to try to undercut the prospective competition by other merchant groups in Veracruz and Guadalajara, consolidate its traditional dominance in the south and southeast as well as in the Manila trade through Acapulco, and gain royal and viceregal concessions to soften the implementation of measures harmful to their monopoly of Mexican trade.[108] The imminent reforms-related loss of the exclusive rights to the import tax known as *derecho de avería* (.06 percent on goods imported through Veracruz), with which the consulado funded the Desagüe contract, most likely also kindled their wish to be quickly released from Desagüe commitments.[109] Patriotic credit from the fulfillment of a project as huge as the Desagüe could be traded for tangible concessions.

This strategy seems to have borne fruit: the man Viceroy Flores named to evaluate the consulado's request to discharge the open trench was none other than Francisco Javier de Gamboa, a longtime ally of the corporation. To urge the release of the consulado from its contractual obligations, Gamboa used Costanzó's assessment that such deviation from the technical terms of the contract was actually an improvement upon Aylmer's design, since the strong current was exactly what permitted the sweeping effect, which was still necessary as excavation of the trench sides needed to continue beyond Bóveda Real.[110] In 1789, despite his own displeasure at the flaws in the trench, the new viceroy, the second Count of Revillagigedo promised to "contribute to the just remuneration" of the consulado's services "to the welfare of the Fatherland and the state" by sending the king the names of individual merchants who had distinguished themselves in the Desagüe in case the sovereign "sees fit to reward them."[111]

Instead of the merchants, it would be a viceroy who would promote the canalization of the Desagüe and other water flows to improve transportation for the benefit of commerce. During his tenure as viceroy, Revillagigedo asked military engineer Miguel del Corral to study the possibility of constructing canals to carry goods from Oaxaca and other districts to Veracruz. Del Corral drew plans and estimated the cost at 157,551 pesos, but despite their desirability his canals did not prosper because of competing demands on the royal treasury.[112] Looking back on his term while writing his *Instrucción Reservada* (the end-of-term confidential viceregal reports on the state of the realm) in 1794 for his successor, the Marqués de Branciforte, Revillagigedo would insist that "many more advantages could result from canals, or from making certain rivers navigable, which can be done at little cost, than from roads . . . for although its rivers are really few, and not very plentiful, some are valuable because of their origin, direction, and outlet with regards to this capital."

Revillagigedo admitted that the cost of some of the long-distance canals would be immense relative to the funds the crown might wish to expend on them. Moreover, he also recognized that it would not be for a long time yet that "the population or the commerce of these kingdoms would demand such a huge undertaking." But this was not the case for the Desagüe, which was his prime candidate for such canalization out of the capital and which he felt was realistic. Having seen the trench's potential for eventually connecting the capital to the sea, Revillagigedo expressed disappointment at the lack of response from the crown regarding his request for two military engineers permanently assigned to such projects.[113] If the crown would not support this idea, some consulado might.

Indeed, by the 1790s, some urbanites would bestow expectations of great deeds from the consulado of the City of Mexico in transport infrastructure, mining, and of course the Desagüe. Verses composed in 1796 (and possibly commissioned by a merchant from the capital after the formation of the long-dreaded Veracruz and Guadalaja consulados) to flatter both Charles IV—the king "that all nations envy"—and Viceroy Branciforte broadcast how this viceroy had "succeeded in engaging the patriotic zeal of the consulado" in his enlightened campaign to bring progress to the realm:

*Si defensas al Reyno le previno*
*si a este, y al Rey aumenta los caudales*
*con los arbitrios de un talento fino,*
*no menos precavió riesgos y males*
*a esta Ciudad con obras que imaginó,*
*que se han de celebrar en sus Anales:*
*tal es la del Desagüe en donde aprueba,*
*lo que su Juez Conservador promueva.*[114]

The verses also proclaimed that the consulado's efforts in road construction supported "that long-meditated vast project to build roads that are of such interest to the King and the State."[115] Maybe so, but this was a rather strange way of showing support, since when the merchants of the City of Mexico finally became interested in remedying the pitiful state of long-distance transportation, their evident motivation was to undercut the Veracruz consulado. Of the two possible routes for the Mexico-Veracruz link, it lobbied for the one to Orizaba because that route bolstered its own dominance in the east and south.[116] With the 1793–95 construction of the road to Toluca, the camino de tierra adentro remained a future possibility, since although Toluca was a major grain-producing region, it offered only a round-about way to the north, more easily reached via Cuautitlan—that is, through the Desagüe.

We know now that the consulado used its donations, loans, or public works designed to woo concessions from the crown at the time of commercial reforms.[117] It is nonetheless possible that without the crisis of colonial rule that would ensue in 1808, the consulado might have pursued this navigation idea as a project unto itself. However, to accept this potential as evidence of the corporation's inherent interest in developing fluvial communications to further its internal markets, not to mention the good of the realm, requires a great deal of good faith. In this light, the consulado's public works contracts might be seen as the colonial precedent for the large, publicly funded and allegedly publicly minded construction projects promoted by alliances of private firms and government officials that prevailed in the modern era and made fortunes for individuals.

As it was, and interpretations aside, the hypothetical canal along the Desagüe to connect to the existing route to tierra adentro through Cuautitlan never went any further than individual imaginations and the uncontrollable flattery of verses penned by inspired letrados. If the crown itself was unresponsive to this potential, so too would be the capital's elites, merchant or otherwise, for the remainder of Spanish rule. There was a common denominator to the vociferous patriotic Enlightened demeanor among quite assorted sectors of the elite: it was discursive, not substantial.

The alignments of savants and technicians had implications for the kind and content of knowledge that its various actors marshaled in the Desagüe. The refutation of Vitruvius, the "Prince of Architecture," and the alignment with experimental science opened the possibility for the exploration of hydrostatic and hydrodynamic puzzles drawn from the observation of water. Hypotheses and experimentation might create an autonomous corpus of scientific knowledge; savants might engage with European science other than merely applying, imitating, or rejecting it. With hydromechanics, Joaquín Velázquez de León shook the foundations of the scientific paradigm that had justified the material form centered around the open trench of the Desagüe. But none of this translated also into a shattering of the other old paradigm: that the Desagüe existed to safeguard the city and nothing else. After Velázquez de León, few still believed that dramatic gradients were necessary in the Desagüe bed, but none went beyond this. Ironically, however, whenever he was cited thenceforth with regard to this finding, it was in the same spirit in which people had cited Vitruvius before: as coming from the "sabio americano," an undebatable and final authority to which little could or needed to be added.

The context in which these people intervened cannot be extricated from what they said about the Desagüe. The fact is that they were operating within the specific conditions created by the consulado with the extended powers it enjoyed in the project. So far as the corporation itself was concerned, this control militated against work in the Desagüe jurisdiction, against the articulation of the project to any economically generative activity, and against the delicate balance among hacienda, township, and drainage needs for manpower that the alcaldes mayores understood and had an interest in maintaining. The corporation would "finish" the trench, but its method of doing so wasted human, environmental, and technological resources, and few suggested they should act otherwise. Those who did tended to be metropolitan agents themselves—the military engineers, not the creole savants and letrados.

Colonial subjects correctly understood the engineers' mission to be threatening to them. Because in the second half of the eighteenth century there were more of them, and because the Desagüe was one of several public works projects in New Spain, engineers were brought into contact and conflict with many more sectors of society, and more visibly so, than before. As Aylmer's somewhat naïve expectations about the force of the 1767 contract illustrates, engineers would often still remain insulated from intimate familiarity with the interests of local networks and how these might interfere with the execution of their projects. This would turn out to be a key weakness in their work, since the successful completion of their projects, on schedule and according to the original plans, hinged upon contracting out to these local networks. At the same time, however, the sting of slights inflicted in these exchanges may have contributed not only to local self-assertion against the foreigners but also to a keen sense of the need to keep abreast of the cultural changes beyond the sea, if only for self-preservation.

Much as they emulated the form of the new scientific and technological culture of military engineers and other agents of the European Enlightenment, the savants and technicians who intervened in Desagüe matters during this period would not—or could not—partake of its substance. That is, the underlying interrogating demeanor was not there and neither was the inextricable link between techno-scientific "advances" and the usages of human and nonhuman resources consonant with them (which of course did not necessarily mean these usages were more humane). As members of a creole lettered and social elite, local players had little to gain and much to lose from rearranging the social relations prevailing in the Desagüe. Since the consulado did not turn over the works until 1789, this extended surrender of the privileges of the officers of the crown in favor of the merchants of the City of Mexico could very well have been a factor precipitating the implementation of the intendancy reforms that sought to sever the patronage and client relationships that entangled royal officials with the wealthy local elites and the learned with the state.

By the time the consulado returned the Desagüe proper to the oidor superindendants, all creole technicians and savants who had something to do or say in the project had been made acutely aware of the new standards of state employment and patronage. If they could do nothing about their American birth, they could about their culture. In coping with the new circumstances, people like Alzate, Castera, and Velázquez de León adopted a variety of strategies, combining resistance, emulation, and conviction in varying degrees in different aspects of their interventions. However much or however little the drainage itself or the kinds of knowledge and practices technicians and savants coalesced around it would change from this point on, the fact is that even without radical transformations of its content, the Desagüe was once again actively helping to deepen the urban elites' colonization of the region, sinking thicker roots into its soil and water and increasingly segregating the fluid interaction both between these elements and between residents of the hinterlands and the biota.

CHAPTER SEVEN

# Toward "Waterless and Dry" Ground

When the consulado was released from its obligations in 1789 following its completion of the open trench the previous year, the command of both the Desagüe proper and the ancillary structures in its district was unified again under the oidor superintendants. This event coincided with the phasing in of the reforms articulated in the royal *Ordenanzas de Intendentes*, which aimed to break the hold of colonial elites—particularly the entrenched American consulados—on local administration, commerce, and production, and to reorient all these activities toward the recovery of Spain itself. This growing assertion of metropolitan control would sweep over the Desagüe too, opening up a second wave of potential challenges to the status quo as new approaches to management were instituted and the creation of new technological and scientific institutions multiplied the sources and kinds of expertise available to the project.

After the opening of the Real Academia de San Carlos in 1785, and more so with the opening of the Seminario de Minería in 1792, it became possible to consult mathematicians and technicians in these institutions about problems in the Desagüe.[1] This opened the door for these institutions to broaden their concerns to include more hydrodynamics and hydraulic engineering in the curriculum, both with an eye to solving the specific problems encountered in the project. These institutions were thus well positioned to provide a home for the increasingly detailed expertise about the basin that both fed and resulted from the expansion of the jurisdiction of the Desagüe that occurred after the return of the oidores. Until this point in time, hydrologic and hydraulic knowledge had been held—dispersed and unsystematized—in the Desagüe records, in the heads of the people associated with it, and in the memory of the on-site wardens and foremen of the project and its rural neighbors. The potential for a symbiotic relationship between the new institutions and one of the most important engineering projects in New Spain was huge. Long absent from the Desagüe, the practices used in mining—the most dynamic sector of the economy, where capitalist forms of resource use and relations of production prevailed—promised to transform technology and work in the project, while having professionals in the modern sciences so near at hand could result in the production, or at least the application, of new knowledge about the mechanics of fluids.[2]

For elites, there was also a practical need for new solutions in the Desagüe. With the open trench completed, great hopes were riding on the possibility of finally extending the Desagüe into a general drainage. If this were accomplished, "the soil of Mexico, which is now moist, would in time become waterless and dry," wrote Viceroy Revillagigedo in 1790, and this would be an outcome "clearly convenient to the health of the basin's inhabitants." Evacuating the water of the basin would consolidate the ground, providing buildings with "better foundations and greater endurance and magnificence." Most contemporaries in Revillagigedo's class and many a historian thereafter would wholeheartedly agree with him that by extending the Desagüe's drainage capabilities, "a City such as Mexico, which can well be worth one hundred million pesos, would be buying its perpetual security, beauty and decorum very cheaply for two or three."[3]

A different scenario far less beautiful and decorous was already unfolding underneath these discussions. Despite the fact that the crown's legal protection of the peasantry's customary access to watery and dry commons seemed on the whole to be surviving the Bourbon onslaught, in the Desagüe district the intervention of the project's regulations and structures were silently changing the practical meaning of this protected access. The effect of the fifty-vara rule and other dispositions was compounded by a more irreversible fact—that whatever its troubles, the Desagüe was gradually favoring the shrinkage of the lakes. For example, as can be seen in the 1775 boundary map in Figure 7.1, at that time the township *términos* (legal boundaries) of Zumpango and Teoloyuca still contained a tule-gathering area within Lake Zumpango.[4] By the nineteenth century, however, the edges of this lake at the peak of its season had retreated to some three kilometers from Teoloyuca.[5] Although to ascertain how much this retreat progressed and when would require field surveys, it is reasonable to assume that was happening during the period covered in this chapter and beyond.

With the phantom of urban flooding still lurking despite this shrinkage and the trench, city and viceregal authorities did pursue the goal of a general desiccation following the conclusion of the consulado's contract. They significantly expanded the Desagüe, and these additions allowed it to tap deeper into the basin. The way in which the expansion of the drainage was designed and undertaken tells us much about how deep the reforms actually reached, who was doing the colonizing that either informed or accompanied the whole project, and for what ends.

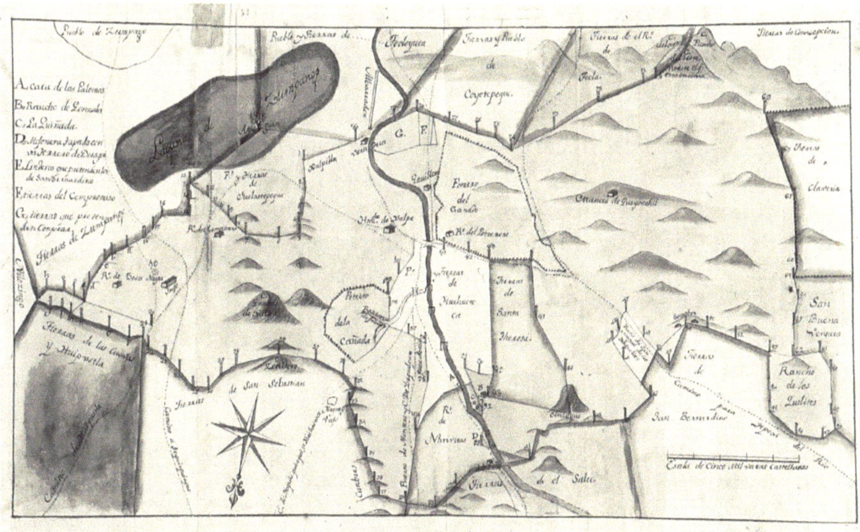

*Figure 7.1.* The haciendas, ranchos, and townships surrounding the Desagüe in 1775. Teoloyuca's lands extend into Lake Zumpango.
SOURCE: AGN, Tierras, vol. 287, exp. 6, f. 30. Reprinted by permission.

Even with the involvement of the most salient exponents of the reformist ideology and scientific culture, there were clear limits to how far the conceptualization of the city's relationship to land, water, and economic life in and around it would go from that which had informed the design of Martínez's Desagüe.

## NEW DIRECTIONS WITH A NEW ACADEMY

As the Desagüe proper and the consulado parted ways, the intendancy reforms threatened to leave the project rudderless while the storm surrounding their implementation quieted down. A French import, the intendancy system was designed to short-circuit the ties between American elites and officials that hindered the crown's access to resources in its richest territories. As "regional executives," intendants were supposed to assert the control of the mother country by intervening in their regions' political, economic, and ideological affairs.[6] First deployed in Spain in 1718, made permanent there in 1749, and sent shortly thereafter to "marginal" colonies, these officials finally arrived in Peru in 1784 and New Spain in May 1787. An alliance of merchant elites and the viceregal government had delayed the implementation of these measures, proposed back in 1768 by José de Gálvez, and would eventually also undermine their substance.[7] But in the meantime, the existence of intendants had the potential to rock the Desagüe. Although gone from the works, the consulado's man Gamboa would ensure this did not come to pass.

Under Article 28 of the Ordenanzas de Intendentes, both the ways and means (*propios y arbitrios*) of the City of Mexico and the Desagüe superintendancy would cease to be under audiencia supervision and come directly under the purview of the intendant general.[8] This event coincided with the consulado's request to turn over the works. Viceroy Flores, another friend to the consulado, immediately commissioned Gamboa to report to him on the state of the Desagüe. Characteristically, Gamboa wrote back with great admiration for the consulado's oeuvre.[9] He also took the opportunity to advocate the nullification of Article 28:

> Once Article 28 is revoked, things must to be returned to their previous state, and it is the prerogative of Your Excellency in his viceregal authority to name the Juez del Desagüe conjoining to it also the superintendancy of Propios y arbitrios of the City, as was done before Ordenanza de Intendentes . . . and especially because the City is so indebted that it will hardly be able to achieve the financial wellbeing in which the Illustrious Gentleman Don Domingo de Trespalacios left it.[10]

It was equally important, Gamboa argued, that oidores remain at the helm of the drainage because of the "great harm and detriment" the Desagüe would suffer if it were assigned "a regidor or other private individual lacking in knowledge of the matter." The Desagüe "by its nature demands not merely authority and integrity, and expertise, but also experience, health and readiness," Gamboa continued, and coincidentally these were qualities that were "notoriously conjoined in said gentleman Don Cosme de Mier," the prominent oidor. With or without Article 28 of the Ordenanzas, Mier became superintendant and trickled down the reforms to the Desagüe administration—with a dropper. This means that while eager to endow the Desagüe with all the accoutrements of "utility" so dear to Bourbon culture, Mier was less than anxious to overhaul either the aims of the project or the social relations imbricated within it.

Nothing was better suited for realizing these desires than the institution created precisely to root out the "useless" excesses of the American baroque and replace them with the rational and pristine aesthetics of neoclassicism—the Real Academia de Bellas Artes de San Carlos. Born out of a Mexican arts school but modeled after the Real Academia de Bellas Artes de San Fernando in Madrid, founded in 1752, the Real Academia de San Carlos came into being in 1783, during the consulado's tenure in the Desagüe. The consulado, the mining guild, and the city council each provided one councilor to a governing junta that included other eminent denizens of the city and oversaw the affairs of the academy on behalf of its protectors, the king and the viceroy. As in the Desagüe, the audiencia played a key role as it provided the president of this governing body. As a result of this leadership composition, the Academia de San Carlos reflected the sometimes conflicting priorities of different sectors of the colonial elite and the tensions that existed between American territories and Spain's aspirations to be a true metropole to them during this delicate stage in Bourbon reformism.[11] And the professional activities of its directors, as well as its lack of subordination to the Madrid academy, simultaneously rendered the institution permeable to concerns in the Americas themselves. The Desagüe became one of these concerns for a few years beginning in 1795 because Superintendant Mier made it so.

Shortly after the conclusion of the open trench, it became evident that the potential for flooding was as present as ever. In the district of the Desagüe itself, the structures that supported the proper functioning of the trench as an evacuator of excess water had been neglected during the consulado contract. For example, the silting pool of Coyotepec had not been cleaned out, and as a result it had accumulated so much sediment that it was often hindering the free flow of the Cuautitlan River into the Desagüe. In 1792 and 1795 the city experienced floods again—not catastrophic, but clearly due to the failures of structures designed to retain the Cuautitlan River. On inspecting the site in August 1795, master architect Ignacio Castera determined that the river's channel lay under approximately eight meters of silt, and that cleaning all this out was imperative, since when impeded from flowing into the Desagüe the river burst through its diversion dam and went straight for Lakes Zumpango and San Cristóbal.[12] In response to this situation, in November Mier proposed a plan to drain Lakes Zumpango and San Cristóbal directly into the open trench excavation via a tunnel. The goal of this plan was to open up storage areas for floodwater that would otherwise enter Lake Texcoco from either the north or from the Cuautitlan River if it should escape its routing to the open trench of Huehuetoca. It would also open the way for eventually extending the Desagüe all the way to Lake Texcoco (see Map 3).[13]

In the commission the audiencia named to study Mier's proposal there were two members of the Academia—Antonio Velázquez, chair of architecture, and Diego de Guadalajara, chair of mathematics. Also commissioned were Ignacio Castera, and military engineers Pedro Ponce and Miguel de Costanzó, Guadalajara's predecessor in the Academia. Tapping the Academia for help in the Desagüe at this point had potential implications, particularly in the case of Guadalajara. A creole, Guadalajara had studied mathematics with both Antonio León y Gama and Joaquín Velázquez de León,[14] who in turn were part of a growing circle of academics and autodidacts spreading Newtonian science in the viceroyalty in the second third of the eighteenth century.[15] During 1777, Guadalajara had

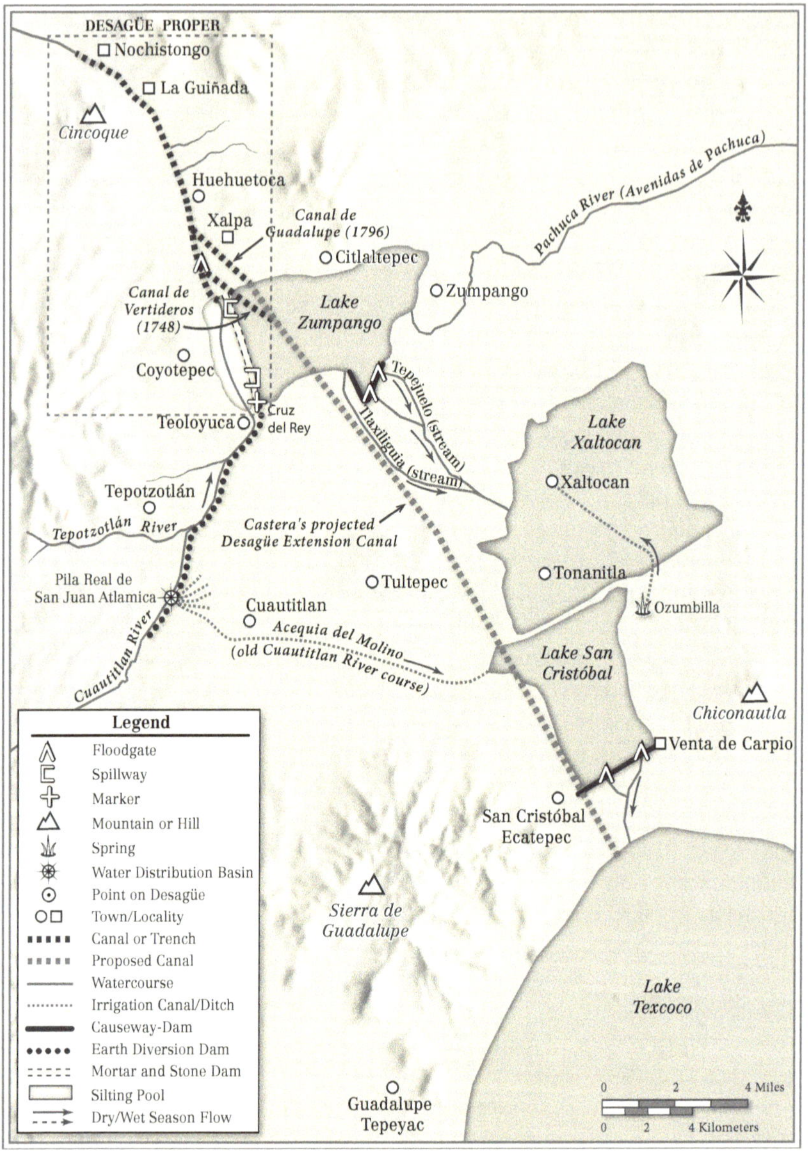


*Map* 3. The Desagüe from Mier's reforms to the end of Spanish rule.

SOURCES AND NOTES: (a) AGN, Desagüe, vol. 14, exp. 5, ff. 13, 20v; (b) AGN, Desagüe, vol. 30, exp. 4; and (c) Tacuba and Cuautitlan together are elsewhere quoted at 3,053 varas.
CREDIT: Gerry Krieg.

published a periodical entitled *Advertencias y reflexiones conducentes al buen uso de los relojes y otros instrumentos matemáticos, físicos y mecánicos*. The inaugural issue promised to cover precision instruments used in physics, such as the aerometer, and in mathematics, such as the "English pantometer" that was particularly useful for trigonometry.[16] It is unknown if all these were ever published (only five issues are extant and they discuss clocks), but Guadalajara was clearly skilled enough to serve as instrument specialist in Alejandro Malaspina's expedition.[17] Like Velázquez de León, this academician also had expertise in mining and "subterranean architecture." Having access to such diverse but relevant realms of knowledge, he was thus uniquely placed to develop the Desagüe's links to mechanical understandings of nature, particularly water, and to mining technology, both largely absent from it for a long time.

Using Velázquez de León's 1774 survey data, the commission was split over the feasibility of an underground drainage for these lakes, given the historical experience with Enrico Martínez's tunnel. They cautioned that due to the potential difficulties of an underground excavation through the inconsistent and clayey terrain, canals could well turn out be the only way to drain the two lakes. Each lake could be desiccated by its own canal; these canals would enter the lakes to the depth necessary to remove all water except that needed for local livestock. Inside the lakes, the canals would be flanked on each side with earthwork levees to keep silting out of their channels. The two canals would join into one at a point to be determined and then the single canal would run parallel to the Cuautitlan River for 1,660 meters, to connect with it at Vertideros, where the waters it carried would be led to the Desagüe open trench. The entire length would be 13,031 meters; the width, 3.3 meters. The depth would vary according to the terrain. All this would require the removal of 714,000 cubic varas of soil and rock. In addition, they concluded that the course of the Cuautitlan River ought to be straightened. A new survey would be needed, but in the meantime they estimated that the entire project would cost 195,000 pesos.[18]

But Mier wanted his tunnel and felt that a mining expert could back him up. He asked Guadalajara to see if he could find solutions to the objections of the commission. Guadalajara obliged: yes, a tunnel to drain only Lake Zumpango by taking water from the Vertideros or spillway of Lake Zumpango and the Bóveda Real section of the Desagüe open trench was possible, he explained with the help of the map shown here in Figure 7.2.

Content, Mier had Guadalajara draw a preliminary plan where the tunnel would be 6,640 meters long and have 198 shafts or adits (*lumbreras*) (see Figure 7.3). He countered his colleagues concerns: to deal with the

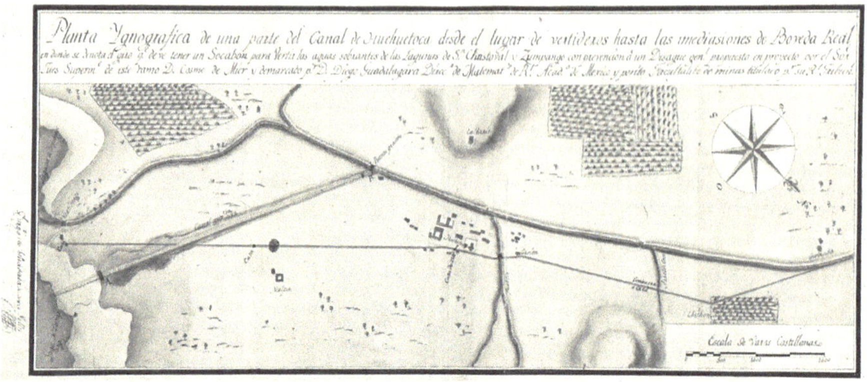

*Figure* 7.2. Mier's tunnel to drain Lake Zumpango, as imagined by Diego de Guadalajara in 1796. It was to tap the lake at Vertideros (on the left) and take its "excess" water to the Desagüe at the Bóveda Real.
SOURCE: AGN, Desagüe, vol. 29, exp. 10, f. 92. Reprinted by permission.

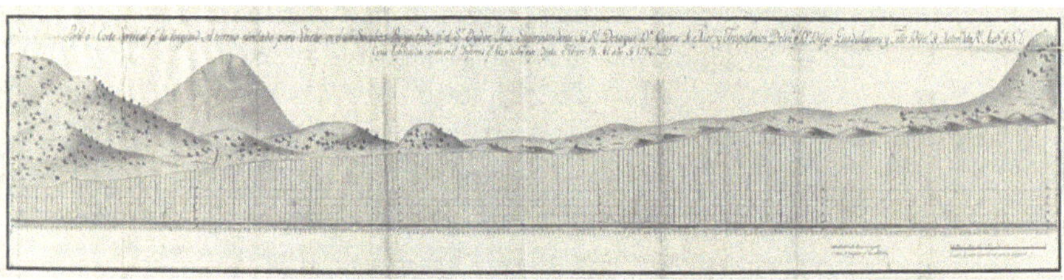

*Figure* 7.3. Profile along the length of Diego de Guadalajara's 1796 tunnel project to drain Lake Zumpango, with scales in *pulgadas* for heights and *varas* for length.
SOURCE: AGN, Desagüe, vol. 29, exp. 10, f. 91. Reprinted by permission.

loose terrain and the constant infiltration of groundwater could flood the shafts, Guadalajara proposed shafts of different functions could address this problem—vertical ones would provide excavation guides and ventilation. Between each of these, shallower adits at 45-degree angles would accommodate either "the simple hollow spiral machine of Archimedes, or hoisting buckets." These drainage machines were "quite simple and of little cost, and exert a powerful effect in shallow depths," according to him. Between these measures and having crews working from both ends of the tunnel, Guadalajara felt "the waters shall be defeated and the excavation will proceed with violence."[19] Tapping Lake Zumpango with a tunnel instead of a canal would cost forty-two thousand pesos in all, the lower price tag being due mainly to the lesser quantity of earth to be removed.[20]

Like Velázquez de León had done in 1774, Guadalajara used scientific methods to assess the feasibility of this preliminary plan. Initially relying on the water level instrument available at the academy, he sought to replace it with a built-to-order air-bubble precision instrument with telescopic sights to minimize errors in readings from the effects of parallax, refraction, and wind.[21] By using test bores ("*catas de inspección*"), he also gathered information on the quality of the ground along the projected path of the tunnel and the depth at which the groundwater would be met. These data he considered so crucial that he asserted that on them "hangs no less than the progress of the works."[22] On the basis of the data, he determined which segments of the tunnel could be left as bare rock and which required reinforcement.[23] The test-bore data allowed him to correct his plans. Believing the water he found underground to be filtered rainwater rather than the springs he had expected, Guadalajara determined that "the angled shafts" that he had originally proposed would be "pointless."[24]

Guadalajara relied on a second source of knowledge, his mining experience. Until it could be reinforced with masonry, timbering would support the interior of the horizontal tunnel "in the style used inside mines." Timber props placed against the walls and resting on head-pieces ("*zapatas o llaves*") would distribute the load over a larger surface. Half-rounds of ilex "which can be ordered from Pachuca" would be placed atop these props to support the ceiling.[25] This was not the first time that mining technology was invoked in the Desagüe. Besides of course Enrico Martínez's own adaptation of mining techniques in excavating the tunnel through Huehuetoca, in 1743 military engineer Gaspar de Courseulle had wanted to use calibrated explosions to remove conic sections in the trench excavation. Then, in 1755, master architect Manuel Alvarez had proposed using mining techniques and skilled workers from Real del Monte and Pachuca to blast open a canal segment in the rockier parts of Salto de Tula, at the outlet of the Desagüe. Alvarez's empirical method called for setting off four thousand pesos worth of gunpowder blasts (*cuetazos*) and then figuring out the rest of the plan and its costs from there.[26] Armed with Guadalajara's report, Mier went on the offensive again, and on April 21, 1796, he triumphantly inaugurated the project amid the usual fanfare.

The mining-based approach inherent in Guadalajara's plan brought into the project various elements that could destabilize the very nature of work and social arrangements in the Desagüe. After all, as is well known, mining in New Spain was not just operated with wage labor but with relatively highly paid workers at that. Guadalajara envisioned paying wages as well as expenses for the treatment of work injuries. The greater specialization of the tunnel workers would be rewarded with meat rations, "which with a small increase in cost will impress upon them the idea that they are being distinguished from the laborers in the River, and this rivalry will encourage them to work harder, with which that apparent increase will be more than covered."[27] This idea seems to have been implemented, as guarda mayor Francisco Power recorded having purchased eight heads of cattle to reward the tunnel workers.[28] As importantly, the seasonality of labors vanished, for here debris would be continually extracted with a team of two skilled rock drillers (*barreteros*), each with his own shoveler (*palero*) to clear away the debris, which would be loaded by a third worker into sacks, and raised by two other men above with a rope and pulley (*garrucha*).[29] These improvements in the organization and treatment of labor meant that barely one month after groundbreaking 82 percent of the shafts and 65 percent of length of the tunnel had been completed.

This would all appear to indicate that repartimiento had vanished and that from the end of the eighteenth century on work on Desagüe projects would be performed by laborers as free, and possibly as well paid, as those in mining. Not so. While the necessarily experienced barreteros were miners, their assistants were Indians coming in through the usual rotational labor draft. Complaints and requests for exemption from Indian townships as far away as Texcoco indicate that the roundup of coerced labor was in full force by May 1796. Mier, eager for his laurels, ignored them all and bragged to Viceroy Branciforte on August 11, 1796, that he had twenty-seven hundred men already at work from several jurisdictions.[30] Again the guardas mayores played an important role in guaranteeing this. As resistance to the repartimientos increased, guardas mayor Francisco Power used the powers vested earlier in his position to order subdelegados to send the Indians to him tied up in gangs, if necessary.[31]

By contrast, conditions for these repartimiento workers may have improved somewhat: whipping was regulated, and they now received their pay in hand individually, rather than indirectly through the *gobernadores* of the towns whence the tandas came.[32] Individual payment, however, was a real loss for the Indian gobernadores, whose control over their Indians' Desagüe cash wages for tribute diminished. All in all, workers were only marginally better off than their predecessors in the Desagüe, and they were still coerced to abandon their villages and their fields.

It is not surprising, therefore, that as one of Guadalajara's assistants reported, even the technical achievements were a mixed bag, a result of the lack of experience of the workers (none of whom had ever worked in a tunnel), who were "rustic and inexpert." The result was that while the crews did bore through the terrain in a straight line, "it was not possible to avoid defects in the Bed of the Mine, which reached very noticeable heights capable of stopping the flow of the waters, causing them to rise to a level that filled the entire cavity of the tunnel in its longest section."[33] Not that the workers lacked ingenuity. Rather, they employed it mostly for their own survival: because there was more water than expected underground, the workers solved the problem of digging with their feet in the water by raising earth banks above it on which to stand. These "solutions" were completely counterproductive to the intent and design of the tunnel, since these obstructive mounds were not removed.

But it was not inexperienced labor that doomed the tunnel. Before years' end, the vaulting and supports between shafts sixty-six and ninety collapsed and impeded the flow of water in the tunnel. Mier demanded an explanation. Guadalajara was confused. Maybe he was to blame, since he had had neither sufficient timber nor sufficiently trained workers to do what was needed to reinforce the tunnel in the style of mining he had hoped for. But mostly it was the terrain's fault: the terrain had failed to behave as expected—"an exception to the general rule that is prescribed by the most celebrated mineralogists when they speak about tufas, whose characteristics are as the same as those of this treacherous terrain," Guadalajara defended himself. "Even the most informed wit would have been fooled."[34]

The only solution was to turn the whole thing into a canal. Unlike the destruction of Enrico Martínez's tunnel, this conversion took less than one month. Mier's tunnel became the Canal de Guadalupe (seen as a project in Figure 7.4 and as a reality in Figure 7.5). Mier invited Viceroy Branciforte to watch as the water from Lake Zumpango flowed down this canal into the Cuautitlan River just after the bridge in Huehuetoca. Branciforte was apparently pleased and told Mier to plan the next phase—the excavation of the canal to drain Lake San Cristóbal. The fiscal of the audiencia then recommended that the capital finish these two canals "for the conservation of its opulent fortunes, because excessive as the wealth is that is kept in the storages of their houses and that belongs to His Majesty in rents, it would all be lost if by misfortune the waters of the lakes rose."[35]

Again Guadalajara and Velázquez were asked to lead, with Castera as consultant. Excavation began in February 1798, and by May 1799 Mier had his second canal. All was not well, however. After visiting the works in June 1798, Velázquez asked Mier to be excused from the project to desiccate Lakes Zumpango and San Cristóbal altogether. It was all hopelessly flawed, he felt: while having the two canals merge at an acute angle before joining the Cuautitlan River was correct, the water flowing down from Lake Zumpango made a sharp turn and rushed down into Lake San Cristóbal instead of the Desagüe because the canal of Lake San Cristóbal was level, not inclined toward the merger. In addition, both canals were easily obstructed because their sides were steeper than 45 degrees, which

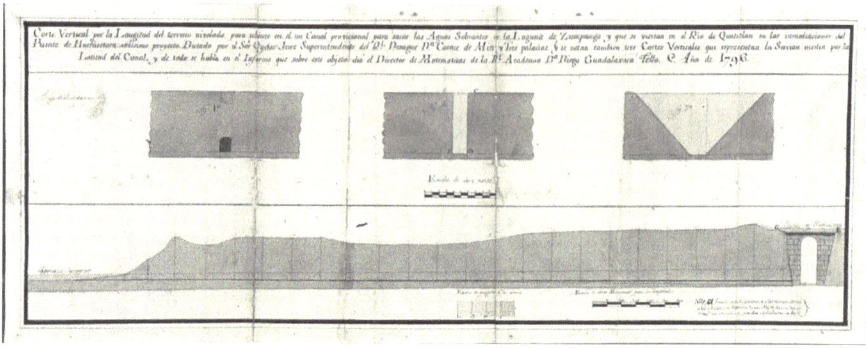

*Figure 7.4.* Guadalajara's response to the collapse of the tunnel to drain Lake Zumpango—the Canal de Guadalupe, surveyed with precision and here pristinely drawn with conventional signs that belied the looseness of the terrain and the probable uselessness of the project.

SOURCE: AGN, Desagüe, vol. 30, exp. 2, f. 58. Reprinted by permission.

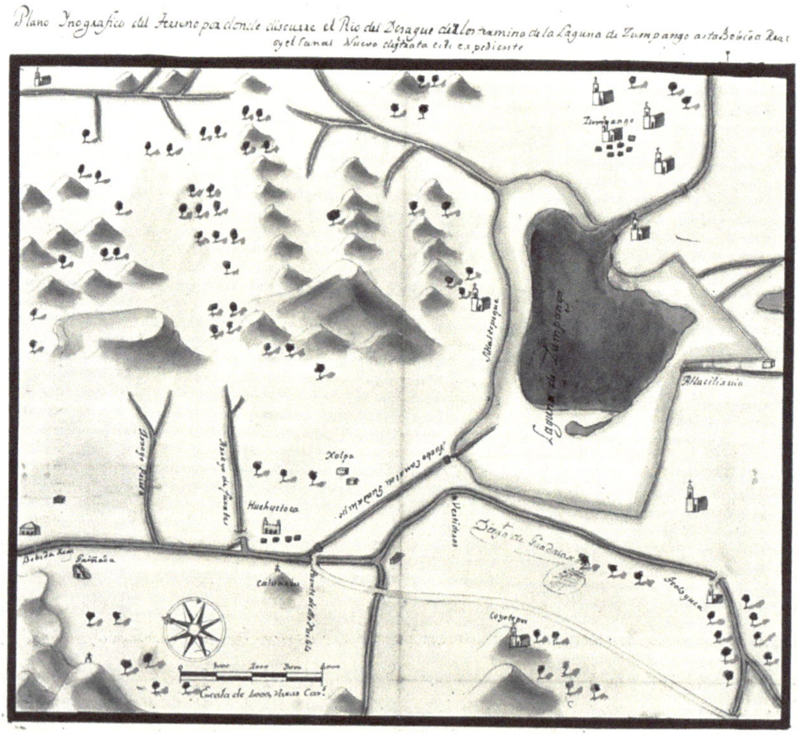


*Figure 7.5.* Mier and Guadalajara's "New Canal of Guadalupe" on the ground. Neither the tunnel that collapsed nor the canal that replaced it extended to greater depths of Lake Zumpango. Note the two margins for the lake—lighter green for the rainy season; dark for the dry.

SOURCE: AGN, Desagüe, vol. 32, exp. 2, f. 241. Reprinted by permission.

promoted crumbling. Velázquez feared that the projects were fated to failure because of careless planning.[36] These criticisms echoed those made twice by military engineer Pedro Ponce—once of the Canal de Guadalupe, and later of the San Cristóbal project. In Ponce's view, the only thing these projects did was "show assiduousness in the protection of the public."[37]

The disappointing results of Mier's projects should not obscure the fact that Desagüe authorities as a whole had eagerly welcomed the foundation of the Academia de San Carlos as a fresh breath of scientific air, calling on its president to send its experts for help in the same way as they had always tried to tap whoever they saw as the best and the brightest. The Academia was equally eager to offer its staff and instruments to help the Desagüe. Whenever technical assistants were needed, Guadalajara brought along its advanced students (*pensionados*). Apparently anxious for the Desagüe to gain from "useful knowledge," in their reports to Mier Guadalajara and his fellow commissioners simultaneously praised the scientific and mathematical approach to the Desagüe's problems that they felt they were using and deplored the lack of *method* and *system* in past Desagüe affairs. To them, "not having formed and followed a fixed and invariable system" in the public work meant that "of the many millions invested in this enterprise more than half have been wasted, for lack of direction and system in hasty, almost fruitless and useless short-term operations and maneuvers."[38] As it poured its expertise into the Desagüe, did the Academia address these problems by also incorporating this call for a scientific approach to the hydraulic project into its curriculum?

Unfortunately, there is no indication that as an institution the Academia took its cue from the Desagüe and its problems or from the lacustrine environment as a whole. The focus of the curriculum remained on aesthetics, drawing, and on areas of science with direct bearing on the construction of solid buildings of "good taste." The founding statutes, approved by the king in November 1784, had called for two mathematics directors to teach the subject "as extensively as possible," yet the Academia would not employ mathematicians or physicists per se as professors of mathematics, but "people well versed in mathematics, preferring always those who to this instruction add the practice of Architecture."[39] Physics was not specifically contemplated in the statutes and although the textbooks did encompass subject areas relevant to the Desagüe, there is no evidence that they were used to actually teach hydromechanics in any great depth.

In practice, mathematics instruction focused on the basics even as Guadalajara created the second mathematics class contemplated in the statutes. Until 1791, the subject had been taught in only one class for all levels. Now, as students graduated from it they were made to teach the beginners' course, leaving the director time to teach geometry and trigonometry to more advanced students. Even the advanced class covered only arithmetic, algebra, geometry, and trigonometry. Students studied this with the *Elementos de matemática*, a multivolume treatise by one of the instructors at the Madrid Academia de San Fernando, Benito Bails, which summarized and simplified in Spanish the works of Bernoulli, d'Alembert, and others.[40]

Some instruction in hydromechanics or hydraulics may have developed at the academy as a means to regulate the construction of water-, wind-, or animal-powered devices. In 1802, the viceroy prohibited the construction of such machines by anyone lacking prior approval of the plans and budget by experts from the Academia and provided that only *peritos* examined by that institution would be entitled to direct the operation of the devices once installed. This was not a response to any Desagüe need, however, but to the public outcry after a hastily built water-powered flour mill in the mining district collapsed, killing or maiming some fifty people.[41] Although there is no evidence that this rule actually resulted in curricular changes embracing more hydromechanics or hydraulics, it is a sign that the institution was not impervious to the demands of the society in which it was implanted. This makes it all the more striking that hydrodynamics and hydrostatics were not developed in the curriculum earlier, when the Desagüe and the Academia crossed paths, especially since the Desagüe projects were putatively at least as much for the "public good" as any grain mill.

The explanation for this is that the school was more in tune with the aesthetic agenda coming from Spain than with local needs, even if these needs were those of patrician elites. As a whole, then, the Academia de San Carlos does not seem to have had the intent to address either the scientific or the technological challenges in the Desagüe or in the area of water management more broadly. At least for the time being, the technicians the Desagüe had traditionally relied on, the social arrangements they secured, and the types of knowledge they relied on the most remained unchallenged.

## PROFITING FROM DRY LAND

Among these Desagüe technicians was master architect Ignacio Castera. Probably pressed by the competition of the academicians in the Desagüe, in 1795 he presented an ambitious plan for the desired general drainage. The plan is significant because at first glance it seems aimed at altering the conceptualization of the Desagüe from merely defensive to economically generative.

Castera's idea for a general drainage was to subdivide the extension of the Desagüe into four discrete legs covering almost the entire lake system. Section One would be a navigable canal located in the southern lake of

Chalco that would drain all but the water needed for the canal in the area of Chalco and Mexicalzingo (the yellow line to the south of the city in Figure 7.6). This would be the first section to be undertaken. The canal for Section Two, to be started after the conclusion of Section One, would extend from Lake Zumpango to Vertideros. The aim of this and subsequent sections would be solely to desiccate. Section Three, next, would go from the end of the canal in Lake Zumpango southeastwardly to Lake Xaltocan, draining that area, while Section Four would finally reach from the isle of Tonanitla in Lake Xaltocan to Lake Texcoco. Castera argued that Section One would "result in freeing the lands that the lagoons occupy between Chalco and Mexicalzingo to enable using them in most fertile and beautiful agricultural haciendas with year-round springs to irrigate them." This might seem to be an effort to connect Desagüe design to productivity, but the rhetoric about freeing up land for agriculture turns out to have been just that—a discursive resource for promoting a plan with a different objective.

While taking advantage of a very real desire for arable or pastoral land, the plan's call for starting in the south instead of the north, where most of the floods came from, had nothing to do with the agricultural potential of drainage per se. Instead, it was designed to harness the desire

*Figure 7.6.* Master architect Ignacio Castera's 1795 "General map of the lakes that surround the city of Mexico," showing two potential routes for the extension of the Desagüe to Lake Texcoco are shown. The eastern one in red is Joaquín Velázquez de León's project; the western, in yellow, military engineer Pedro Ponce's. Note again the significant variation in the surface covered by the lakes in the wet and dry seasons in light and dark green. The "receptacle of the lakes" covered the light area, making the reservation of this surface onerous for its riparian users.

SOURCE: AGN, Fomento Desagüe, vol. 1bis, f. 213. Reprinted by permission.

for profit and land speculation, as the first section of canal would provide the bulk of the capital needed to put the entire project in motion, through the sale of the resulting dry land:

> Once this canal is concluded, the lands can begin to be sold to the neighboring haciendas since they will be most useful to them because of their having the houses and workshops already erected and when they are not able to pay in cash, as is needed, they can be put up for auction, at which time there will be instantly an infinity of buyers due to the proximity to the city that all desire for their country estates.

Castera estimated that the land created by the desiccation in Section One would amount to three hundred caballerías of first-rate land irrigated by springs, which, at three thousand pesos per caballería, would net 900,000 in no time because of the desirability of the location.[42] This sum would jump-start the second leg of the canal system, which would free up land slightly harder to sell due to its distance from the city and its lesser quality. Nevertheless, the desiccation of Lakes Zumpango, Xaltocan, and San Cristóbal could yield another 450,000 pesos from the sale of another three hundred caballerías of new land. Lastly, the fourth leg would dry up less desirable land in Lake Texcoco, which would sell for only one thousand pesos per caballería because of its salinity. In all, the entire sale would net 1,650,000 pesos.

The amount of space Castera dedicated in his written proposal to each aspect of the project is telling. While fashionably physiocratic allusions to "the enterprise of working the fields" and its connection to the "wealth and stability of states" were lightly peppered over the document, explanations of flooding in the city, which at the time came from Lake Texcoco's silting, were slathered everywhere. This imbalance supports the interpretation that it was the latter issue, and not opening up arable land for its own sake, that motivated its author and, presumably, the officials who might approve it.[43] It was, after all, an effort to sell a juicy project to potential patrons or clients. In this sense, Castera's behavior was no different from that of modern-day development and architectural firms.

The proposal was fundamentally a financial scheme to ensure the execution of the works that the architect was proposing as a remedy to the flooding problem in the urban hub. The agropastoral potential alluded to in the report was not translated into design. None of the new canals contemplated irrigation, for example. The economic motivations the elites and their technicians still pursued in the desiccation had to do with rents and the exchange value of land. Quite simply, land covered by water had no exchange value: it could not be sold, it would not appreciate, it could not be rented, it could not be taxed. Free of water forever, it could generate income in all these ways, relative to its various potentials as real estate. What was important was that the land be created and acquire a market value, not that it be made to produce. This is why the water to be evacuated had no value whatsoever, not even when it was as capable of irrigating crops—as was the case with the fresh water in the southern lakes.

This plan as a whole stayed on paper. But in 1804, Castera would win over Viceroy Iturrigaray's favor for what amounted to a revival of Enrico Martínez's idea to prolong the Desagüe to the northwestern tip of Lake Texcoco through Lake San Cristóbal—again, for flood protection. From the Vertideros to there, the whole canal would have measured almost thirty-two kilometers and cost six hundred thousand pesos. Given its huge labor demands, by the time the insurrections of 1810 forced the government to divert all resources to counterinsurgency, the project would be little more than a shallow ditch in the plains between Lake Texcoco and Lake San Cristóbal.[44] Thus, the last of the new Desagüe projects under Spanish rule still only treated the potential agropastoral yield of the drainage as little more than a discursive crutch to ensure approval for different goals altogether.

But perhaps there was a new conceptualization lurking somewhere nonetheless before the last gasp of empire? In 1808, with Castera still acting as the Desagüe's main technician, Superintendant Francisco Robledo asked him for a list and map of rural establishments bordering Lake San Cristóbal. Robledo wanted to be able to levy a fee on the indigenous townships, haciendas, and ranchos that had supposedly gained from Mier's drainage canal, charging them sums commensurate to the value of the land they had allegedly gained, defined as "the value of the improvements, which is the difference in worth of the land when it went from flooded to dry."[45] The response of the hacendados, rancheros, and townships is instructive.

Manuel de Arzipreste, owner of the hacienda San Mateo, stated that far from gaining land from the San Cristóbal auxiliary canal, he had actually incurred losses. He and other hacendados were barred from cultivating the fifty varas from the edge of the auxiliary canals (which was a general prohibition in the Desagüe jurisdiction). In addition, San Mateo and Santa Inés haciendas were now inhibited from draining irrigation remnants into the San Cristóbal canal or the Tepejuelo (a small stream connecting Lake San Cristóbal and Lake Xaltocan to Lake Zumpango), a right they had been conceded in 1801 and without which Arzipreste claimed his entire hacienda would become a lake.[46] For their part, the Indian townships of Ozumbilla and San Cristóbal protested that "even when [the lake] is free from all the water they will not gain from it at all nor can they cultivate therein because of the large quantities of salt residue which inhibits any results."[47] In addition to these types of objections there was the thorny issue of the original land titles that had to be presented for the correct estimation of the difference between the pre- and the postdesiccated holdings and the sums owed by each. Indian townships did not seem to have a problem producing the documents, but fiscal Sagarzurieta of the audiencia pointed out that most hacendados would not be able to produce legitimate land titles. Significantly, it was this problem of titles and not the protests over the alleged land gains that created obstacles for superintendant Robledo's payment plan.[48] The alleged beneficiaries seem to have felt anything but gratitude about the supposed boost to their productive capacity. They viewed Mier's canals as totally alien to their interests. For the moment, changes in the conceptualization of the Desagüe would have to come from elsewhere.

## WHERE THE MINERS ARE

As it turns out, these canals of Mier's and the new phase of Desagüe expansion happened to correspond in time to the development of another enlightened institution concerned with many of the same natural phenomena that were of relevance in the Desagüe—the Real Seminario de Minería. Mining had been one of the original sources of imagination and know-how in Martínez's time. In addition, mining shared with hydraulic engineering the manipulation of the static and dynamic properties of fluids (including gases) to power machines, multiply forces, drain flooded tunnels, ventilate galleries, and so on. When the Seminario opened in 1792, these areas of physics were included in the curriculum. It was a situation full of possibilities for the Desagüe and for the science of water: a site for the production of knowledge and another for its application were within reach of each other. The institution could use the project for experimentation, and the project could gain valuable knowledge and insight from the institution. Tracing what happened to this potential takes us back to the threads linking the consulado, Joaquín Velázquez de León, and the revival of the general drainage idea.

Velázquez de León, in fact, was in large part responsible for laying the groundwork for the potential for mining to transmit into the Desagüe the manner in which it coordinated labor, organization, and technology. This had occurred nearly two decades before the Seminario's opening. Velázquez de León was in 1773 and 1774 compiling his report on the feasibility of extending the Desagüe for the consulado. Besides this, he was also working on two other reports. One was his *Descripción histórica y topográfica del valle, las lagunas y ciudad de México, de las inundaciones que ha padecido y principalmente de las grandes obras que se han hecho y aún se están haciendo para preservarla de esta calamidad a que está expuesta*. This was a manuscript on the history of the Desagüe and the basin commissioned to him by the viceroy, which he was developing on the basis of both his observations and surveys and Desagüe documents. Although he did not complete it, even in its fragmentary state the text conveys a detailed understanding of the hydrology, geology, flora, fauna, and evolution of the basin.[49] The other report was a proposal based on his work on a junta convoked by Viceroy Bucareli acting on a royal cédula from November 12, 1773, which asked for solutions for the multiple problems in Mexican mining and the organization of a "formal and unified" body to oversee this vital industry "in imitation of the consulados"—that is, as an autonomous, self-regulating corporation with its own ordinances.[50] Besides relevant government officials, this junta brought together delegates from the main mining districts.

The resulting report, the *Representación que a nombre de la minería de esta Nueva España hacen al Rey nuestro Señor los Apoderados de ella, D. Juan Lucas de Lassaga y D. Joaquín Velázquez de León*, listed the lack of a central body for the industry, scarce capital, and inadequate and unscientific methods among miners as mining's main liabilities. Among other things, the report rejected Gamboa's proposal that the industry be governed by a company led by the consulado, recommending instead the creation of a governing mining tribunal controlled by miners themselves and endowed with its own bank. It also called for the establishment of a school of mines with a four-year theoretical and experimental program covering mathematics, physics, chemistry, mineralogy, and metallurgy to be followed with a two-year residency at a mine.[51] After reviewing this report, the crown supported its main points, hoping that the mining school would soon produce "expert, well-versed individuals with practical knowledge of Geometry and subterranean and Hydraulic Architecture, and also of Machinery and the aspects of the arts of Carpentry, Blacksmithing and Building that are utilized in Mining."[52]

The plan to reform mining had powerful opponents in New Spain, and although the tribunal was created in 1777 with Velázquez de León as director, the consulado and its tenacious allies torpedoed its functions once the merchants failed to gain control of its finances.[53] As a result of this and other factors, the reformers only managed to fulfill two tasks. They created a mining code based on Lassaga's and Velázquez de León's *Representación* (the *Reales Ordenanzas para la Dirección y Gobierno del Importante Cuerpo de Minería de Nueva España y de su Real Tribunal General* of 1783), which became the basis for modern mining regulation throughout the Americas, and founded the Seminario de Minería, the continent's first specialized scientific engineering school.

The human and instrumental resources that would be brought together in the Seminario became available long before Mier's Desagüe projects of the 1790s got under way. By September 1788—that is, as the consulado was "finishing" the open trench—Fausto de Elhuyar, a Basque mineralogist trained at the Freiberg School of Mines and named by the king to serve as the first director general of mining in Mexico, had arrived with a first cohort of scientists recruited in Germany: Fredrick Sonneschmidt, Francisco Fischer, and Ignaz Lindner von Lindenthal. These men strove to improve mining practices by implementing what they believed to be a new and better silver-refining method by amalgamation.[54] Elhuyar counted on them to found the Seminario on the reputed German model.[55] Even this the consulado tried to torpedo, alleging that the utility of the mining school was not demonstrated and then delaying handing over to Elhuyar the documents needed for its foundation.[56] Despite all this, by January 1790 Elhuyar had presented a detailed plan for the curriculum and regime of the Seminario, which opened in 1792.[57] Recruitment of faculty continued in Europe, resulting in the arrival of Andrés Manuel del Río for mineralogy (another Freiberg alumnus who had had continued training in Paris, London, and Vienna), Francisco Antonio Bataller for experimental physics, and Andrés José Rodríguez for mathematics.

Although the performance of the Seminario was halting at first, for lack of sufficient texts, faculty, and other factors, a huge qualitative leap had taken place. This was an institution where teaching and research went hand in hand, and experimentation was the order of the day as the faculty responded to the stimuli of the land. As a result of this work, del Río, for example, discovered "erythronium" (vanadium) from a mineral brought to him from Pachuca and published extensively on the results of his research. Significantly, as a result of his earlier study of European drainage machines, detailed in his *Memoria sobre una máquina con la cual se extraen las aguas de las minas de Shemnitz* (1788), he was able to successfully adapt the Hungarian model to solve the problem of flooding of Pachuca mine shafts.[58] Bataller, in turn, wrote his five-volume *Principios de física matemática y experimental* as a practical work of Newtonian physics designed expressly to address the problems faced in mining in New Spain. It differed from other works in physics by authors in New Spain in that it was written in the vernacular and did not engage in theological or philosophical polemics, despite the fact that its author had a clerical background.[59] An entire volume of this work dealt with hydrostatics and hydrodynamics, including chapters of clear relevance to the Desagüe: "On the movement and velocity of fluids upon exiting the vessels in which they are," "On the collision or percussion of fluids and the resistance of intermediates," and "On the application of the collision or force of water to the movement of machines."[60]

In other words, during Mier's superintendancy there was, for the first time, an institution and a cohort of scientists with considerable capabilities in the very sciences relevant to the Desagüe. Given these capabilities and the Seminario's responsiveness to concrete problems, unmatched anywhere in the continent at the time, it would be reasonable to find the Seminario and its professors in the Desagüe.[61] But they were not. Why?

The most obvious explanation is timing. The Seminario was still in its formative years during Mier's project, with many professors reporting an inability to fully impart their lessons due to lack of equipment and texts, inadequate student preparation, or inappropriate location in temporary buildings.[62] Still, even if the school was still in the process of acquiring its laboratory space, equipment, and funding for experimentation, the faculty was as available for consultation as Guadalajara and Antonio Velázquez were at the Academia, while the requirement that students work on actual problems in the industry could easily be made to include the Desagüe. Moreover, as Mier was aware, his plans to drain Lakes Zumpango and San Cristóbal with a tunnel depended on inputs from mining, not just for the process of drilling and reinforcing the shafts and gallery, but for the ever-present problem of draining away water seeping into the excavations.

What is probably a better, if still partial, answer for the lack of connection between the Seminario and the Desagüe is the unfortunate alignment of private local interests that was the context for both the Desagüe and the Seminario. Cosme de Mier y Trespalacios, a colleague of Gamboa on the audiencia, was, alas, no better disposed toward the school than the consulado.[63] As oidor, Mier took potshots at the institution, advocating its weakening by slashing its funds, on the basis that it had too many students and an excessive budget that would quickly dissipate the funds of the miners.[64] Mier's otherwise puzzling abstention from using the resources of the only institution capable of meeting the needs of the public work under his care can be understood better in this light.

Conversely, uninvited to consider the problems of the lacustrine landscape on which it had landed, the Seminario seems to have failed to address on its own initiative the specific problems of the Desagüe or of hydraulic engineering outside of mining through its curriculum or through field work. Bataller's manuscript addressed aspects of physics relevant to problems specific to mining, such as drainage of adits and tunnels and using the properties of fluids as force multipliers in mining equipment (hydrostatics). It did not cover topics such as the hydraulic jump, friction and the shape of canals, the velocity of water in different sections of an inclined canal, and so on that would be pertinent in the design of works to manage the basin's rivers and lakes (hydrodynamics). Perhaps this was due to the fact that the director and the faculty were overwhelmingly foreign, and mostly recent arrivals at that, who lacked experience with major floods in the city and felt no need to address them.

## TOWARD A "REAL SEMINARIO DE HIDRÁULICA"?

It mattered that the Desagüe did not really have an institutional "home" where its problems could be the basis for experimentation, as those in mining were for the Seminario, and where the strands of scientific and technical knowledge it had generated here and there could coalesce. The cross-fertilizing effect of combining, under an institutional umbrella, scientific inquiry and a practical need for fuller understandings of nature by gathering, centralizing, developing, and deploying scientific and technical expertise is well demonstrated by examples from the many shores of early modernity, from the Iberoamerican efforts of the Real Jardín Botánico explored by Daniela Bleichmar to the Venetian Arsenal's "republican" conjoining of realms of expertise as diverse as shipbuilding, forestry, lumbering, and more detailed by Karl Appuhn.[65] In hydraulics, as L. E. Harris has argued, such a function was served by the Dutch polder boards, which built up that republic's accomplishments in the science and technology of water by not only administering waterworks but also by financing study and experimentation.[66] If the Real Seminario de Minería would not serve an analogous function in the context of the Desagüe because of its focus on mining and the alignment of interests against this confluence, what about the creation of an institution dedicated specifically to the science of water—a "Real Seminario de Hidráulica"?

Since the Bourbons were avidly pursuing scientific and technological advancement of the *patria* relative to Spain's competitors, the necessary motivation clearly existed. So did the potential to materialize it specifically in the realm of hydraulics and communications, as demonstrated by the 1774 creation in Spain of a Corps of Engineers of Roads and Canals. But such an institution never came to exist in New Spain, be it under royal or local sponsorship. Lacking an institutional home supporting hydraulic investigation within the Desagüe context, which together with the mines was the most challenging technological project of the epoch, would make it harder for Mexicans to develop an autonomous science of water comparable to that of the Italians, the Dutch, and the French.

To be sure, some at the time were aware that the lack of an institutional home for a science of water was an unfortunate deficiency relative to their own goals. Back in 1777, echoing the points made in the *Representación*, the fiscal of the audiencia José Antonio de Areche, who was the close ally of José de Gálvez, had in fact compared the Desagüe and other hydraulic works in Mexico to those of France and central and northern Italy. Drawing a contrast with the practice in New Spain, he explained how in Rome Pope Alexander VII had called in physicists ("mathematicians") Cassini and Viviani to work on flood control there. Areche stressed that "in many parts of Italy they endure similar worries [as in

Mexico]," and that Cassini was named "water intendant of the Church state," a post he already enjoyed in his homeland of Bologna.

Areche had felt that the real reason for the deficiencies in the Mexican case was the reliance on technicians who were not scientifically trained, in contrast to the practice in European states, which despite having "Architects and Building Masters of far greater abilities" than those in America, rarely resorted to them for "extraordinary endeavors" such as the Desagüe.[67] Areche described the example of European states' useful employment of "great mathematicians" such as Domenica Guglielmini, who directed large hydraulic public works that distributed water and diverted rivers in various parts of Italy, Louis XIV's Jean-Felix Picard, who was mobilized to survey and design Versailles' water supply, and Philippe de La Hire, made "intendant of waters, bridges, roads, and causeways."[68]

For Areche, the only remedy for Mexico in the long run was to establish a scientific educational institution and to create a directorship of hydraulic works that would enjoy exclusive jurisdiction over technological matters, including the critical original act of conceptualization, depriving contractor or civil administrators of any say whatsoever in this. "Henceforth," he stated, "the interests of the state" required that the study of mathematics be institutionalized and promoted, and that Desagüe directors be "truly versed in mechanical and hydraulic geometry." Ideally too, they should "have knowledge of the country and of the metallic mining works which are the most analogous to those which occur in the Desagüe." It would be this director's job, said Areche, "to conceptualize, give form to, and present the plan for the works, and then direct its approved execution." He probably had Joaquín Velázquez de León in mind for the directorship.[69] But nothing happened: the consulado was still enjoying its open trench contract, the oidores were still superintendants of everything that lay outside this contract, and the Academia and then the Seminario responded feebly—only when asked and only in practical terms—to the Desagüe.

As the eighteenth century drew to a close, urban elites' romance with the Academia's most prized human resources faded, and the academia grew ever-more distant from the Desagüe. At the same time, the Seminario remained as unconcerned as ever, and military engineers grew busier with more urgent imperial tasks. By 1801, it was clear that the academicians and military engineers were acting as episodic consultants at best.[70] If an opportunity to found a Real Seminario dedicated to hydraulics—a key field given its centrality to various transport and productive activities—ever existed, it was now extinguished.

The ebbing of the reformist tide had direct consequences for the Desagüe. As the professors receded, the men who had been in the Desagüe all along became more important than ever. This included the master architects, of course, but also the often unnamed men who actually ran the works on a daily basis.

The career of creole master architect Ignacio Castera exemplifies the eventual triumph of the local status quo over imported scientific expertise. Castera had been named maestro mayor of the Desagüe in 1781, with a salary of two hundred pesos a year. When the Real Academia de San Carlos was founded in 1785, all guild-trained masters were required to recertify through the institution by presenting drawings for a major public building. The academicians saw all *académicos de mérito*, as these recertified men were called, as inherently inferior to those trained in the institution and required that they remedy their purported deficiency by attending class as often as possible.[71] Although Castera passed his recertification on the basis of his established career as an architect before the founding of the institution, military engineer Miguel de Costanzó—then the Academia's professor of mathematics—believed him to be utterly inept, suspecting Castera of paying a better draughtsman to do his plans for the recertification exam.[72]

The bias Castera faced was in many ways a mirror of the tension between the peninsular "agents" of the Bourbons (including the Seminario professors) and the creole elites over the resources of the colony. The tension was manifested in the Desagüe. Whenever the viceroy named commissions of technicians to examine the state of Desagüe structures, it was common for the viceroy and audiencia to name a military engineer or academician and the city its maestro mayor to more loyally represent them and report back to them alone. In the 1788 inspection of the open trench, for example, city officials sent Ignacio Castera to represent them, while the viceroy chose Costanzó.[73]

Castera participated in the 1796 commission to assess the feasibility of a tunnel drainage for Lake Zumpango. To his credit, he was completely against the idea, but Mier basically ignored him and replaced him with the academicians. When Guadalajara's tunnel failed, Castera felt vindicated. Again, when in 1801 he proposed an 18,470-peso budget to repair and clean the canals of Zumpango and San Cristóbal and to build a parapet along sections of the open trench of Huehuetoca to keep the soil out of the trench (see Figure 8.1 in Chapter 8), he was rebuffed by military engineer José Cortés, Antonio Velázquez, and Guadalajara: Castera "deserved the most rigorous disapproval" for inflating the estimate and recommending work that was completely futile given the friability of the terrain.[74] And yet, the master architect endured in his post, drafting further plans and budgets for work in the Desagüe complex. He outlived his competitors in the project long enough to be honored with the role of guide to Viceroy Iturrigaray on his visit to the works in 1805 and to be asked to build a twenty-nine-kilometer-long canal all the way from Huehuetoca to Texcoco in 1808.[75]

As the Academia, the Seminario, and the king's military engineers became less and less active in the Desagüe, the lines of technological tension their presence caused dissipated. Basically, by the turn of the new century there was no one to fight with, and the men who had been there all along, like Castera and the on-site guardas, came to the fore. In other words, what continuity of knowledge and practice existed in the Desagüe was provided not by the Bourbon's scientists and academicians, but by creole master architects, on-site supervisors, and the Indians who maintained sections of works. And, of course, lawyers and letrados. This may not have led to the creation of a Mexican science of water, but it did ensure the continuity of forms and contents of knowledge about the hydrology of the basin and how to manage it that could very well have been pushed closer to oblivion had the Bourbons founded a "Real Seminario de Hidráulica."

## A "NUEVO ORDEN ECONÓMICO"

What did change under the Bourbons was the style of governance in the Desagüe. This was even more the case during the two decades ending in 1810, when the revolutionary upheavals would paralyze the project. The clearest indicator of this change in the style of governance was the manner in which the work was quantified, reported, and analyzed. This was now dramatically different from that reflected in the long, narrative reports of previous epochs, more closely than ever resembling the tabulated forms of exposition in military engineers' reports and plans. At the same time, the actual social arrangements that governed who did the work itself in the Desagüe proper, its district, and its ever-widening jurisdiction remained what they had been since the early seventeenth century. If anything, the chasm between form and substance widened, and the more this happened, the deeper the resentment the new Desagüe impositions generated.

In an effort to systematize Desagüe reporting, in February 1796 Mier ordered the wardens to begin sending him weekly reports on both construction and maintenance tasks. He wanted charts, not narratives: the information about the state of the diverse components of the Desagüe complex was to be conveyed in a systematic and succinct form. A typical report on the cleaning of the bed of the Cuautitlan River, for example, had four columns listing the name of the haciendas, ranchos, or Indian townships responsible for the segment; the number of laborers working that week; the number of linear varas completed, and the number of linear varas remaining in each party's assigned channel segment. The guarda mayor was now also required to compile systematized data on expenses within the Desagüe proper—that is, the part paid for with Desagüe funds. For this, he would create a table with the tasks in rows and the resources expended in columns (divided into the number of overseers, workers, days each category worked on any given task, and money for all wages), and totals tallied bottom and right of the categories. Below this, further information on payments to individuals, such as the smith, was itemized (Figure 7.7).[76]

Reporting in this fashion, tabulating data also permitted its more plastic usage and removed the impressionistic elements of the narrative reports of yore. Using information on these two charts, for example, the superintendant could quickly assess the productivity of labor, or whether the method of distribution of project segments to private parties was efficient. Although we do not know exactly how Mier himself used the information, his hope "that the reports and my replies instruct and create an economic governance" pointed to his drive to impose greater accountability and stricter enforcement of maintenance obligations.[77]

These reports were actually part of Mier's own small-scale emulation of the "economic government" in the similarly titled imperial-level

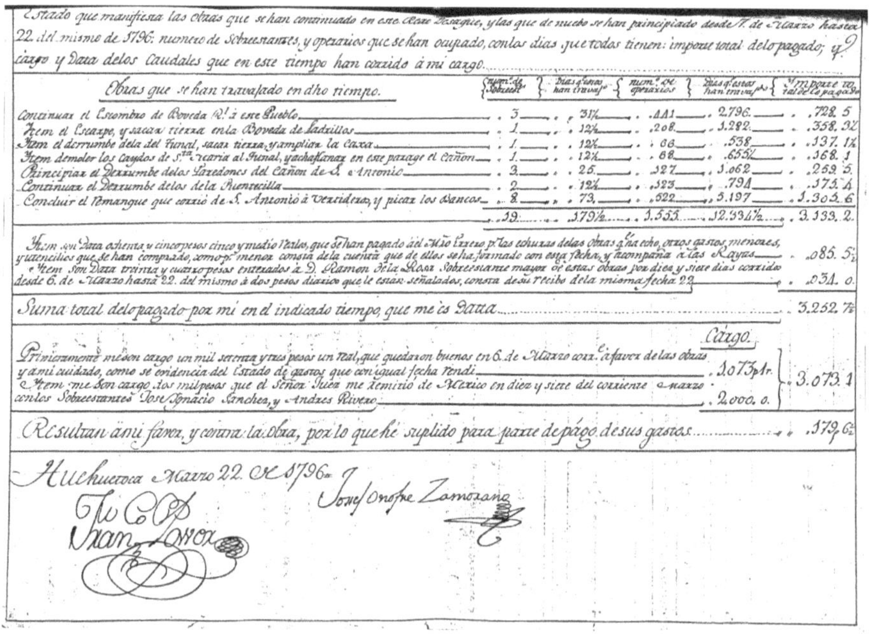

*Figure 7.7.* Guarda mayor Francisco Power collecting and transmitting data from his subordinates in tabular form, according to Mier's *Orden económico.*
SOURCE: AGN, Desagüe, vol. 30, exp. 4. Reprinted by permission.

reform programs of *proyectistas* such as José del Campillo y Cossío, Gaspar Melchor de Jovellanos, Gerónimo de Uztáriz, and Bernardo Ward. In 1798, as he launched his plan for the new canal to drain Lake San Cristóbal, Mier's "Orden Económico de las Obras" included detailed technical and organizational provisions. It instructed the wardens how to draw the register to be followed by the excavation; when and how they should open sluicegates along the canal, and which of its portions they needed to reinforce with struts of mortar and stone. It aspired to an efficient usage of manpower, ordering wardens to provide workers with shacks free of cost; require work in the water only while the sun provided warmth, and end excessive physical violence ("only" whips were allowed to prod laborers along). A special mounted guarda was to prevent drunkenness among the crews by keeping the pulque vendors away. The sick would be tended to and the dead buried. Indians from any jurisdiction who appeared to suffer from a contagious disease, such as smallpox, which was then in epidemic, would be dismissed together with their entire tanda, and replaced with workers from a different jurisdiction altogether. The instructions governed everything, in other words, even bodily functions: when they went to defecate, Indians were to be watched, lest they run away.[78]

Mier's orden económico suffered the same fate as the proyectistas' own programs, afflicted by a "wealth of analysis and poverty of execution."[79] That is, the systematically collected, presented, and interpreted data moving among guardas and superintendants would produce little change in the aims, the scientific implications, and, above all, the social arrangements of the Desagüe. Nothing illustrates this more than the fact that the enlightened reformist mood was simultaneous with the intensification of repartimiento as a method of labor recruitment and organization in the Desagüe from 1762 to 1789 and from 1795 on.

Although the indigenous population had grown along with the demands for labor and the trough of depopulation lay in the deep past, the impact of the labor drafts was in some regards much as it had been during the early decades of the project—that is, detrimental to indigenous communities and their forms of social organization. Disease among populations subject to the Desagüe, for example, reverberated into late in the colony, working its way through the northwest quadrant of the basin in the late 1790s. Smallpox was mentioned, which coincides with Florescano's reports of the disease for 1793 and 1797 in the basin at large.[80] Most of the nineteen workers sent to Mier from two hamlets in the San Cristóbal Ecatepec jurisdiction had been ill, for instance.[81] Undoubtedly, in requiring crucial human and material resources from susceptible communities, the Desagüe under the orden económico regime aggravated the conditions that invited and surrounded disease epidemics, in addition to being impacted by them.

## BECOMING UPPITY

As in the early period, indigenes did what they could to escape the Desagüe draft. Petitions continued to be the path of least resistance to secure a release from Desagüe duty. But now the context was different: because the indigenous population had stabilized and even increased, no official—or letrado, as we have seen—expressed concerns such as those aired in the early seventeenth century about the Indians' probable extinction as a result of Desagüe labor (see Table 3). In addition, the more rigid Bourbon regime had narrowed the room for negotiation. So in 1798, for instance, when the Indian governor of eight townships in Cempoala used the same arguments that Indians from Papalotipac and Tlacuylultepeque had deployed successfully in 1631—that they were from far away in tierra caliente and would be going to tierra fría at great harm to their health—he did not get far. This despite the fact that the Cempoalans were fourteen leagues away from Huehuetoca, where the two hundred laborers the Desagüe demanded would "be in a strange country" without food because their mothers or wives would not be there to cook for them, and the district's nine hundred tributarios were already overworked with agriculture, church repair, and other obligations.[82] Mier stood his ground

*Table 3.* Tributary counts and rates of population growth for the basin of Mexico and Cuautitlan province, 1720–1800.

| Year | Tributaries basin of Mexico | % growth basin of Mexico | Tributaries Cuautitlan province | % growth Cuautitlan province |
|---|---|---|---|---|
| 1570 | | | 10,600 | |
| 1643 | | | 1,193 | |
| 1688 | | | 1,866 | |
| 1720 | 34,179 | | 3,377 | |
| 1725 | 35,552 | 0.8 | 3,860 | 2.9 |
| 1730 | 38,962 | 1.9 | 4,000 | 0.7 |
| 1745 | 33,548 | -0.9 | 3,054 | -1.6 |
| 1750 | 37,226 | 2.1 | 3,151 | 0.6 |
| 1765 | 32,757 | -0.8 | 2,725 | 0.9 |
| 1775 | 39,261 | 2. | 3,094 | 1.3 |
| 1780 | 40,909 | 0.8 | 3,091 | 0 |
| 1785 | 39,532 | -0.7 | 2,898 | -1.3 |
| 1790 | 40,686 | 0.6 | 3,479 | 4.0 |
| 1800 | 48,756 | 2.2 | 3,978 | 1.4 |

SOURCES: Arij Ouweneel, "Análisis exploratorio de la serie tributarios del Anahuac," in *Ciclos interrumpidos. Ensayos sobre historia rural mexicana, siglos XVIII–XIX* (Zinacantepec, Mexico: El Colegio Mexiquense, 1998), 209–60, tables 14 and 16; and also Peter Gerhard, *Geografía histórica de la Nueva España* (Mexico City: UNAM, 1986), 130–31.

because he needed the Cempoalans to replace Xilotepec workers who had been quarantined because of contagious fevers.[83]

From 1796 on, townships near the Desagüe were under greater pressure from the Desagüe than those from distant jurisdictions like Cempoala as a result of the double effect of the drainage's direct and indirect appropriation of labor. They used petitions as a tactic to resist Mier. In 1803, for instance, Indians from five townships in the jurisdiction of Cuautitlan asked to be exempt because they were already overtaxed and could not bear it. Being neighbors of the drainage, they not only had to contribute labor as maintenance duties in other structures of the Desagüe, but they also had to send people to the saltpeter works and provide unpaid road duties on the Camino Real, which passed through their town on its way to the mines in the north, including serving as porters to passing troops and the king's post, or leading prisoners to court. In addition, the overseers were so brutal, they claimed, that they refused to excuse them "even for those natural bodily functions."[84]

The less flexible Mier became with his drafts, the more villagers were pushed beyond petitioning. Two years into the new canal projects, for example, a district magistrate (subdelegado) had to reluctantly chase after Indian workers from Santa María Tecagete, again in the jurisdiction of Cempoala. Mier had ordered him to act "with all necessary rigor" to ensure the delivery of workers, so he tied them up and sent them to the works practically as convicts.[85] But rigor did not always work. The Indian Lorenzo Baca and other villagers who were supposed to join this gang were playing cards as the subdelegado arrived. When the magistrate ordered them to "get up in the name of the King," Baca darted for the door armed with a knife and a stick, shouting at the officials to "get out of the way, *carajo*," because he would "obey neither King nor Alcalde."[86] Baca and two others managed to flee. The women of their households promptly followed them, so that for all intents and purposes what started as resistance became flight.

Flight began to take a collective form when Indians fled as gangs while they were being led to the Desagüe.[87] By the early nineteenth century, Indians would become "uppity" (*insolentados*) and even revolted against subdelegados who tried to take people away. In response, magistrates tried responded by pouncing on the men as they left Sunday mass, tying their hands and forcefully taking them to the Desagüe.[88] But some subdelegados preferred to ignore viceregal orders to collect their area's Indians for Desagüe duty over being forced to react to the refusal of Indian governors to yield their men.[89] This late colonial local form of "obedezco pero no cumplo" is understandable, since distant orders could upset the delicate local balances that wise district magistrates chose to respect.

Although they did not work in the Desagüe themselves, indigenous authorities also bore the pressures of the public work, and this had as much of a ripple effect on their communities as the actions of commoners. Indian gobernadores and alcaldes were responsible for selecting the individuals in their townships going to the Desagüe as part of repartimiento summons. Measures from the mid-seventeenth century imposing penalties of imprisonment and fines of up to two hundred pesos upon governors who failed in this task were still in effect.[90] Every year, when superintendants inspected structures, such as the Cuautitlan River diversion dam, that were under the care of Indian townships, ranchos, and haciendas, governors and alcaldes would be in attendance along with the mayordomos of haciendas to answer questions and to receive maintenance orders and any specific instructions about how to carry out their tasks. They were then responsible for coordinating the execution of these orders.

The role of gobernadores of townships in dealings with the Desagüe thus paralleled what was happening to Indian gobernadores elsewhere as intermediaries between the two *repúblicas*: success in contesting Hispanic impositions was key to both the township's and his own welfare. This is well illustrated in the Cempoala case, where the Indian notables who drafted the petitions for reprieve angled for favorable responses by warning that "not even the Governors could survive in their posts because the task of assembling the tributarios [for Desagüe duty] would be insufferable."[91] Over time, these exertions probably contributed to the weakening of the legitimacy of Indian governors within their jurisdictions that was already under way.

Different forms of resistance had different effects. While collective acts such as petitions, protests, and riots tended to affirm the township as an entity, flight did not: by the eighteenth century, once they left for the mines, haciendas, or towns, people tended not to come back. In the long run, unless compensated by other factors, this could do nothing but erode the viability of the Indian townships. Paradoxically, ending the reliance on repartimiento could very well have "weakened the fabric of the república de indios" by reducing the importance of the gobernador and village structures in labor procurement.[92] Fundamentally, "public" works undermined Indian villages under both repartimiento and wage-labor conditions because either way this was energy that was not circulating back into the indigenous economy and was thus a net loss, particularly when in the case of the drainage the project was so wholly antagonistic to the manner in which the townships obtained their livelihoods.

The fact that Desagüe labor drafts corroded indigenous townships was clearly manifest to colonists and authorities at all times. But despite this awareness, and despite the fact that indigenous townships were by definition collective and official crown policy was to favor collective interests over private ones, indigenous townships rarely won out in conflicts involving the Desagüe. Quite the contrary: while in cases of conflicts between haciendas as private interests and Indian townships as collective interests the courts often ruled consistently with the official crown policy of favoring the collective, when it came to weighing the interests of the Desagüe against those of Indian townships, the interests of the drainage were nearly always deemed more important. This is because the Desagüe transcended the private-collective divide: although in it were clearly vested the private interests of the urban elites, the oidores who made its decisions were officials charged with defending the collective interests of the crown and the city.

These priorities are evident in decisions and policies directly and indirectly pertaining to the Desagüe. Denials of repartimiento exemptions were common examples of the former, as when in 1796 Texcoco Indians wanted to be exempt from the Desagüe draft because they were already burdened with repairing the dam of Acolman and cleaning the irrigation canal in their district on the east side of the basin. Superintendant Mier denied their petition, claiming that the Acolman dam served only that district and not the capital, which "in no way benefited from this canal being clean." The Texcocans, he concluded, should be forced to send their people to the Desagüe just like everybody else.[93]

Far subtler manifestations of the absolute superiority of Desagüe interests appear in the record from the 1720s on. The audiencia, through the Desagüe superintendants and even guardas mayores, began interfering in the annual election of gobernadores and other officials of the república de indios. All these officials started to manipulate the success of candidates willing and able to assure Desagüe duties, particularly in the township of Teoloyuca, being as it was that with the single largest tract of maintenance duties on the Cuautitlan River diversion dam (see Table 2 in Chapter 5). In 1730, for instance, upon a report from the guarda mayor that the Teoloyucans had elected a "plotting and restless" Indian as gobernador, the superintendant confidentially asked the attorney general of the audiencia to nullify the election and instate a candidate known to be dutiful toward maintenance duties.[94] The audiencia official obliged. Similar interventions where superintendants prevailed to ensure gobernadores in Teoloyuca who would be "capable of issuing the appropriate orders for the care of the Desagüe's River" took place in 1732, 1734, and 1768.[95]

## *PLUS ÇA CHANGE . . .*

The Desagüe's substantial reliance on indigenes thus included but transcended the issue of repartimiento labor. In fact, had the revolution and the collapse of Spanish rule not intervened in the Desagüe's history, the coerced indigenous labor drafts could very well have become extinct in the works, as that seems to have been the dynamic as the new century began. Despite his refusal to give it up altogether, in 1804 Mier came to the conclusion that repartimiento labor was a problem because it was both untrainable and less subject to discipline or control than he thought ideal. He thus set out to modernize this aspect of the Desagüe, or so he seems to have thought he was doing. He combined a new plan for labor recruitment and training with a reasoned proposal for the transformation of the projected extension of the Desagüe to Lake Texcoco into an economically dynamic and forward-looking project: a navigable canal. To meet the labor demands of this endeavor without repartimiento, he proposed a sui generis factory town with a stable, resident, well-fed, and trainable workforce made up of convicts. Everyone would gain not just because these workers would be unpaid, and therefore less of a burden on the treasury, but also because the tight and exclusive control imposed by their residing in a new jail to be built on Desagüe premises or nearby. Mier thought convict labor would improve work in the Desagüe, as it would be "on-going in that the Inmates are escorted and subjected to severe penalties if they desert, whereas the Indians of the jurisdiction suffer no punishment if they flee."[96]

Updated with concerns for efficiency, the human condition, and "public utility," which were hallmarks of the epoch, yet harking back to coercion as a way of recruiting and keeping labor, the project was promoted as a vast improvement on the past. Mier and his contemporaries thought this was better than repartimiento labor because the workers' inability to flee and long sentences created the potential to continually improve their skills. Conversely, the expense incurred in training Indians would not be lost at the end of each tanda.[97]

The Desagüe would benefit also because repartimiento labor had the added problem that it was still polyglot, which not only delayed the communication of instructions but also raised costs, as more foremen than necessary had to be hired on the basis of their knowledge of languages. Indians would be spared hardships and unnecessary deaths. The convicts would learn, exercise, eat well, be given access to hygiene for their good health, and receive the consolation of family visits. A yearly budget for installations for four hundred convicts was estimated at 17,270 pesos, which represented a savings of 3,170 pesos with respect to the two hundred men employed on average at the Desagüe, as Mier explained in another of his characteristic charts. The plan went ahead, with Romero de Terreros donating the independent second patio of his Hacienda de Xalpa for the presidio, and Castera drawing up the plans and the 4,750-peso remodeling estimate.[98] Although the navigable canal did not materialize, the convict labor plan did, since later maps show the jail (on lands of the

Hacienda de Xalpa) and Orozco y Berra used its ruins for the topographical measurements in his 1864 *Memoria para la carta hidrográfica.*

Mier thus introduced the form of the gobierno económico in his standardized and systematized reporting system and work rules, but vitiated its substance by remaining loyal to repartimiento and to the old conceptualization. Thus, much as had happened in the realm of scientific culture and technological communications, imitation went only to a point. In absorbing many of the executive traits the intendants were supposed to enjoy, the last enlightened oidor superintendants did much to infuse the reformist spirit of their age into the drainage administration. What they did not do, however, is change anything in the purpose of the Desagüe or its nonexistent link to productive activity.

When it came to any proposal that threatened either the coerced character of the labor system or the privatization of a large part of the costs of Desagüe maintenance, the economic order had an answer succinct and clear: "no." We already saw how the calls of the military engineers for more capital investment or more efficient recruitment and usage of labor were dashed against the established arrangements in the Desagüe. At the beginning of the new century, Mier demonstrated that the officers of the high court were fully aware of these contradictions, but were also too entangled in the social relations that brought them about to be able to solve them. As reformists, the oidores were unwilling to undo arrangements that shifted the costs of saving the wealth of the elites onto the backs of someone else, and that is why new technologies and new ways of organization in the Desagüe were few and far between until long after their rule. While seriously interested in "modernizing" and infusing enlightened culture into the Desagüe, they could obviously not be revolutionaries. Neither could most of their fellow statesmen and letrados after the oidores were long gone. The Desagüe would remain the guarantor of propertied and rentier priorities in the city even as it became a vehicle for the transformation of the social meaning of water and land as fixed and segregated entities, capable of being quantified and bought and sold as any other commodity and subject to speculation and accumulation.

Over its long history the Desagüe embodies the characteristics and legacy of Spanish rule. The Habsburgs had created a regime ruled by landed aristocrats, mercantile capitalists, and a vast lettered bureaucracy that not only did not require the extinction but even relied on the persistence of the peasantry. This basic reality continued with the Bourbons by inertia—there was no social sector sufficiently powerful to dislodge it, much as the enlightened advisors of the dynasty tried to will it into preponderance. There were, of course, huge areas within the Spanish empire where mining and estate elites were fast becoming one of the few "recognizably capitalist societies" in the world.[99] In the Desagüe—as in much of the indigenous spine of the New World—the consulado, Mier, and the officials of the independent republic might insufflate as much discursive modernity as they wished, and might even manage to change aspects of the material realities, but the indigenous peasantry was there to stay and, for the moment being, help make predatory capitalist transformations the stuff of dreams.

CHAPTER EIGHT

# Deep Colonizing

In 1804, the intellectually influential Prussian savant Alexander von Humboldt visited New Spain, inspected the Desagüe, and surveyed much of the drainage's documentation. While impressed by the Desagüe's engineering daring, Humboldt was critical of the many imperfections of the project as he found it, barely fifteen years after the completion of the open trench. It was overdimensioned and ill executed, he regretted. But he reserved his most rueful tones for the conceptualization he believed had spawned it—the manner in which Mexicans in the capital valued water and land, which in his view did not stress productivity.

When the audiencia's fiscal del crimen, Francisco Robledo, took over the superintendancy of the Desagüe in 1807, Humboldt's unfavorable appraisal still applied. The trench, in the imperfect state the consulado had left it, had suffered the effects of time, with chunks of tepetate crumbling in and obstructing the water's path.[1] Mier's canals lay uselessly silted up (see Figure 8.1); Castera's general desiccation ones were stalled. The Desagüe as a whole was still far from fulfilling its mission even as discontent rumbled ever more loudly through the countryside, threatening its prospects for completion the foreseeable future.[2] The dream of ordered dry land among the City of Mexico's well-to-do was about to turn into a nightmare of blood-soaked revolt and repression—and more flooding.

The Desagüe suffered the effects of rebellion and war from 1808 to the end of Spanish rule in 1821. Particularly detrimental was the entry of Hidalgo's rebels into the area in 1811 and the aftermath. As a strategic point on the royal road communicating the capital with the north and as a prime supplier of foods to the city, Cuautitlan and its environs were key to both the royalist defenders of the capital and to the insurgents. Having reached the township of Huehuetoca on April 4, Hidalgo's followers gathered adherents among local populations, sacking the Hacienda de Xalpa.[3] This alone would have weakened the Desagüe, since its administrators had developed a symbiotic relationship with the proprietors, the Condes

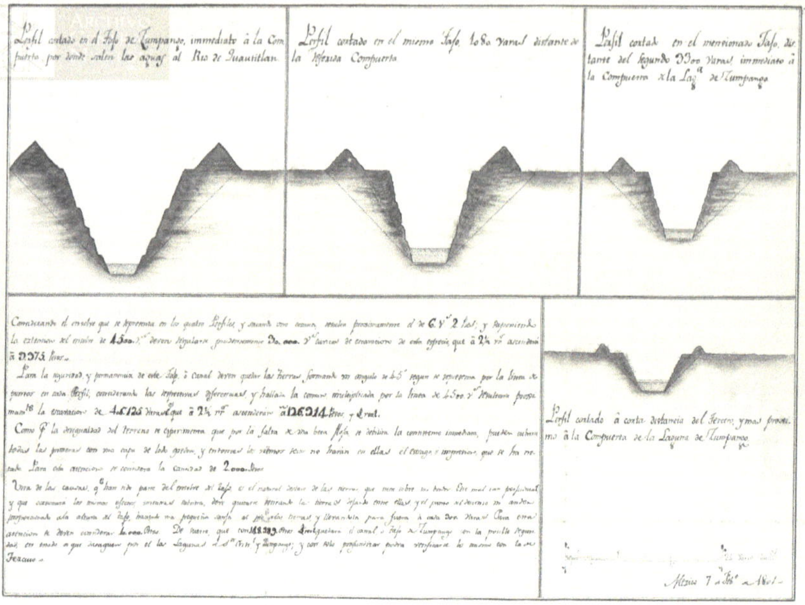

*Figure 8.1.* Four sectional cuts of the Canal de Guadalupe five years after completion, showing original slope inferior to angle of repose of the soil (450) and accumulated silting: the canal was useless.
SOURCE: AGN, Desagüe, vol. 35, exp. 1, f. 18. Reprinted by permission.

de Regla. But the fact that both rebels and royalists literally devoured their way through local resources and absorbed the energies of local men brought all work on the project to a halt, since it depended heavily on them. The rebels' retreat north brought no respite, as the royal troops pursuing them were billeted in the Casa del Desagüe throughout 1812, laying it and the tools it housed to waste.[4]

One might read the events leading up to the revolution—including those involving the Desagüe—as part of the epoch-changing struggle between colonies and metropole. While quite frequent, this way of reading events overlooks something important: the most fundamental dynamic in which the Desagüe was enmeshed was not that of colonizing state versus colonized territory, but of colonizing classes versus colonized classes. Colonization understood in this manner had begun with the conquest (and with the construction of the Desagüe) and it would continue long after the revolution.

This, of course, is something that for all his insights Humboldt could not see. Humboldt's diagnosis of the Desagüe and its ills, despite being accurate in some respects (perhaps *because* it was accurate in some respects), would provide the conceptual scaffolding upon which later scholars would construct a variety of misconceptions about the Desagüe specifically and colonialism and "development" in Latin America more generally. These misconceptions—that the drainage was universally beneficial, flooding universally problematic, and liberal progress universally desirable—have persisted to this day, skewing the thinking of historians in two ways. First, picking up on Humboldt's critique of the design and execution of the project and usually unaware that they were viewing the Desagüe through the lens of urban elites and technicians, authors limited their technical assessments to discussing whether and why the drainage project worked or not. Second, because they identified with Humboldt's historical analysis, writers did not question the basis for it, even when critical of the project's costs.

Now that the foregoing chapters have provided the tools with which to do it, it is time to make a balance sheet for the Desagüe, dismantling the scaffolding that writers have built upon Humboldt's foundations and examining more keenly what lay behind the misconceptions it has supported. The point of this exercise is not only to better understand the dynamics unfolding in Mexico around the Desagüe but also to submit the prevalent understanding of an Atlantic realm divided into "nations" and "colonies" to a similar interrogation. If this division, too, is understood historically, as a paradigm created by particular writers at a particular point in time in an effort to explain, legitimate, or advocate for historically specific *outcomes*, might removing it not reveal a far more contingent Atlantic early modernity?

## MOCTEZUMA'S REVENGE

Paradoxically, Humboldt's own analysis enables us to begin dispelling the misconceptions standing in the way of understanding the relationship of the Desagüe to colonization and of using desiccation and drainage works elsewhere as sites to reconsider colonization's centrality within the Europe itself. His analysis of the Desagüe had two components: one equivalent to what this book has called design, which determined its hydraulic functioning; the other, to conceptualization. A hydraulic evaluation of the drainage in the context of what is now known is necessary to determine the Desagüe's full material implications. With this in place, the historical implications of how the project was conceptualized, and the analytic benefits of abandoning the Humboldtian groove, will be clearer.

To Humboldt, the Desagüe had been a truly heroic piece of engineering. Although its execution had been flawed, Martínez's tunnel design had been brilliant, visionary even. The open trench idea, in contrast, was a hydraulic design and execution disaster in every possible way. Misshapen in width, slope of sides and bed, and height, the puzzling lack of calculations of the amount of water that it needed to evacuate had been compensated for with absurd gigantism. At their peak, the waters it was supposed to evacuate could wet a maximum cross-section of twenty meters square, but the actual cross section over much of the trench hovered between eighteen hundred and three thousand meters square.[5] To him, such overdimensioning was fatal: the Desagüe simply did not work.

In reality, despite its defects, the colonial Desagüe was not entirely unsuccessful from a hydraulic point of view. It did contribute to the ongoing process of desiccation of the lakes. However, the fact remains that even as they were increasingly disappearing, the lakes were still flooding the city from time to time. Urban flooding persisted despite the Desagüe in part because its causes multiplied and became more complex. Water from the southern lakes and rivers, ground compaction, loss of ground moisture, and the natural and landowner-promoted silting of the wetlands increasingly competed with the maligned northern waters as the city's enemies.

In any given flood, much of the water came from sources other than the detested Cuautitlan River. In fact, although this river and the Desagüe's failure had caused the catastrophic flood of 1629, it had been clear since then that the Desagüe could not protect the capital against "waters it received from the South, East and West."[6] In addition to this, one of the new and important causes of postconquest flooding was the expansion of the built area. Over the colony's first century, the urbanized area doubled from 2.7 to 5.4 square kilometers, reaching 6.6 square kilometers by the end of the seventeenth century. Under the weight of the buildings, the clayey ground underneath compressed, effectively lowering the city dangerously closer to lake level. As early as the 1630s, Andrés de San Miguel and others had noticed the compaction of the ground, but attributed it solely to the weight of city buildings and even of the water itself.[7] They were right about sinking, of course, but not quite about its full causes.

According to geophysicist Marcos Mazari Menzer, some of the sinking was due to the fact that away from the twelve-meter-thick platform built by Tenochtitlan's engineers in the core area of the city occupied by the viceregal palace, the cathedral, and the archbishop's palace, the underlying clay layer was not nearly as well compressed or preconsolidated, and the thickness of the in-fill varied considerably. The thickness of the in-fill decreased significantly to the north of this core area and rapidly to its south, so that outlying areas were more vulnerable.[8] Sinking over time has followed this pattern of uneven preconsolidation, sometimes visible in a single structure, as was the case with the viceregal palace—now the Palacio Nacional—which tended to sag in its northeast corner where the in-fill underneath was less thick. This unevenness resulted from the postconquest relocation of seventy-five thousand cubic meters of rock from the above-ground portion of the Templo Mayor to thicken the foundations of the palace and the cathedral.[9] Seventeenth-century sinking was of course only the beginning, from the modern-day point of view, but it was clearly a visible phenomenon at the time.

Another problem, paradoxically, was the loss of moisture content in the soil, although it is hard to determine how much of this was happening in the colonial era as a result of the gradual diminution of the area covered by lakes. Until the nineteenth century, the only one to connect the Desagüe, flooding, water-table levels, and the sinking of the city, and to advocate for preserving rather than evacuating the lakes, was the vituperated Adrian Boot. In 1620, when he reiterated before the cabildo his 1615 project to manage the basin's hydrology in the manner of the Low Countries, Boot called for keeping as much water as necessary in the lake system, "because removing the water entirely would be to harm said city as the earth is flabby, the buildings heavy, and it will not be able to sustain the foundations without the aid of water."[10] In this, too, he would prove prescient: this is exactly the diagnosis of modern specialists, from sewer engineer Roberto Gayol in 1925 to soil mechanics expert Raúl J. Marsal and his disciples, some of whom have advocated recharging the aquifer and others even restoring the lakes, motivated by a 1965 proposal by Nabor Carrillo.[11]

The loss of moisture, the uneven artificial in-fill, and the way both acted on the underlying clayey layer boded ill for all living and building in the city over the colonial period. As can be seen nowadays, after more than a century of accelerated depletion of the water table due to pumping and the loss of infiltration surface as cement and asphalt have covered the ground, the city has not sunk evenly; as a result, buildings have tilted or cracked, and the direction of gravitational flow in canals has changed. As Mazari humorously put it, *this* is the real revenge of Moctezuma, and not the digestive commotion endured by foreigners who drink Mexico's tap water.

This differential sinking may have been at play when, between 1781 and 1795, during the flurry of Bourbon urban reforms, military engineers Nicolás de Lafora and Miguel de Costanzó headed the project to replace the old clay tubing and open canals with paved streets and covered sewage canals (*atarjeas*) under the sidewalks. The Tenochca king's revenge clearly fell on their efforts too, which were repaid with flooding in the city streets and the derision of creole savants and technicians. Master architects, whom royal policy had been increasingly marginalizing vis-à-vis military engineers and academic architects, gloated that against their advice, Costanzó in particular had erred in choosing the direction of the slope of the streets and canals, which instead of draining eastward toward the sluice at the albarradón de San Lázaro so they could be emptied into an increasingly fetid Lake Texcoco were now draining westward back into the traza.[12] Both Cosme de Mier y Trespalacios and the then-aging José Antonio de Alzate y Ramírez felt that it had been a mistake to replace the old system of canals because not only were these water conduits more capacious, but also streets with which they were associated were slanted toward the middle to allow water to drain into their central gutters.[13] In this "disastrous enterprise," according to Alzate, the maestros were no less guilty than the engineers for eagerly going along with it for their own contracting profit. "One would need a miracle," he said, for the constricted new covered sewage conduits to carry both the liquid wastes and the seasonal downpours that even the more capacious old canals could not handle. "As soon as the Maestros and their collaborator the Engineer demonstrate that this is possible," proclaimed Alzate:

> I will spend my days burning what book has been printed on the Mathematical Sciences, and I will shout from the Squares and the Terraces that our recent Architects are greater magicians than those of the Pharaoh because without the help of the Devil (save that which prevails in personal interests) they overcome what the Luciferian charmers of the stupid Pharaoh could not have.[14]

Clearly an infuriatingly poor design was one of the causes of flooded streets. It is telling, however, that the most flooded streets (half of Plateros Street and most of its continuation, San Francisco, which is present-day Madero) seem to have lain well to the west of the contours of the twelve-meter-thick pre-Hispanic fill-in platform. Inside this ancient preconsolidated area, the built area has sunk 6 meters. Outside it, where the thickness of the preconsolidated platform abruptly decreases by half, modern subsidence is between 7.5 and 8 meters.[15] This suggests that the buildings, streets, and canals built outside the area where the thickest platform had already precompressed the clay before the arrival of the Spaniards were subject to ongoing compression over the colonial period. In any event, although given the scientific possibilities of the epoch, nobody could factor the sinking into the gradients of the drainages they built, ground compaction seems to have already been disrupting the anticipated gravity flows of these works, as a precursor of how it currently wreaks havoc in Mexico's Sistema de Drenaje Profundo.[16]

From the 1740s on, for reasons that will be discussed below, the actions of hacendados, rancheros, and even a few Indians further complicated the causes of flooding. Haciendas and ranchos that practiced flood fertilizing (*enlamado*) near the city could create havoc even in the dry season, as after having directed river flows onto their fields to deliver moisture and nutrients to them, they all released the water at once when the time to prepare for planting approached.[17] In the littoral of Lake Texcoco, which was brackish, this was also done to wash salts from the soil and render it arable. The city's north-bound causeway of Guadalupe seems to have been particularly vulnerable to flooding from these practices, since it traversed low-lying ground. It could become impassable, which was a problem for mule-driven provisioning of the city. Its neighbors were an array of landholders, including Indians from Santiago Tlatelolco and San Miguel Nonoalco, who engaged in these practices to create pastures (*potreros*) to rent to the mule-train and livestock drivers plying the causeway.[18]

In addition, throughout the lacustrine littoral, landholders put the city at greater flood risk as they little by little expanded their holdings into the wetlands and lakes. They did this during the dry season, creating ditches, low levees, and raised borders on the margins where the water had evaporated or been absorbed. In each subsequent rainy season, while the city risked flooding as the water thus constrained had nowhere to go but up, these barriers trapped eroded and nutrient-laden materials carried down the slopes by runoff while keeping the lake away, thus effortlessly enlarging landholdings over time and increasing the productivity of crops.[19]

This brings us to siltation. Eighteenth-century Mexicans knew well that sediments carried by runoff raised the lakebeds and that this gradually and dangerously diminished the difference between the level of city streets and the water surface in Lake Mexico and Lake Texcoco.[20] As the maestros noted, as the level of Lake Texcoco on the other side of the albarradón de San Lázaro increased, neither the covered or above-ground drains of the city could discharge their water through its sluicegates because the water on the other side impeded it.[21] Explaining the connections among erosion, siltation, and flooding, Superintendant Rodríguez del Toro observed how "the rains drag to the bottom slime, sand, and dead vegetation that silt up the lake," and this

> has greatly increased after this Kingdom was conquered because with the use of ploughs (of which the Indians had no knowledge since to plant their crops they only dug small holes), the soil is loosened and the rains drag it with greater ease and in larger amounts to the lower parts; and this with unbelievable speed if the hills and slopes are ploughed as has been the case with those that surround this Valley; so that it has been a constant tradition since the beginning of the conquest to prohibit the cultivation of said Hills.

This prohibition had been "invalidated by greed and the connivance of justices." An hacienda administrator showed Rodríguez del Toro a hill that had formerly produced wheat but which at the time had been denuded by the rains of "all substance, leaving it bare to the bone," with the sediment having eroded down the slope and ended up in the lake. The only thing that grew on these higher lands thereafter were some grasses for cattle, but when these dried up they were burnt and the ashes too wound up as silt in the lake.[22] Not everything that bedeviled the basin can be pinned on the Desagüe, of course. It was the entire constellation of things that Hispanic and Hispanicized elites did in the basin (drainage and desiccation, cropping on erosion-prone slopes, grazing livestock and plowing fragile soils, controlling seasonal fluctuations in lake levels, putting the city in a very vulnerable place) that allowed flooding to continue (and other problems to occur). What follows explains how it was this whole constellation of practices bound with their class content that constituted deep colonizing.

This denuded landscape, with its enduringly threatened city at the bottom and the signs of what must have seemed to him wasted opportunities for progress everywhere, cannot have pleased Humboldt. In the *Essai Politique*, he ruefully wrote that in aiming purely at desiccation and not considering irrigation functions for the Desagüe, Mexicans had "killed the germ of fertility" in a large part of the basin, and that in neglecting to give it a navigational use they had failed to do something more grand and useful.

Without mentioning them as such, Humboldt was manifestly critical of the rentier priorities of the city that had chosen desiccation as a means to safeguard its accumulated wealth. He clearly wished they had acted consistently with the notions of utility prevalent among enlightened circles of his epoch, where the function of "public works" was to ensure progress as defined by greater production, greater circulation, and greater mastery over nature. But this did not shake his fundamental agreement with the elites in the City of Mexico's understanding of flooding as a universal problem and their right and drive to master their environs. Humboldt's own conceptualization of the hydraulic situation of the City of Mexico both legitimated the essence of what the elites in the City of Mexico had done—try to protect their capital—and outlined a program for their future.

But not everyone conceptualized the basin in the same way. It is hard to imagine rural populations caring much about the fate of the city, its inhabitants, or the idea of progress. It is equally difficult to picture the ghost of Moctezuma doing anything but dance while watching floods endure, sinking accelerate, and infertility creep up the slopes to plague the heirs of his victors so far into the future.

## CLASS CONFLICT AND COLLABORATION

The indigenous commoners who lived on after Moctezuma, however, had little to celebrate. This is because over the course of its long history all the Desagüe's actors participated in a protracted and contradictory process of colonization in the basin. The public work accompanied this process and often made it possible. This colonization was deep rather than extensive. In the shadow of the works, colonization included, but was not limited to, indigenes losing units of land and irrigating water to Hispanics. Likewise, colonization included, but was much more than, a "colonization of the imaginary," to use Gruzinski's acute phrase; it was much more than an adjustment of psychological and symbolic structures to cope with changed relations of power. This was deep colonization because the Desagüe was transforming the broad, fluid relationships of indigenes with their lacustrine resources and seasonality into more fixed and simplified ones. As it did so, projecting onto the countryside the rentier priorities of the urban mercantile, bureaucratic, and propertied classes, the agrarian landowning classes were busy projecting theirs too, seeking also to fix and simplify water, ecosystems, and human communities into mere inputs of production. The colonizing was thus carried out not by "Spaniards," but by classes, each with specific interests, which happened to be made up largely of Hispanics—or Spaniards, in the colonial legal sense of the term. With some acting through the Desagüe and others independently of it (and even unwittingly against the welfare of their luxurious capital), and deliberately or not, all sectors of the elite were colonizing everything in the hinterland, including the very social definitions of what things were, where they were supposed to be, and how they were to be used.

To unpack the role of the Desagüe in these regional relationships of power over land, water, and living organisms, it is necessary to think of these elements as more than unifunctional elements, inputs in production severed from any ecosystem and distinct from each other. If instead of being treated as inputs, "water" and "land" are understood as having both a use value and an exchange value in different proportions for different classes, then both water and land constitute complete ecosystems replete with extractable biomass and elements (moisture, soil, mechanical energy from the flow of water) that can be used as inputs for the production of goods for either subsistence or the market.[23] Seen in this fashion, the Desagüe becomes a technological object through which all Hispanic elites in the basin colonized the countryside.

Urban elites consciously designed the Desagüe for this purpose, but this does not mean at all that they used the Desagüe to colonize their rural and lacustrine environs by literally occupying it. Rather, their Desagüe allowed them to colonize by subjecting water, land, and life forms in the region to their needs and definitions. Conversely, agrocommercial landowners (also mostly Hispanic) did colonize water, land, and people by literally occupying them. As a class, they had nothing to do with the conscious design of the Desagüe, but they profited from it nonetheless. Nobody in the elites conceptualized the Desagüe as Humboldt would, but the capitalist colonization of nature in the basin crept on nonetheless.

As a whole, this colonization was manifested not only in the visible patterns of land or water acquisition as private holdings, but also in the penetration of private property relations—and a new regime of value over all resources as a whole—right into the heart of indigenous life. Communities of indigenes as well as the wealthier families among them took to these novelties out of need or desire, in some places and at some times more than others. The regime of value inherent to capitalist forms of production that Iberian colonists transmitted to America created a peculiar situation in places with settled indigenous societies, such as the basin of Mexico. Here, while both Hispanic agroproducers and indigenes used water, land, and work as inputs to produce things for the market (the place where these things realized an "exchange value") among indigenes the predominant use for these resources was for the production or extraction of "use values," or things they needed for their own sustenance.

However, in the basin much "land" became "water" during the rainy season even late into the colonial era, creating wetland ecosystems upon whose ebb and flow the indigenous peasant economy had been predicated. The activities of agriculture, hunting, gathering, fishing, and exchange this class carried out all required the ongoing interplay of water and land, so throughout the colonial era communities tried to protect both this interplay and their ability to extract values from it. For the agrocommercial landed class, by contrast, only permanently dry land was suitable for crops, which then received moisture from rain, irrigation, or artificial inundation. Both of these agricultural regimes were affected by the Desagüe—the more it interfered with the hydrology of its district or with the rights of access, the more the Desagüe irked its neighbors.

Water was of course a central element through which the city related to and was in conflict with the hinterland. But, in approving the transformation of the Desagüe into a single open trench back in the 1630s, authorities had turned water into an even more critical element in these relations for the rest of the colonial era because they chose to rely on water to sweep the debris from the channel. Because the most dependable source of water was the Cuautitlan River, this choice had real costs for irrigators along the river, whose access to the water was impinged upon when the Desagüe was given one-third of the entire river's water.

Naturally, in constricting access to the Cuautitlan, the Desagüe curtailed the full usufruct of land in its district. In times of drought, this could be a disaster, as it was during the consulado's tenure, which coincided with prolonged dry conditions in the basin.[24] But there was more. The fifty-vara rule that from 1724 on forbade cultivation along the margins of any river or lake in the Desagüe district also impinged upon this full usufruct. While merely unpleasant for the hacendados, this exclusion from the most fertile soils was painful for indigenous townships already under pressure of increasing populations.[25]

The Desagüe's disturbances of the Cuautitlan River—the diversion of its water and the yearly removal of decomposed plants and sediments from its channel—eliminated or compromised many of the river's former benefits. Because the diverted Cuautitlan River was prevented from inundating its floodplain and thus from delivering moisture and nutrients, groundwater levels, soil moisture content, and biological activity in the soil were all reduced, and this in turn impacted wild and cultivated vegetation.[26] The channeling of the river also greatly diminished the ability of both the river and its adjacent lakes to serve as spawning habitats for fish.[27] By the end of the eighteenth century, therefore, the river provided very little besides water for irrigation and for the Desagüe. Everyone was affected by the reduction of the river's "ecosystem services."[28] But the indigenous peasantry was disproportionately affected by the effects of this biological impoverishment on food webs because of their need to supplement their sustenance with the gathered, hunted, and fished species dependent on these webs. Although all rural producers in the Desagüe district could form common cause against the Desagüe to protect access to the Cuautitlan River's irrigation water, indigenes' dependence on seasonal inundation and water's use values made their specific needs different from those of Hispanic landowners, and that ultimately broke down these alliances. No Hispanic landowner in his right mind would fail to take advantage of the fact that the Desagüe's actions on water and land as ecosystems promoted the colonization of the countryside by people like him.

But the way the Desagüe assisted the colonization of the hinterland went much deeper than this. As I discussed in the Introduction, the crown protected the peasantry's customary rights to watery and dry commons. As the Desagüe evolved, the crown faced the contradictory position of promoting the Desagüe on behalf of its imperial city *and* protecting the land and water endowments of the indigenous peasantry living partly off the lakes that the project sought to desiccate. In accordance with the latter responsibility, viceregal orders had repeatedly protected indigenous communities' access to Lakes Zumpango, Xaltocan, and San Cristóbal (in 1586, 1589, 1594, 1607, and 1667).[29] As the physical structures and regulations of the crown's Desagüe interfered with the ability of rural communities to interact with the aquatic and wetland ecosystems of the northwest quadrant of the basin, however, this stressed their ability to grow and extract foods and resources from the ecosystems, to communicate with each other, and to reinforce intercommunal dependencies through exchange. The Desagüe's intervention in the realities of these lakes had been changing the practical meaning of this protected access, through its rules and the way its structures favored the permanent retreat of the waters. "Access" and customary rights on wetlands and lakebeds no longer meant contiguity to them. Overall, the Desagüe's regulations and changes to hydrology meant that villages became increasingly isolated socially and physically, both from the source of their gathered goods and from each other.

This, in turn, affected regional power relations among Indian and Hispanic landholders and the Desagüe. Increasingly, the interests of the hacendados dovetailed with those of the Desagüe's superintendants, weakening the position of indigenous townships relative to both. This was manifest in two linked processes, the tension over water and land rights and the more subtle one over ecosystemic usage of water and land. A good example of the former took place in 1746, when the Xaltocan Indians, whose chinampas were discussed in Chapter 1, came into conflict with the Jesuit owners of the Hacienda de Santa Lucía over the freshwater of the Ozumbilla spring. The Jesuits had built a ditch to divert this water to their estate and away from the Xaltocanmeca chinampa zone. When the Indians marched to destroy this ditch with "drum, bugle and banner and making much noise," Superintendant Trespalacios, himself a landowner in the area, sided with the Jesuits.[30]

Episodes of conflict of this type were in reality efflorescences of the more persistent ecosystemic tension that was flowing continually underneath, which pitted peasant communities and productive systems, oriented toward subsistence more than the market, and Hispanic ones, oriented largely toward markets. The Xaltocanmecas' concern over the Ozumbilla was tied to a broad set of practices underpinning their communal economy, which required that they manage the water level in their lake in their "industrious and skillful" manner, not just for the chinampas but also for the purposes of sustaining populations of fish that provided them with food.[31] The specific chemical, physical, and biological impact of their manipulations of the lake with the freshwater from Ozumbilla allowed them and their neighbors in Tonanitla to raise fish in their lake.[32] Xaltocan was not the only township vying to pull the regional power balance in favor of indigenous productive systems. The townships of Zumpango and Teoloyuca also defended their customary rights to similar ecosystem services offered by Lake Zumpango while it was full—they still gathered tule and other plants and insect larvae, fished and hunted in it—and Teoloyuca was particularly tenacious in defending its ability to plant on the lakebed when the water receded. In 1771, 1777, 1781, and 1784 it had combined lawsuits against the Desagüe with defiant reoccupations of its lakebed using embankments and ditches. This resistance only increased from 1785 on and during Mier's superintendancy, which so affected this lake.[33]

By the time Mier took over the works, meanwhile, the indigenous population was growing, stressing the limited fundos legales of the townships in the crowded basin. However, the impact of demographic growth can be overestimated when considered in relation to land and water as static entities—as arable or irrigating units—and as simple inputs in agricultural production. What makes the 4 percent growth of the indigenous population in the district of the Desagüe that happened after the harvest and epidemic crisis of 1785 (see Table 3) significant is not the simple addition of mouths to feed on an unchanging fundo legal, but the fact that the relatively modest indigenous population increases coincided with three Desagüe actions. The hardened rules of Trespalacios, the consulado, and Mier that restricted access to and usufruct of water and land, the gradual fixing of what and where these two elements were, and finally the way all this altered the ecology of wetlands, lakes, and soils, *all* militated against the fluid relationship between water and land that sustained indigenous life and made the production of adequate food amounts even more difficult in the face of demographic growth.

All this makes the increasing push-back of indigenes against the Desagüe over the course of the eighteenth century much more than a struggle for more generic "land" and "water" for increased populations. It was a struggle to protect their ability to continue their way of life, which depended on combining the growing of crops in different soils and moisture conditions with the gathering and harvesting of various plants and animals from the surrounding ecosystem and allowed them greater flexibility with regard to how they engaged value regimes. In order to practice this way of life, indigenes had to keep these practices and material conditions broadly within the reach of the techniques and forms of social organization of work they had developed to capture the resources of their surroundings—enough water for floodwater or canal irrigation, intercommunal collaboration for drainage and irrigation maintenance, lake levels suitable for canoe transport, specific plant and animal balances for net and decoy designs to work, survival of and access to good earth-binding grass species, and so on. The correlation between population growth and water and land conflicts in the shadow of the Desagüe should not be considered a mechanical one. The fact that water and soil had both ecosystemic use values and exchange values for communities means that their struggles were multifaceted and dynamic, not merely over acreage or irrigation units. Religion, communal organization and integrity, town finances, cultural identity, and the entire fabric of social life were wrapped up in how lands, water, and the liminal space between them were used to meet food and subsistence needs.

Indigenous hydraulic devices sustained this social life by redirecting water to maximize biological productivity predicated upon seasonal fluctuations. Conversely, created to end these fluctuations, the Desagüe's multiple structures altered the region's hydrology more deeply than those of its indigenous predecessors, impacting the productivity of regional ecosystems in ways that disproportionately affected indigenous communities. The Desagüe's hydrological and ecological impact should thus be seen as part of the (contested, it is true) *longue durée* colonization process in which Hispanic urban and landed elites deepened and extended their control over the territory and its indigenous peasantry.[34]

This colonizing process started to become more intense in the first half of the eighteenth century. In several regions of Mexico, particularly those close to mining centers or the capital, the commercial cultivation of wheat had begun to overtake livestock rearing as the axis of Hispanic rural production, driving hacendados and rancheros to secure both arable land and the water required to irrigate it and triggering increasing conflicts with a recovering indigenous population over both land and water.[35] By the 1770s, haciendas, ranchos, and even some Indian townships on the lacustrine littoral had been responding to these conditions by claiming wetlands as dry ground, and they continued to do so. They did this by building levees or other barriers far into the lakebeds during the dry season. This was done all along the transport canal artery (acequia real) to Mexicalzingo, Chalco, and Xochimilco.[36] It was particularly pervasive around Lake Texcoco, where one of the greatest grain-producing enterprises in the entire basin, the Hacienda de Chapingo (now the Agriculture School), alone had by 1770 "usurped more than half a league, which at present is cultivated, and was formerly known to be part of the Lagoon."[37] This was a cheap way of enlarging holdings, since there was no purchase or grant involved. In fact, entire haciendas were apparently created in this manner. The Hacienda de los Dolores in Mexicalzingo, which yielded "considerable maize harvests," rose entirely out of the former lakebed.[38]

But even though this evident drive on the part of hacendados, rancheros, and some Indian townships and caciques to increase agricultural output was made possible in part by the Desagüe, it remained completely separate from the Desagüe's official purpose. Indeed, it was explicitly treated as antithetical to the Desagüe: both Trespalacios and Rodríguez del Toro ordered the destruction of any structure or ditch preventing the water of any existing lake or wetland from fully extending over their beds. When this proved impracticable, the superintendants prohibited any future retention walls.[39] In other words, while landholding groups were carrying out this de-facto colonization in part thanks to the Desagüe—which although useless against episodic urban flooding was nonetheless contributing to the gradual desiccation of the wetlands and the occupation of the land beneath—the Desagüe's technicians and superintendants continued to conceptualize it as a project to safeguard the city alone because that was the consensus among their urban employers.

## AN UNINTENTIONAL COLONIZATION BY DRAINAGE

It will by now be clear that class perspectives and conflict were multifaceted in the basin of Mexico, and that they penetrated deep into each group's actions upon nature. Although often taken to apply only to the relationship between the oppressed and ruling elites, class conflict had in fact been present within and among all classes, much in the same way it is today. This manifested itself in three related paradoxes in the Desagüe. First, the aristocratic, mercantile, bureaucratic, and lettered elites resident in the capital did not pursue colonization deliberately, and they were unaware of the Desagüe's power as a tool for facilitating the process. Second, no sector of the elite pushed for a reconceptualization of the Desagüe that would exploit its potential to be hitched to economically generative functions, in apparent denial of their own long-term interests. Third, indigenes both resisted and participated in their own colonization and that of the terrain where they lived. The simultaneous existence of these three dynamics is why in this section the term "unintentional" qualifies "colonization" as a process.

There is no evidence that any sector of the elites was deliberately pursuing the colonization of the city's hinterland in the manner described in this chapter nor that they were aware of the way in which the Desagüe assisted this process. This does not mean that nobody speculated about other potential benefits of a general desiccation (as we saw, when Mier resuscitated the long-postponed general drainage idea with his projects for Lakes Zumpango and San Cristóbal, the effervescence pushed master architect Ignacio Castera to draft a proposal of his own). It only means that this did not amount to the urban elite or hacienda owners having a clear and conscious vision of the Desagüe as a means to transform the hinterland and human relationships with the water, land, and natural environment. Therefore, the transformations that did happen resulted from processes and dynamics going on despite the creole elite's or the crown's designs and purposes, not because of them.

Unpopular among rural enterprises, belated attempts to give the Desagüe, its auxiliary works and projects for its extension an aspect of economic utility were, as it turns out, a purely discursive veneer. Nothing in the actual conceptualization, design, and construction of Desagüe structures over the entire colonial period suggests that economically generative considerations were more than this. This contrasts with the canalization and desiccation plans that Europeans had been implementing during this time, which articulated the urban and rural through commerce and agriculture. In the 1750s, court ministers in Spain itself commissioned the naval officer and scientist Antonio de Ulloa to study these canals abroad and to import specialized hydraulic engineers to create comparable grids in Castile and Aragón in part to promote the regions' agriculture and commercial integration. Along with the hegemonic drive that propelled Bourbon Madrid to step up canal construction, it must be said that the resulting designs of the Canal de Castilla and the Canal Imperial de Aragón reflected multiple goals, balancing at least in theory the needs for irrigation water in fields adjacent to the canal with the minimum level of water required for adequate navigation conditions.[40]

This contrast illustrates not only the priorities of the colonial elite in the City of Mexico, but also those of the Bourbons with respect to the Americas, where the infrastructure that mattered most to royal officials was that which benefited the colonial link, such as mining, ports, forts, and administrative offices. However little individual Hispanic landholders may have gained as a result of late-colonial additions to the Desagüe, these land and financial schemes could not but aid their collective concentration of land in the hinterland of the City of Mexico. This is illustrated by an example from 1808, when Desagüe superintendant Robledo tried to get riparian landholders to pay for the San Cristóbal auxiliary canal works in proportion to the land gained from the lake's desiccation, as gauged by original land titles. Most hacendados could not document land they had taken from either the wetlands or the indigenes, so Robledo revived the ancient *composición*, where illegal land holdings were legitimated by payment to the state, now through the Desagüe authority. Because composiciones were expensive, only the better-off landowners could afford them, at the expense of indigenous lands.[41]

As it was, those in charge of the Desagüe were not motivated by a desire to turn the hinterland into any particular kind of landscape. The intended purpose of the project and of the rules that underpinned its functioning was to safeguard wealth and rents in the city itself. The *effects* of the material objects and the regulations of the Desagüe on the ground, however, were the result of the context in which the project unfolded. That is, the Desagüe contributed to a process of colonization in the hinterland by the city and by a class interested in owning land and water as private property and inputs in production, but furthering these particular interests was not why city officials and elites—even those members of this class who were simultaneously landowners and decision makers in the city, such as Trespalacios himself—started and continued the project.

But the Hispanic colonization of the countryside went further, affecting human relationships to water, land, and ecosystems. When the Desagüe began, indigenes had a relationship to water and land that was as fluid as their seasonal variation. While they certainly treated water and land as distinct elements that intervened as inputs in the production of crops, they also used the flowing and ebbing lakes, wetlands, and shores where water and land were inextricable from each other as a complete input—when they used them as nurseries for larvae, or fish pens, for example—or as ecosystems from which they extracted foods, medicines, and materials.

Desagüe regulations and structures threatened this fluidity, by fixing the definitions and physical locations of "land" and "water." Fixing had begun in earnest as a result of orders that declared the beds of lakes at their point of rainy-season cresting to be "receptacles for the lakes," and it advanced thanks to subsequent measures. By 1784, the audiencia had prohibited cultivation in these fluid areas and also ordered the destruction of all embankments and ditches within the lakebed that impeded the expansion of water over them.[42] The fity-vara rule had been extended in 1770 to all water margins in the jurisdictions of San Cristóbal, Texcoco, Coatepec, Chalco, Xochimilco, and Mexicalzingo,[43] and in 1808, to the margins of the new canals draining Lakes Zumpango and San Cristóbal.[44] Combined, the fixing of definitions and the restrictions on use militated against the ability of the peasantry to sustain a mode of life predicated on communal ownership, on the seasonally fluid interaction of water and land in agriculture, and most especially on the use values of these elements and the life forms in them.

Conversely, these two actions militated in favor of private relations of property and production in the countryside. Land, water, and the things that were produced with them could not become distinct, quantifiable things liable to taking on an exchange value mobilized through the market so long as they remained fluid and in communal control. It was not possible to sell or buy units of land-water, only units of land and units of water, as reflected in the measuring systems imported by the Europeans, which had no units of "land-that-becomes-water" and no units of "water-that-evaporates-leaving-land." Once divorced from each other, land and water were much more amenable to their transformation into measurable inputs in the production of crops and goods. The Desagüe assisted this.

But this help from the drainage would have counted for little without the willing or unwilling participation of the indigenous in their own colonization. Ever since Charles Gibson's magisterial work on how the Spanish appropriated preexisting Mexica social, political, and economic structures and relationships to erect their own rule over the Valley of Mexico, scholars have revealed how these kinds of appropriations worked throughout the Americas. As we have seen in this book, maestros and others also appropriated specific indigenous organizational and instrumental technologies and redeployed them into the entirely nonindigenous object that was the Desagüe. But what made all appropriations possible in the first place was the fact that willingly or not indigenes provided valuable knowledge of regional topography, materials, and historic hydrology, actually fulfilled their obligations to maintain the Cuautitlan River diversion dam and other earthworks entirely on their own, and supplied the township crews that would work together in the Desagüe proper. This appropriation made the Desagüe an object by which the urban elites drew nutrients from indigenous and rural society and in so doing spread their class roots deep into the hinterland, facilitating the penetration of a different regime of resource use.

At the same time, this Desagüe-assisted colonization process penetrated deep into the ecosystem and into indigenous society. On the one hand, not all indigenous townships lost and not all Hispanics gained. Among indigenes, class formation progressed through the colonial era, making some families grow in wealth as a result of their appropriation of communal resources and others not. This in itself is neither new nor particularly interesting. What gives it significance is that indigenous communities or individuals rarely—if ever—resisted the opportunity to gain more land, even at the expense of the lakes or of their traditional relationship to them, when the apparent benefits outweighed the losses, and that as the better-positioned members of communities began to see these benefits more tangibly, their actions promoted the commoditization of land, water, and biotic resources within the fundos legales. Commoners in the Desagüe district's townships saw their ability to use common resources in the wetland ecosystems for the extraction of use values erode over time. The companion to this process was the growth of exchange values for these resources at the expense of use values within the communities. Colonization in this context therefore meant not the simple domination of one people by another, but rather the deeper transformation of class usages and valuations of land, water, and ecosystems accompanied by greater class stratification among indigenes.

To untangle the three paradoxes outlined above we need to conceptualize colonization as a class-based process. This is both because "class" cross-cuts ethnicity or national identity, and because of the unexpected ways that a class's "interests" are expressed vis-à-vis other classes. Often understood as a process whereby an ethne or "nation" expands over a given "empty" or populated space, colonization can therefore also be understood as a process carried out by classes. In the basin of Mexico, the Desagüe assisted the transformation of class relations among people and between them and their environments because it helped Hispanic commercial individual landownership at the expense of indigenous communal landholding while also pushing along processes of class transformation within indigenous society itself and in its relationship and usage of land, water, and ecosystems. In adapting to changing economic circumstances and definitions of water, land, and ecosystems, indigenes increasingly and often gleefully embraced individual private property and private gain, thus partially accepting the segregation of water from land. This did not mean a wholesale Hispanization or abandonment of communal obligations and loyalties, of course. It only means that the Desagüe simultaneously sustained continuity in indigenous life by depending so much on indigenous townships and militated against it by assisting processes of transformation that were happening independently of the intentions behind the drainage. The Desagüe regulations that fixed what and where land and water were and what was to be the human relationship with them certainly promoted these transformations. But we have no evidence that this was intentional and conscious. The urban elites and the crown did not pursue the Desagüe in order to free up inputs for production any more than they did so to change the indigenous peasantry into a class of dispossessed proletarians. If this was happening anyway, it was because neither indigenous society nor the elites were homogeneous, and this heterogeneity is what is behind the different simultaneous and often contradictory directions these groups took.

By the end of the colonial period, the Desagüe had helped shape the relationship between the indigenous peasantry and Hispanic landowners and between the city and the countryside so much that the most insignificant aspects of its design showed what these relationships were. An urban elite made up of merchants, bureaucrats of various kinds, and even some city-dwelling landowners used the public work to safeguard the sources of their wealth and rents, a priority that came before any other. And yet, because of the choices these very same people and their technicians made in it, the Desagüe wound up both eroding the indigenous peasantry *and* providing a vehicle for its survival. However tense, the coexistence and survival of both of these classes and their respective priorities is what distinguishes much of Latin America from both Anglo-America and Europe. This is the essence of the region's "uneven and combined development," whose features used to be considered signs of "underdevelopment," a term rejected here not because it is démodé, but because the author shares none of the longings that underpinned the term.[45]

Having assisted the loss of acreage, units of water, and the broader relationship to the lacustrine system whereby indigenous communities might still extract from it both use and exchange values, the Desagüe had, by the end of Spanish rule, left this deep colonization incomplete. This was the result of the limited motivations of the urban elites, the relative lack of incentive for landowners or others to push them aside and of the interruption of the project in the first half of the nineteenth century. As the open trench filled with silt, the sluicegates rotted, and nobody enforced maintenance obligations on the Cuautitlan River diversion dam, the fifty-vara rule, or any of the other ways the Desagüe had impinged upon hinterland populations, rural power relations with the Desagüe and among Desagüe neighbors were up for grabs. When the urban elites finally got around to finishing the Desagüe, it seemed as though Humboldt's wistful lament was not applicable any more.

Had Humboldt's fantasy been fulfilled, it might have been an even greater disaster, for it would have brought even more "progress." The fact that it was not, however, was both the effect and the cause of the backwardness of the elites. Whether this is a "good" or "bad" result depends on the alignments of the viewer: for some, this outcome might be beneficial as it allowed the peasantry—and therefore an alternative manner of using water and land—to survive to within the range of present-day living memory. This makes it possible to rescue from it what is useful, to restore or create sustainable practices instead of the predatory ones that characterize capitalism, particularly in "emergent" nations.

## KNOWLEDGE AND ITS MAKERS

The Desagüe was far more than a node of socioenvironmental struggle. Although by Humboldt's visit it was still very imperfect and certainly incapable of preventing floods in the capital, all along the Desagüe had been a site of tremendous technological experimentation, natural observation, and production and accumulation of knowledge. It was a factory of expertise. This fact does not square with the enduring notion that Spanish colonialism inhibited innovation in technology, science, and other realms traditionally linked to economic growth and thus condemned the continent to "routinism" and underdevelopment.[46] But neither does it undermine it. Instead, it invites a different narrative and analytical approach than has been commonly applied in the historiography of science and technology in the Spanish empire and its successor republics.

The basic narrative of knowledge production in science and technology in the Iberoamerican realm goes something like this: in the sixteenth century, most of the letrados that Castilian universities increasingly produced made do with indirect monetary supports such as tax exemptions because the crown could only afford to sponsor two of these schools (Granada and El Escorial).[47] In the early seventeenth century, the state was the most important patron for scientific activity and nontextual production (instruments, notably), funding activities and authorship through the Casa de Contratación, the military engineering and artillery academies, the Real Farmacia, and other institutions that catered to a variety of imperial needs in navigation, fortifications, infrastructure, weaponry, commerce, the health of royal lineage, the salubrity of urban life, and so on. A Golden Age in all cultural production ensued, science included. However, in science it was not innovative because there was also censorship.

Unfortunately, as the empire's fortunes declined, so too did this direct patronage. Then, in the second half of the eighteenth century things picked up again thanks to the Bourbons' resumption of sponsorship of institutions and interest in useful knowledge. Science and technology, however, suffered from this interrupted patronage. While accurate in its essentials, this narrative arc is incomplete.

Among its significant merits, the historiography of science and technology in the Spanish empire and its successor republics that produced this narrative has equalized the role of the "periphery" with that of the "core" in the history of science and the production of knowledge.[48] To accomplish even more, however, it needs to take two further analytical steps. One is to reconsider what constitutes patronage and how it related to knowledge production. There is agreement that the market for books and texts was fairly restricted to the letrados, and that since producers and consumers were basically the same sociological group, the production of knowledge depended on patronage, whose decline therefore led to a decline in scientific, technological, artistic, and literary works. If however, following Angel Rama, we include the remunerated leisure facilitated by many kinds of church- or state-provided incomes as patronage, then in fact full-time and part-time patronage abounded in the Iberian world at all times. The "problem" of patronage thus practically dissolves, and so does the answer to the "problem" of a decline in the production of scientific and technological knowledge.

The question then becomes more about what those patronized did with their time. As it turns out, authors generally preferred to orient their literary production less toward the physical sciences or technology than toward religion and the theory and craft of monarchy.[49] The scant scientific or technological authorship specifically dedicated to the phenomena in the Desagüe confirms this pattern, as what was written about them was insightful but imitative overall.

The second analytic step that is necessary is to treat formal and tacit knowledge and the people who worked with them in fluid relationship with each other, as scholars of other regions have begun to do. The tacit realm has been explored more by Latin Americanist and Iberianist ethnohistorians, historical archaeologists, and anthropologists than by historians, who have preferred the formal realm. Unfortunately, this means that the latter mostly tell the story of knowledge production and expertise among the far from numerous lettered elites, as though this production could exist without its tacit relatives and as though the only medium for the storage and transmission of knowledge were texts. Both steps are necessary for a fuller comprehension of the Desagüe and of other objects or processes where knowledge was appropriated, developed, mobilized, stored, and transmitted. They take the analysis on a path where the questions about knowledge become who produced it, from and for what, and how, not whether it confirms, refutes, or just nuances the notion that the Spanish empire's intellectual life was poor and imitative.

So if not the letrado caste, then who provided the know-how the empire needed to physically emerge, grow, and endure even in the midst of competition and "decline"? Colonization is a demanding business requiring hard work and ingenuity, after all. Most of those who provided these two things in and out of the Desagüe were not in the lettered caste and did not use their time for literary authorship, in or out of the academies or patronage circles. As this book has shown, in the Desagüe—that important thing that allowed an imperial capital to consolidate—most of those who held, recombined, tinkered with, and redeployed knowledge were craftsmen, indigenous peasants and workers, practicing engineers, foremen, and wardens. Of course the oidores, friars, academicians and savants, and others who qualify as patronized letrados contributed to this process, but they were neither numerically nor qualitatively more important than their social inferiors who performed *oficios mecánicos*.

These social underlings lodged and passed on their knowledge less through formal writing than through work procedures, oral communication, the quotidian Desagüe record, and the structures themselves. This lack of formal texts on the science and technology of water is perhaps the reason why historians have generally abstained from delving into areas such as hydromechanics (hydrostatics and hydrodynamics) and hydrology. While there was no autonomously developed "Mexican" hydromechanics—not even a single magnificent new discovery in these areas of physics to entice a modern scholar—maestros, friars, guardas, sobreestantes, engineers, indigenous townships, and Hispanic landholders did nonetheless synthesize and amass a substantial body of hydrological and hydraulic knowledge. This is visible through their largely non-letrado products and their work—their dams, ditches, earth levees, canals, cutwaters, stoplog sluices, wood and metal-tipped tools for grass pulling and river cleaning, decisions to open sluicegates, underground inspections, and other ordinary objects and actions as well as in the highly codified plans of the engineers.

Objects and processes like these were the media through which the majority of the population expressed itself—and still does. They have the additional benefit of being largely unselfconscious and therefore vulnerable to exposing their hidden truths. Unfortunately, because they are not luxury goods or collectibles, most of these items—along with the tacit knowledge and the oral and gestural culture that often sustains them—do not survive well over time. Getting at them requires "reading" small and large objects and physical systems—even the textual or visual representations of them when they are not extant—as historians, archaeologists, and historical archaeologists of environments, landscapes, and material culture have done. This means treating the physical qualities of objects and their content as registers of practices and knowledge just like one would treat the appearance of a name or event in a document as a register of presences and relationships.[50] Reading both material remains and textual sources in this fashion, Chandra Mukerji, for instance, found a dynamic of appropriation of local tacit knowledge by the much lauded French hydraulic engineers in the Canal du Midi.[51] The tacit knowledge of Mukerji's laundresses is matched by the tacit knowledge that was absent from treatises or any other letrado product ("formal writing," she calls it), but was common as weeds among indigenous townships and rural producers in the Desagüe district. Without tacit knowledge, there would have been no drainage and possibly no Canal du Midi either, at least not in that form or for that price tag, large as it was.

The ability of Mukerji's technicians to detect and use the technically savvy Languedoc laundresses was as crucial to the French Bourbons as Charles Gibson's first bureaucrats and settlers' eye for preexisting indigenous social structures were to the Habsburgs, and as the Desagüe's technicians who took from the peasantry's mastery of the Cuautitlan River basin were to the imperial City of Mexico. In other words, the master architects, guardas, sobreestantes, and others actively working in the Desagüe who dexterously appropriated and redeployed practices of varied origins were much like their much lauded counterparts elsewhere: all channeled material and organizational devices developed by indigenes to assist the colonizing projects of the various classes, groups, dynasties, or alliances who employed them; all produced uniquely complex—and for their epoch tremendously intrepid—technological objects by recombining existing knowledge and practices to meet a challenge. Since this, incidentally, is what innovation consists of, then the Desagüe's history demonstrates that colonialism in New Spain did not discourage technological innovation and experimentation at all—in fact, this activity was going on not *despite* Spanish rule but *because* of it, since early modern colonialism itself was a monumental and complex experiment. This would suggest that the Iberian world more broadly actually embraced the activities that are commonly associated with the production of knowledge—observation, tinkering, innovation, reflection—because it needed to, and that the majority of those directly engaged in these activities did not make books but things.

Indeed, the Desagüe helps explain why and how the empire the Habsburgs built managed to dominate for as long as it did, and then sustain its rule over vast and distant lands even as its competitors struggled to elbow their way to hegemony: by sagely mobilizing its most renewable resource—its population. The Habsburgs unleashed common people throughout the Atlantic world to pursue their most basic desires and needs by searching for, experimenting with, and combining matter, species, energy, and knowledge to channel or produce wealth. Simultaneously, the crown restrained the colonists' most ravenous cravings with state and religious interventions whose considerable effectiveness the intrepid Bourbons tried and failed to surpass. In the Desagüe, such commoners included humanist and enlightened savants, friars, master architects, military engineers, academicians, illiterate foremen, and of course countless indigenous laborers and peasants. Simultaneously exploitative and preservative of indigenous populations around it, the Desagüe was not just an engineering feat, it was also imperial rulership at its most ingenious and dynamic.

In the Desagüe and in the farthest corner of this empire, various actors appropriated and reworked strands of useful information from others, collectively producing tacit knowledge. In and around the drainage works, people embedded this knowledge in things and diffused it orally and in informal writing. Inspections, deliberations, mundane conversations, orders, instructions, reprimands, complaints—these events and others in the Desagüe district were media for the transmission of information and knowledge between individuals and groups. Knowledge also diffused in written form, in the notaries' record of these oral transactions, in the dispatches that flew back and forth among the guardas and between the guardas mayores and the superintendents, in the legal complaints the Desagüe's neighbors lodged about the uses and misuses of water, land and productive infrastructure in the region, and so on. Much as Desagüe underlings and the inhabitants and workers in its district might thus take information from their social betters, the current of knowledge flowing from the bottom up was far stronger.

Having access to the record of and often participating in the same inspection teams as these people, the savants, letrados, and bureaucrats themselves were also party to this knowledge. Those with exposure to the various strands of written science—San Miguel, Iriarte, Alzate, Velázquez de León, for instance—then made syntheses of their own, according to their alignments and needs. In doing so, those groups closely dependent on a broadly defined patronage eagerly adapted the culture of Europeans amidst them. Even after Andrés de San Miguel and the men of his time, the most frequent demeanor of the letrados, savants, and academicians who did think about the Desagüe during and after the emergence of a new, post-Galilean science was to apply this science's tools, not to generate their own. In the physical sciences and mathematics relevant to the Desagüe, their approach was certainly erudite, but it did not interrogate nature. As military engineers became more persistent presences, and as it became clear their culture and training earned them royal favor (not to mention income), these sectors increasingly adapted the enlightened *language* of science, reason, technological advancement, and cartographical communications to their circumstances. In the process, so far as the Desagüe was concerned, they mutated its content, using it to reinforce practices that helped sustain their social privilege and hindered the emergence of alternative uses for energy and matter.

The absence of institutional specialization around the Desagüe capable of housing and breeding hydraulic engineering and hydromechanics as pursuits linked to economically generative activities is rooted in the Desagüe's class content. It was once suggested that one of the reasons that hydraulic technology and hydrostatics developed well in the Netherlands was because of the existence of the Polder Boards. This institution not only coordinated the maintenance and financing of the land reclamation system of dams and pumps, but also promoted and paid for research. As a result, hydrostatics and dam technology flourished there and supported Dutch economic power.[52] The lack of such specialization diminished the possibilities for the Desagüe to serve analogous functions in Mexico. But if lawyers and intellectuals were so crucial in the Mexican drainage, it was because no other social sector took their place to make a specialized techno-scientific corps a priority. Despite its magnificence, the Desagüe and its successor would, until the mid-twentieth century, continue to coalesce around it lettered people whose intellectual work and authorship would be derivative of imported knowledge. This would change but also be constrained in new, peculiar ways with the social energy unleashed by the 1910 revolution and then channeled by the postrevolutionary regime and its institutions. We might rue this failure to hit such developmental milestones. To determine whether this mattered at all to the workers and technicians whose story is told here, to the Desagüe itself, which showcased the ingenious and knowledgable solutions to practical problems this particular piece of the empire demanded from all, Hispanic or indigenes, or indeed to modern Mexico, it will be worthwhile to pursue this Whig lament a bit further.

## BRAUDELIAN BOURGEOIS BETRAYAL IN COMPARATIVE COLONIAL PERSPECTIVE

In most of the literature on the subject of development in Spain and its American colonies, backwardness in scientific and technological capabilities has gone analytically hand-in-hand with the failure to fully achieve the capitalist social relations and structures that occurred in England, other northern European nation-states, and eventually also the United States. Most of this scholarship has been concerned with *how* these putative failures evolved; in contrast, French *Annales* historian Fernand Braudel, in his influential book on the Mediterranean world at the height of Spanish power, worried about *why* they did. To answer, he reached for class behavior, famously reproaching sixteenth-century Spanish entrepreneurs for their act of "betrayal" against their nation's capitalist development with their rent-seeking behavior that diverted wealth away from productive investment in Spain's nascent industries.[53] Were the conservative motivations that drove urban elites to build and keep the Desagüe as they did an American example of this Braudelian *trahison de la bourgeoisie*, making the drainage project one spectacular instance of betrayal of the potential for full-throttled capitalist "development"?

The answer, in a word, is yes. As we have seen throughout this book, the Desagüe is indeed an expression of the "betrayal" of the Mexican "bourgeoisie" in that the mercantile and landholding elites had no interest in using the Desagüe to develop "national" markets out of their "regional" ones, to maximize commercial agropastoral production, or to pursue any other economically generative purpose. Instead of taking the Desagüe under their wing and through it mobilizing capital to advance and revolutionize the means of production, these elites imbued it with rentier priorities, allowing it to be used to conserve wealth in the capital.

But the reason why it is important to deal with this question is not that the answer supports—yet one more time—the standing characterization of backwardness. It is unfortunately necessary to deal with it because the Desagüe has, since Humboldt, played the role of villain in a wishful story of national progress and development that still reigns despite the evident crisis of nation, progress, and development in Latin America and beyond, and because considering this question with class and nature at the analytical center actually permits a fuller consideration of what colonization is, where it has happened, who have been its actors, what have been its outcomes, and—for all these facets—why.

To help answer these questions, this book has linked the phenomenon of colonization in the basin of Mexico to an understanding of the trajectory of capitalist development as something that depends centrally on the capability and desire of propertied classes to alter whether and how the peasantry will relate to land, water, and biota and how all are incorporated into the privatization and commoditization process through which elites amass wealth and power. It has explained why and how the Desagüe, even though it was not deliberately designed to promote this process, assisted it nonetheless because it was unfolding independently of the drainage. It has also discussed how the Desagüe, in the context of privatization and commoditization, contributed to the region's uneven and combined development—the phenomenon characterized by the sustained presence of both capitalist and peasant classes in the same space, each unable or unwilling to defeat the other's control and usage of the same resources.

It will now be productive to consider the class and ecosystemic content of colonization more broadly, by comparing the Desagüe to contemporary desiccation and canalization projects elsewhere. While such works were being carried out in Asia and French America, here only European projects will be considered for the sake of reducing the number of variables and to tack close to the standard contrasts the literature prefers. Two such projects stand out because they were begun at roughly the same epoch as the Desagüe and because they were and are seen by their promoters and historians as key to turning "wastes" into "productive" lands. We turn first to the English Great Level.

Two decades after Enrico Martínez finished building his Desagüe, an alliance of English private investors, the state, and privatizing landlords began the drainage of the fenlands in many parts of England. Some of their main targets were in the wetlands of Lincolnshire and Cambridgeshire fens, where this alliance pushed against commoners and the peasantry with large-scale drainage projects that would transform fenland ecology and class relations. As in the Desagüe, foreign and domestic technicians were hired and their expertise mobilized and combined to survey, design, and execute ditches, canals, diversion dams, and other structures and devices to suck the fens dry. The expertise, the technology, and the material form they took belonged much to the same universe as that deployed in the Desagüe.

Also like their Mexican counterparts, fenlanders developed a way of life adapted to a landscape characterized by seasonal flooding. Peasant farmers and poor commoners drew both sustenance and marketable things from the wetlands, which were commons: they ate and sold cheese made from milk their animals produced while grazing rich summer fenland pastures; used and sold peat as fuel; gathered reeds, willow branches, and other plants to make traps, decoys, baskets, and so on; and they used these tools to gather fish, fowl, and crops for consumption and market, as well as to pay dues to the manor.[54] More pastoral than agricultural, their economy depended on the rise and recession of water over land that nourished the soil with silt, kept the water table sufficiently high for plants to reach, and prevented the peat from compacting and losing its anaerobic balance to corrosive oxygen.[55] The entire fabric of social and cultural life was predicated upon their capacity to continue extracting use and exchange values from this fluid setting.

Drainage efforts were an assault on this fenlander way of life. Fenlanders could respond in various ways, along a spectrum from complete rejection to complete acceptance, and each response would have different meanings. They could adapt the techniques, tools, forms of organization of work, and even the often communal property and access regimes to social, economic, or environmental changes around them. As in the Mexican case, this adaptation *by* the indigenous populations of these wetlands can simultaneously be read as a colonization *of* fenlander peasant and commoner society by other classes, in the form of the penetration of private property relations within communities, an increased focus on commodity production for regional markets, a reduced reliance on hunting, gathering, and common pasture, and so on. Among both fenlander and Mexican lacustrine indigenes this adaptation/colonization could very well have taken place without drainage. But the fact is that the drainages were there.

What was different about these two drainages was the conceptualization, and that is why the results were different. The class coalition behind the fenland drainage designed it *deliberately* as a vehicle for their "bourgeois" colonization of the countryside. They sought not just to split and fix land and water to better control them, but to systematically assault fenlander control of land and water so as to better commoditize land, water, and labor, and mobilize all three as separate inputs to produce things with an exchange value, capable of multiplying capitals—an economically generative plan, in other words, despite the speculative side-deals that took place in its shadows. The Mexican Desagüe also sought to fix what and where land and water were supposed to be, but this only *assisted* their transformation into quantifiable inputs and commodities and the process of bourgeois colonization of the countryside; it was not done deliberately and self-consciously for this.

Of course, there was as much conflict over the colonization of the fens as there was in the Desagüe district. In fact, until the bourgeoisie had steam power to finish the job, fenlander resistance to the overall process—as well as the Civil War, the demise of the courtier alliance behind the effort, and Parliament take-over—marked the ebb and flow of the class struggle over the drainage and colonization of the English wetlands and their people.[56] Even so, the Great Level projects launched in the 1630s were aggressive enough to ensure the sudden loss of two-thirds of the seasonally flooded common pastures to the gentry and the crown's allies, who fixed it as permanent land for commercial agriculture.[57] This did not happen as swiftly in the basin of Mexico and certainly not through the Desagüe because the class coalition behind the Mexican drainage had a markedly different balance, not the least because the crown was a major player, and it had an interest in keeping the endowed indigenous peasantry very much alive as a class, not wipe them out. Unlike their counterparts in New Spain, the English peasant farmer and commoner indigenes of the fenlands faced an uphill battle because they did not enjoy the protection of the state and because they faced a rather more unified front of different sectors of the elites. The bourgeois in their parts were certainly not guilty of *trahison* to the capitalist outcome historians have praised so much.

As in New Spain's Desagüe, in France too the state took the lead in desiccation. In 1559, Henri IV launched the Association pour le dessèchement des marais et lacs de France to drain marshlands. The alliance of classes supporting this venture is different from the Mexican case, as evidenced not just by the involvement of Dutch capitalists and technicians, but by the multiple objectives of the association's drainage projects: to improve "public" health, intensify commercial agricultural production on large estates, develop transport and communications, and increase treasury income. Again, a key process was to fix what and where land and water were supposed to be. All marshlands were targeted: those belonging to the king, the church, the nobility, and the third estate. The entrepreneur was allowed to expropriate and destroy any physical structures that impeded desiccation—locks, mills, embankments, dams, and so on.[58] The colonizing success of these bourgeois-crown alliances was uneven, lying somewhat in-between that achieved by the deliberateness of the Great Level and the lack of intention in the Desagüe. In many places, such as the mouth of the Rhone and on the Atlantic, there was significant resistance by commoners and peasants protecting common rights to wetlands. Like their counterparts in Mexico, and in the fens, indigenes of the Camargue and the marais poitevin sabotaged the physical devices (drainage canals, especially) that the Dutch engineers of Henri IV and his successors deployed to separate water from land.

Examined closer to the ground, so to speak, the tensions of colonization actually look quite similar in all three cases. In addition to being contemporary projects of Europeans on both sides of the Atlantic, all three feature indigenous inhabitants of a particularly fluid type of ecosystem dealing with the efforts of foreigners to incorporate, colonize, and reshape their region and its customs. The significance of this is lost because, to quote French environmental historian Jean-Michel Derex, historians have been "prisoners of a reading of the landscape inherited from the eighteenth century."[59] Wetlands become interesting to historians only when they were targeted for extinction through desiccation, but rarely in and of themselves with all the social and ecosystemic relations that developed within them and between them, the regions, and the "nation." They are studied as "lands to be conquered and exploited so as to release from them the fruits of human activity as soon as possible."[60] In looking at marshy environments and their inhabitants in this fashion, "historians have done nothing but take the side of political and scientific elites of their epoch."[61] The first step in peeling back these analytical constraints legated by liberal historiography is to abandon the assumption that the indigenous peasants and commoners of these regions felt any allegiance whatsoever to the nation-states that the English, French, or Spanish monarchs and their class allies were creating at the time, thanks in no small part to large-scale hydraulic works such as the Great Level. Indigenous notions of *país*, *pays*, or *countrey* had to do with locality, not an abstract supraregional entity.[62]

Rather than a distinct phenomenon, therefore, the New World colonization that the Desagüe assisted might more productively be seen as comparable to that which was unfolding in Europe at the same time. The main obstacle to conceiving of both New World and Old World colonizing processes as part of a larger early modern phenomenon is the presence of that disfiguring entity, the nation-state, which projected onto the past artificially separates them.

Mexican, fen, marais, or bonifica drainages were all useful for a class colonization of the countryside that assisted the penetration of the capitalist mode of production, the city, and the state into the countryside's water, land, and people, but they achieved different degrees of success according to which sectors of the elites were doing the drainage and how forcefully. The resulting divergence between "colonial" regimes and "colonial" social orders in the far-flung territories from those in Europe was far more contingent than is often admitted. That is, the outcomes were determined by the class composition and capabilities of the foreigners, the greater or lesser ability of the indigenous peasantry and commoners to resist efforts to transform the definition, ownership, and usage of land, water, and biota, and finally whether the crown supported these transformations or not. In other words, not by the invisible hand of progress, but by the far-from-predictable outcome of the class struggle, as quaintly démodé as that may sound.

Tracking not only the forms of struggle over ecosystemic resources, but also the environmental implications of the particular aims of the class alliances that underpinned the Spanish monarchy and the colonial order in Mexico, as opposed to colonizing alliances in northern Europe, in turn can help explain the features that characterize uneven and combined development in Latin America.

In the end, the "bourgeois betrayal" of this process in the basin of Mexico turns out to be a self-betrayal. The ruling elites' failure to develop an autochthonous science of water, to infuse hydraulic engineering with the technology available in Mexico's most capitalist social and economic sector—mining—and to abandon rentier motivations and embed economically generative goals in the Desagüe was worse for them than it was for the "nation." Squeezed out from the top of a new global order, and continually challenged from below, these elites reluctantly emerged from the collapse of Spanish rule unable to truly possess and rule over the territories they inherited.

But they were not the only ones responsible for the region's uneven and combined development. In fact, much as they may have wished for a different reality, if Mexican elites were still using repartimiento in the Desagüe at the end of the colonial period it was because it was not just up to them to imagine an alternative—more enlightened, perhaps—use for the human energy that went into it. If drafted, coerced, and rotational indigenous labor persisted it was also because an Indian peasantry survived to provide it. Although the constraints that operated on the enlightened classes in New Spain were not fundamentally the indigenous peasantry's doing, as the liberal tradition was wont to claim (as we have seen, the savants and institutions themselves helped set the limits of their reach), the survival of the Mexican peasant communities certainly helped frustrate the peculiar take of the Mexican elite on the Enlightenment.

If all this seems a terrible waste of opportunities, it is only such in the context of the developmental agendas created by and disproportionately benefiting dominant classes. In the life of a commoner or working person living in any society, it is not clear that it matters whose science develops first or farthest, just as it is not clear at all that the macroeconomic growth and contraction of viceroyalties, regions, or nations really improved or worsened conditions for these sectors.[63] What matters is that technological and scientific practices, as well as the economic and social usages of water, land, biomass, and energy are consonant with the needs of the majority of the population and the survival of the ecosystems necessary for human life. Judging from the impact of the class colonization, however inchoate, that actually did take place over the long term in the hinterland of the imperial City of Mexico, it is as improbable that the elites' achievement of the liberal desiderata would have translated into a lasting improvement in the lives of the majority of urban and rural populations as it is likely that environmental destruction would have been deeper.

If the material, social, and cultural realities that the technicians and workers in this book created in and through the Desagüe appear to "lag" behind those achieved by their counterparts elsewhere in the Atlantic, it is due to the developmentalist lens through which they have been examined. "Failing" to enthrone practices that seem desirably propitious to the idealized capitalist, egalitarian, and democratic futures that have putatively blessed other peoples, the Indian workers and village hydraulicians, on the one hand, and the Hispanic guardas, sobreestantes, master architects, friars, military engineers, bureaucrats, and savants, on the other, helped create and preserve the uneven and combined development of their corner of the New World. That is, they did not actively promote "capitalist predations" that elsewhere transformed property relations and environmentally savvy peasants into disposssessed proletarians, assisting instead the rentier priorities of crown and aristocrats which required the persistence of indigenous peasant access to and possession of land, water, and biota. Viewed in this fashion, the technicians and laborers who populate these pages facilitated the survival of the knowledge, practices, and classes that underpin often more sustainable modes of life in nature, and may provide solutions to current socio-environmental crises. Hardly a failure at all.

# Epilogue

The Desagüe encapsulates and displays the essence of uneven and combined development that characterized the Spanish legacy in those American regions that had once supported populous and complex indigenous societies: the survival of both a battered but by no means defeated peasantry and a weak, fractured, and fearful elite, perpetually desirous of a different reality yet unable to impose it. Throughout the territory, the competition among these and other social sectors to redefine the relationships among themselves was manifest in myriad ways during the decades between independence and the triumph of the oligarchic republic in 1877. In the Desagüe district, the symptoms of this tension were the silted trench, the rotting wood of the sluicegates, and the slack enforcement of the measures and obligations the colonial superintendants had labored to put in place.

Not until the new regime headed by Porfirio Díaz finally succeeded in imposing a nation-state called Mexico on all, orienting production and infrastructure toward the demands of a voraciously expansive Northern Hemisphere capitalist system, did conditions change. Now the various sectors of the elites had the resources to join forces and actually rule, pushing through the infrastructural projects necessary to this purpose, wooing malcontents of all classes, and repressing those who remained recalcitrant. In the basin of Mexico, the motivations that had been embedded in the Desagüe began to shift as land values increased, making entrepreneurial and propertied classes pine for future urban plots then still submerged under water. But this outcome was far from certain when Spanish rule collapsed.

As the Desagüe lay slowly deteriorating, the flood cycle continued, periodically inundating urban streets and buildings. Initially, the Desagüe became a provincial matter; in 1826 it returned to the central government. Either way, the contention over authority and taxation that followed the end of Spanish rule brought a halt to new projects and even regular maintenance of existing works. In 1823, as the federal republic took shape, the liberal politician José María Luis Mora inspected the works and found them to be "in a state of decay."[1] Subsequent rainy seasons, particularly in the early 1830s, sustained a consciousness of the vulnerability of the city, as about half of the open trench was useless, obstructed by up to six meters of silt.[2] In 1832, conservative statesman Lucas Alamán, then a cabinet minister, put complete desiccation of the lakes back on the agenda. In his proposal, new land created by the draining of the lakes would be property of the federation, and existing landholders would compensate the government the modest yearly sum of 5 percent of the improvements resulting from desiccation.[3] Funding was not forthcoming, however, and would be less so as the pendulum swung toward a centralist republic (1836–48) that endured the cost of war with seceding regions and the US invasion. American engineers with the invading armies also issued plans to renew the Desagüe works, as did those employed by Emperor Maximilian, and those of the Restored Republic (1867–77). All deplored the deterioration and insufficiency of existing drainage works for the purposes of general desiccation.[4]

But not everyone everywhere bemoaned these conditions. It is hard to imagine the neighbors of the Desagüe—particularly those along lacustrine shoreline who relied on seasonal flooding for the agricultural, gathering, fishing, and hunting activities underpinning their economies—mourning the apparent demise of the project. Although repartimiento and tribute had been legally extinguished along with Spanish rule, public works could still require the mobilization of labor from neighboring communities, and so as in the colonial period any lull in the Desagüe and other public works was welcome among villages if the work was paid lower than the benefit of keeping it for their own economies. The same was true, most likely, among haciendas and ranchos, as their proprietors need not dispense with their share of work or money for the projects. More importantly, these decades of crisis for the elites were not unkind to small-scale and family economies. They in fact opened a space for rural populations to reconfigure their surroundings as they saw fit throughout Mexico.

In this era of "agrarian decompression," villages and the rural poor in the basin of Mexico, such as the chinamperos of the southern part of the basin, took advantage of the weakness of commercially oriented hacienda-owning elites to affirm their control of water and land.[5] Although scholars have focused on the social confrontations, court cases, and land occupations by rural populations during this time, there was also an important and extremely low-cost theater of communal recovery—the simple and stealthy deployment of technologies of water and land management and use kept alive throughout the colonial era. The peasantry in the basin of Mexico and the Desagüe district could use the same technologies that had been so crucial to the drainage to recover ground they had lost to both the Desagüe and the haciendas.

In addition to the factors summarized in Chapter 8, what made this possible was the fact that through its maintenance obligations, the Desagüe itself had provided a vehicle for the survival of technologies that could, given the right circumstances, become a weapon for its peasant bearers to regain control of water, land, and biota. This is true even though in technological terms, over time, the Desagüe *proper* grew more and more distant from inputs from its rural neighbors, be they indigenous or not. It became increasingly homogenous in the origins of its technicians and technologies. But the Desagüe *district* and its landscape did not. While during the seventeenth century technicians of all kinds had often tapped rural knowledge for Desagüe needs, as the eighteenth century unfolded and metropolitan military engineers and academicians increasingly intervened in the works, the specific weight of autochthonous technological inputs and methods of work decreased. But it did not vanish.

This was because in their struggles with the Desagüe, indigenous townships, haciendas, and ranchos in the northwest quadrant used not just the courts and protest but also nonverbal means to defend their interests. Indigenes in particular just as often withdrew labor; deliberately or not misinterpreted instructions on how to perform work in the Desagüe district or in the Desagüe proper; cut openings in embankments to draw water; and ignored injunctions to abstain from using lakebeds as fields. In short, they overtly and clandestinely intervened in the contest over the fate of the landscape using their knowledge of techniques, materials, work processes, and seasonal fluctuations in ways that kept this knowledge alive and available for other struggles.

As a result, the outcome of the process of class colonization of water and land and the social usage of both that had been advancing in the lacustrine and rural environs of the city alongside the Desagüe was still altogether uncertain and up for grabs during the first half of the nineteenth century. This incomplete class colonization was expressed in many ways, among them the way that the usage and allocation of water and land were decided. Up to 1870, it was the local users—represented in town councils, those of indigenous communities included—who ruled over these matters. Consequently, the allocation and usage of resources was decided by local correlation of forces among smallholders, communities, and hacendados. This was a continuation of a colonial practice that was favorable to the commons; it survived because rural populations wanted it that way and the state and national governments were unable to impose anything to the contrary on an irritable peasantry.[6] This impasse was manifest in the fact that the liberal measures of the midcentury Reforma, such as the Ley Lerdo of 1856, which sought to end corporate ownership and forms of utilization of land and water, did not become a reality until later.

It was only when all sectors of the elites—liberals and conservatives—coalesced under Díaz and secured the resources to carry out these reforms that the colonization of the countryside could resume in practice. Further legislation to partition and privatize "vacant" lands (tierras baldías) enacted between 1885 and 1891, and the very first law making water a federal matter, the *Ley de vías generales de comunicación* of 1888, now had teeth: resistance could be dealt with by the *rurales*, the constabulary that president Benito Juárez had created in 1857 as the executive aid to the liberal reform legislation and that Díaz bolstered with more weapons, training, and recruitment.[7]

The revitalized drainage project was one of the concrete forms that this offensive against communal control over what and where land and water were supposed to be and whether their ecosystems would use values at all took. By the late nineteenth century, drainage was perhaps necessary for a growing city consuming more water and creating more sewage, and Porfirian authorities defended desiccation as a public health measure.[8] More importantly, the outlook was changing as markets for products and land grew and the prospect for greater profit overtook the simple desire to protect wealth among the elites.[9] It was now reasonable to speed the desiccation along with potent steam-driven machinery and new designs, methods of work, and organization. It was also desirable, as the Porfirian elite embraced wholeheartedly the idea—and financial opportunities—of making Mexico "modern."

Long-debated and redesigned, the new desiccation project began in earnest in 1888, when Díaz named a Junta Directiva del Desagüe del Valle de México and signed contracts with two English firms, Reed & Campbell and Wheetman Pearson. The new design was, like the original Real Desagüe, a system made of a canal, a tunnel, and an exit trench, but it tapped directly into Lake Texcoco and traversed the encircling ridge along a path northeast of the colonial Real Desagüe. From Lake Texcoco, a forty-eight-kilometer canal would lead water through Lakes San Cristóbal and Zumpango. Near the town of Zumpango, it would connect to a ten-kilometer-long tunnel and finally exit on the other side through a trench that discharged into the Tequixquiac River. After enormous investments, the former company finished the tunnel in 1894, while Pearson finished the canal in 1900, whereupon lavish festivities inaugurated the works.[10] The system collected stormwater as well as wastewater through a network of pipes throughout the urban area and delivered it to the Gran Canal, while also drawing excess from Lake Texcoco, and conveyed it all to the tunnel of Tequixquiac, from where it flowed out. It was designed to function by gravity alone, making the relative gradients among its parts crucial.

Finally, the basin was well on its way to becoming "salubrious" dry land. The new desiccation project served mainly the tiny propertied and financial elite who could profit from trafficking in new land. Although hardly comparably, the popular classes made gains too, as thousands of urban jobs—from teachers to engineers, doctors to bureaucrats, tram conductors to bricklayers—opened up as a result of the urban explosion that chased after the vanishing water. The optimistic developmentalism of the postrevolutionary regime, particularly from President Lázaro Cárdenas onward, added further incentive to migrate to the capital as industries and construction mushroomed everywhere, also further stressing finite water resources.

All this contributed to the dissemination of the delusion of progress, which was based on a terrible paradox. While aiming to protect the urban core from flooding by eliminating the water around it, authorities and technicians literally undermined the city by sucking its lifewater from underneath—and from basins far beyond. Both of these actions complicated and increased their original problem with flooding, while creating new problems emanating from urban sprawl on newly created lands.

This paradox developed like the proverbial layers of the onion, each one springing from a common heart—the overwhelming concern with urban flooding at the expense of all else—but developing separately next to each other and then rejoining to resurface as a pervasive, pungent efflorescence. The first layer began to grow quite a while before Díaz inaugurated his drainage. In 1853, with the aqueducts no longer sufficing to supply the city's growing population, the city council authorized the excavation of artesian wells to enable the pumping of water from the aquifer beneath the city. Residential and public potable water needs would be drawn from these wells, believed to tap into more salubrious water, while the waste-infested aqueducts would serve industrial and hacienda needs.[11] The first wells perforated by entrepreneurs Sebastián Pané and Aquiles Molteni were quickly followed by dozens of others, and by the end of the century they had proliferated over the entire surface of the city, precisely where the ground was clayey, sedimentary, and compressible.[12] This was the birth of modern subsidence.

Compaction was actually not new: as early as the 1630s people observed that heavy stone buildings sank into the ground, evidently compressing the resilient layers beneath much like the Aztec platform had before the arrival of the Spaniards. What was new was the cause and rate of this compression of the subsoil layers. That is, the uncontrolled removal of water from the highly porous, spongy aquifer under the city grew exponentially apace with the population, forcing the granular materials in this layer closer together faster than ever before.

It took a while for a proper understanding of this water overdraft to sink in among scientists; longer still for the population as a whole, not to mention the ruling classes, to act in consonance with this knowledge. Although in the 1920s engineers thought that subsidence was the direct result of the desiccation project, in 1948 Nabor Carrillo explained that it was in reality caused by this dangerous interaction between pumping and compaction, and that it had been increasing since the 1890s, with buildings and streets sinking at the rate of five centimeters per year.[13] Accelerating further after 1936, when the wells started drafting from deeper into the aquifer, this subsidence left the heart of the city up to nine meters below its original level. This phenomenon forced scientists and engineers to respond.

In the 1960s, Marcos Mazari and Raul Marsal began studying its origins and possible solutions in greater depth, in the process embedding their research and engineering further and farther into Mexican scientific and technological institutions. In contrast to the colonial period, the concrete hydrological and hydraulic challenges the basin throws at its inhabitants did, over the course of the twentieth century, lead to an autonomous and autochthonous development of relevant scientific fields—notably soil mechanics and hydraulic and seismic engineering. As a result, both compaction per se and its interaction with a slew of other variables, such as climatic change, seismicity, and demographics, were increasingly examined, revealing new dangers. Yet understanding among specialists grew at an inverse ratio to its incorporation into government water policy, which like its colonial counterpart prioritized the capital.

As a result, further wells were opened even after Nabor Carrillo's report. Only catastrophes seem to pack sufficient force to impact this course, however temporarily. The 1985 earthquake halted the operation of wells in the center of the city, where as a result the rate of sinking slowed to as little as two centimeters per year in the old traza where Martínez, Boot, and most of the technicians, authorities, and notables discussed in this book had once lived. But, as the wells spread to areas north and south, the sinking rates there reach up to forty centimeters per year.[14]

Subsidence has had perverse effects on the drainage system. Beginning in the 1930s, differential sinking began changing the gradients for which each component was designed, progressively nullifying the benefits of gravity and requiring costly additions of deeper collectors and pumping stations. In the process, it made (and still makes) the city more, not less, vulnerable to flooding, as in addition to the normal runoff and rainfall there was now the failure of the drainage system to contend with. The natural phenomenon of flooding, utterly indifferent to human efforts but quick to take advantage of the lower ground humans created, first revisited the city in 1925 and continued doing so thereafter. In response, by 1937 authorities began to build a new evacuation tunnel to supplement Díaz's tunnel of Tequixquiac—the Nuevo Tunel de Tequixquiac—west of the original. They also extended the Gran Canal further south and added to the collector piping network. In 1961, an auxiliary collector was built west of the Gran Canal to regulate the runoff coming from that direction. This Interceptor del Poniente was expanded northward, discharging through an Emisor Poniente to the Cuautitlan River, Lake Zumpango, and the old trench of the Real Desagüe. Soon continued subsidence demanded a gigantic and far deeper drainage system, the Sistema de Drenaje Profundo, begun in 1967. This complex system has many features, but its main structures are two enormous tunnels running roughly south-north—Interceptor Central and Interceptor Oriente—which collect and convey wastewater and stormwater toward an outlet tunnel—the seventy-two-kilometer-long Emisor Central. The Emisor Central discharges into the Salto River beyond and thus helps irrigate the Valle de Mezquital, in the State of Hidalgo.[15] Hence, the popular joke among *hidalguenses*: "See why we are more kickass than Mexico City folks? They send us shit and we return it as tomatoes" (*Para que vean que somos más chingones que los chilangos—ellos nos mandan mierda y nosotros les devolvemos tomates*).[16]

All these structures also serve to evacuate the water of the rivers mentioned earlier in the book, which have been channeled or driven underground. All are served by further canals and pipes stretching ever outward toward the edges of the urbanized area. Major new structures do not retire the older ones, which continue operating with pumping stations and other supports: shortly, the new Emisor Oriente will begin operations.[17] To no avail, subsidence has continued all along, perpetually reversing gradients, creating fetid and stagnant pools of wastewater mixed with street runoff somewhere in the city every single rainy season.

The second layer of our vegetable metaphor came in the form of paving and building, each road and neighborhood eliminating any infiltration into the soil and dumping what rainwater fell during the wet season swiftly down the drain. Besides rushing the water away, asphalt and cement spread an impermeable layer on formerly absorbent ground, particularly in what is known as the "transition zone"—the terrains between the clayey former lacustrine bed and the mountains—where most infiltration would otherwise occur.[18] The effect of this was to drastically reduce the rate of filtration and thus recharge of the aquifer, exhausting wells soon after they were drilled and pushing more drilling further out.

In 1941, with urban growth galloping out of control and water needs rising commensurately, the federal government approved tapping the basin of the Lerma, to the east of the Valley of Mexico. Then, a system of wells in the area of Teoloyuca, in the old Cuautitlan River basin (the PAI system) opened up to supply the city. When by the 1970s that too proved insufficient, the city began tapping more distant water resources, starting with the Cutzamala basin, in Michoacán. Currently under consideration are projects that would transfer water from further basins—Amacuzac to the south, Oriental to the east, and Tecolutla to the northeast.[19] Since all of these basins have their own agricultural districts where all forms of production take place, much of the cost of this transfer of water will once more be borne by rural populations far beyond the capital. This is a third layer of the odorous bulb—that the more the city consumes imported water and drafts from its own aquifer, the more it needs to evacuate, further stressing the drainage capabilities and therefore feeding back into the natural phenomenon of flooding.

The paradoxical consequences of this process are many, the most notable being persistent official enthusiasm for the model of drafting and draining, despite its evident historical failure and the return of old enemies. These enemies include more than just flooding, which never really ceased visiting the city. The lakes too are rising from the dead. Extreme subsidence in the southern parts of the basin, on the flatlands of Chalco, caused by deep-well pumping from 1985 on, has allowed a New Lake Chalco to form in recent years. As of the publication of this book, its surface area was four hundred hectares. Unlike its extinct colonial namesake, this lake is in urbanized and industrialized areas and vulnerable to a variety of contaminants.[20] It is highly unlikely that the modern equivalents of the lovely wife of the Spanish ambassador to Mexico, Fanny Calderón de la Barca, who in the 1840s delighted in canoe rides down the Canal de La Viga, buying flowers and fresh produce from Indian boatmen along the way, will be spending their leisurely afternoons here any time soon.

In fact, scientists from multiple disciplines, not to mention the workers and engineers who actually run the city's hydraulic and drainage infrastructure, coincide in their diagnosis that the draft-and-drain paradigm is unsustainable. This is not just because the groundwater contained in the aquifer beneath the city is currently not a renewable resource, but because what is left of it is vulnerable to contamination and the metropolitan area depends on it for meeting more than 70 percent of its water needs.

Studies reveal that contamination by sewage, garbage dumps, industrial chemicals and heavy metals, and spills and leaks of hydrocarbon products could happen in three ways. Surface infiltration through the transition zone already spreads throughout the aquifer. Second, in the "lacustrine zone" (the old lakebeds) itself, formerly believed to be protected by the impervious clayey layer between it and the surface, contamination can percolate through the various man-made piercings of this layer. Contaminants can even reach the aquifer directly by flowing around the outside of the deep drainage system tubes that traverse the lacustrine zone or through cracks and leaks in the system caused by differential subsidence. Finally, surface contaminants can flow down fractures that have developed in the clay layers due to subsidence and loss of moisture and which cannot reseal as they are quickly filled with detritus and sediment, unfortunately highly permeable materials.[21] Under current conditions, an earthquake a fraction of the magnitude of the 1985 event could activate all of these infiltration routes, massively contaminating the aquifer fairly quickly and leaving the vast majority of the twenty million inhabitants of the metropolitan area who lack the resources to elude these conditions in a sticky situation.

The urban sustainability crisis of the City of Mexico—as that of the entire planet—has generated its own growth industry. Entrepreneurs and construction and architectural design firms have offered "solutions" that tap into long traditions of idealization of the ancient Tenochtitlan's relationship with its lacustrine setting while offering little substantially different from the juicy development contracts that fed the crisis in the first place. Fortunately, as a result of this history but in contrast to their colonial counterparts, members of the scientific community in Mexico have over the course of the previous century pioneered many of the disciplines that throw light on the phenomena this pump-and-drain paradigm has unleashed. Geophysicists, hydrologists, soil mechanics experts, urban planners, and a whole slew of technical professionals have studied the problems and devised potential solutions to cut through this paradigm. Critical of the fatal water importing and pumping, they have proposed collecting seasonal rainfall and reinjecting it into the aquifer, thorough recycling of water, a reorganization of the stages of use, and the dispersion of productive activities now concentrated in the basin to outlying areas.

Unfortunately, the laws of nature and physics inexorably apply to these scientists too: the inertia of the draft-and-drain paradigm prevails because there is no greater social force to dislodge it from its path. This is eerily reminiscent of the colonial period, when the weakness of class interests and actions that might have either stopped the Desagüe altogether or redirected it toward multiple and generative goals meant that critiques of its shape or functioning could never gain traction. Drilling artesian wells continues, now reaching farther down than ever. Siphoning of water from adjoining regions continues. Migration to the capital continues also, as governments finish demolishing the last remains of "ecological autonomy" among rural populations,[22] and the closure of opportunities in the United States deepen the crisis for the Mexican poor. Building, paving, and developing continue because it is good for business, as usual. And so do flooding, the sinking of the city, and the exhaustion of that finite resource, fresh water. The dream of dry land may be well on its way to becoming a nightmare.

# Notes

## PREFACE

1. Alison Gopnik, Andrew N. Meltzoff, and Patricia K. Kuhl, *The Scientist in the Crib: What Early Learning Tells Us About the Mind* (New York: Harper Perennial, 2001).

2. "The New Cultural History Comes to Old Mexico," in *Writing Mexican History* (Stanford: Stanford University Press, 2012), 223–64.

3. Ted Steinberg, "Down to Earth: Nature, Agency, and Power in History," *The American Historical Review* 107, no. 3 (June 2002): 798–820. As done in Angel Palerm and Eric Wolf, *Agricultura y civilización en Mesoamérica* (Mexico City: SEP, 1972); Elinor G. K. Melville, *A Plague of Sheep: Environmental Consequences of the Conquest of Mexico* (Cambridge and New York: Cambridge University Press, 1994); Jacinta Palerm Viqueira, "Sistemas hidráulicos y organización social: La polémica y los sistemas de riego del Acolhuacan septentrional," *Mexican Studies / Estudios Mexicanos* 11, no. 2 (summer 1995): 163–78; Michael C. Meyer, *Water in the Hispanic Southwest* (Tucson: University of Arizona Press, 1984; 1996); Cynthia Radding, *Wandering Peoples: Colonialism, Ethnic Spaces, and Ecological Frontiers in Northwestern Mexico, 1700–1850* (Durham, NC: Duke University Press, 1997); Christian Brannstrom, ed., *Territories, Commodities and Knowledges* (London: Institute for the Study of the Americas, 2004).

4. David Edgerton, "De l'innovation aux usages. Dix thèses éclectiques sur l'histoire des techniques," *Annales EHSS* 4–5 (July–October 1998): 815–37.

5. Eric H. Ash, *Power, Knowledge, and Expertise in Elizabethan England* (Baltimore, MD: Johns Hopkins University Press, 2004), 8–17.

6. The term "class" in this book refers to the social groups' overall relationship to capital, land, water, and the resources necessary to produce things (the "means of production") and the differences among specific sectors within classes. The term "elites" is used to encompass the propertied classes (urban and rural landowners, merchants, and capitalized entrepreneurs) and the bureaucratic and lettered castes (who do not own the means of production and therefore are not classes per se).

7. Kenneth Pomeranz, "Introduction," in *The Environment and World History*, ed. Edmund Burke and Kenneth Pomeranz (Berkeley: University of California Press, 2009).

8. Vera S. Candiani, "The Desagüe Reconsidered: Environmental Dimensions of Class Conflict in Colonial Mexico," *The Hispanic American Historical Review* 92, no. 1 (February 2012): 5–39.

## NOTE ON THE COLONIAL REGIME

1. Tulio Halperín Donghi, "Backward Looks and Forward Glimpses from a Quincentennial Vantage Point," *Journal of Latin American Studies* 24 (1992): 219–34.

2. Colin M. MacLachlan, *Spain's Empire in the New World: The Role of Ideas in Institutional and Social Change* (Berkeley: University of California Press, 1991), 28. On the evolution of custom as a basis of the colonial legal regime, see Guillermo Flores Margadant, "La *consuetudo contra legem* en el derecho indiano a la luz del *ius commune*," *Anuario Mexicano de Historia del Derecho* (1990) 169–88.

3. Guillermo Flores Margadant, "El agua a la luz del derecho novohispano. Triunfo de realismo y flexibilidad," *Anuario Mexicano de Historia del Derecho*, Sección de Estudios, I (1989), 113–46.

4. Delfina E. López Sarrelangue, "Las tierras comunales indígenas de la Nueva España en el siglo XVI," *Estudios de Historia Novohispana* 1 (January 1966).

5. One vara is .83 meters.

6. See Woodrow Borah, *Justice by Insurance: The General Indian Court of Colonial Mexico and the Legal Aides of the Half-Real* (Berkeley: University of California Press, 1983), 136–38. On indigenous landholding over time, see Mariano Galván Rivera, *Ordenanzas de tierras y aguas, ó sea Formulario geométrico-judicial para la designación, establecimiento, mensura, amojonamiento y deslinde de las poblaciones*, 5th ed. (Paris: Librería de Rosa y Buret, 1868), 188–206; George McBride, *The Land Systems of Mexico* (New York: American Geographical Society of New York, 1923); Wistano Luis Orozco, *Los ejidos de los pueblos* (Mexico City: Ed. El Caballito, 1975); Francisco de Solano, "El régimen de tierras y la significación de la composición de 1591," *Revista de la Facultad de Derecho de México* 26, nos. 101–2 (1976): 649–70; Enrique Florescano, "La formación de los trabajadores en la época colonial, 1521–1750," in *La clase obrera en la historia de México. De la colonia al imperio*, ed. Carmen Valcarce (Mexico City: Siglo XXI, 1980), 9–124; Robert J. Knowlton and Lucrecia Orensanz, "El ejido mexicano en el siglo XIX," *Historia Mexicana* 48, no. 1 (July–September 1998): 71–96.

7. Estimates vary. See Hanns J. Prem, "Spanish Colonization and Indian Property in Central Mexico, 1521–1620," *Annals of the Association of American Geographers* 82, no. 3 (September 1992): 444–59. See also Charles Gibson, *The Aztecs Under Spanish Rule* (Stanford: Stanford University Press, 1964), 277–80; Karl Butzer and Elizabeth K. Butzer, "Transfer of the Mediterranean Livestock Economy to New Spain: Adaptation and Ecological Consequences," in *Global Land Use Change: A Perspective from the Columbian Encounter*, ed. B. L. Turner II et al. (Madrid: CSIC, 1995), 151–93.

8. For the state of the agrarian question, see Rani T. Alexander, "Introduction: Haciendas and Agrarian Change in Rural Mesoamerica," and Eric Van Young, "Beyond the Hacienda: Agrarian Relations and Socioeconomic Change in Rural Mesoamerica," both in *Ethnohistory Special Issue: Beyond the Hacienda:*

*Agrarian Relations and Socioeconomic Change in Rural Mesoamerica* 50, no. 1 (winter 2003): 3–14 and 231–45, respectively.

9. See details in J. H. Elliott, *Imperial Spain: 1469–1716* (London: Edward Arnold, 1963); Charles Gibson, *Spain in America* (New York: Harper Collins, 1966); Mark A. Burkholder and D. S. Chandler, *From Impotence to Authority: The Spanish Crown and the American Audiencias, 1687–1808* (Columbia: University of Missouri Press, 1977).

10. Louisa Schell Hoberman, "City Planning in Spanish Colonial Government: The Response of Mexico City to the Problem of Floods, 1607–1637" (PhD diss., Columbia University, 1972).

## INTRODUCTION

1. Participants and ceremony in AGI, México, leg. 328, exp. 15, f. 2v; AGN, Desagüe, vol. 25, exp. 2, f. 4.

2. Richard Kagan, *Urban Images of the Hispanic World, 1493–1793* (New Haven, CT, and London: Yale University Press, 2000), 26–28.

3. This makes it difficult to ascertain how far the Desagüe actually extended as a physical object at any given point in time. Therefore, in this book the term "Desagüe proper" refers only to the feeder canal that led water into the tunnel, the tunnel itself, its outlet, and the spillway with sluices that controlled the flow of water into the feeder canal. "Desagüe district" refers to both the structures that were articulated to this central device, mainly the Cuautitlan River diversion dam, the causeways or levees that controlled Lakes Zumpango, Xaltocan, and San Cristóbal, and the geographical space that these structures linked to the project.

4. John Tutino, *Making a New World: Founding Capitalism in the Bajío and Spanish North America* (Durham, NC: Duke University Press, 2011).

5. For Humboldt's complete description of the Desagüe, see his *Essai Politique Sur le Royaume de la Nouvelle-Espagne*, 5 vols. (Paris: F. Schoell, 1811), II: 194–261. On this inspiration, see Richard Weiner, "Mexican Reception of 'Political Essay,'" in *Alexander von Humboldt: From the Americas to the Cosmos*, ed. Raymond Erickson et al. (New York: Bildner Center for Western Hemisphere Studies, 2004), 415–26.

6. *Memoria para la carta hidrográfica del valle de México* (Mexico City: Imprenta de A. Boix, 1864), 132.

7. *Atlas Pintoresco e Histórico de los Estados Unidos Mexicanos* (Mexico City: Debray Sucesores, 1885). On García Cubas's cartography, see especially Raymond B. Craib, *Cartographic Mexico: A History of State Fixations and Fugitive Landscapes* (Durham, NC: Duke University Press, 2004).

8. See Wiebe E. Bijker, Thomas P. Hughes, and Trevor J. Pinch, eds., *The Social Construction of Technological Systems: New Directions in the Sociology and History of Technology* (Cambridge, MA: MIT Press, 1987); and a critique in Langdon Winner, "Upon Opening the Black Box and Finding it Empty: Social Constructivism and the Philosophy of Technology," *Science Technology & Human Values* 18, no. 3 (summer 1993): 362–78.

9. See Robert Redfield, Ralph Linton, and Melville J. Hertskovitz, "Memoran-

dum for the Study of Acculturation," *American Anthropologist* 38, no. 1 (January–March 1936), 149–52; Fernando Ortiz, *Contrapunteo cubano del tabaco y el azúcar* (1947; reprint, Caracas: Biblioteca Ayacucho, 1978). For a critique, see Enrique Rodríguez-Alegría, "Narratives of Conquest, Colonialism, and Cutting-Edge Technology," *American Anthropologist* 110 (2008): 33–43.

10. An exception is Magdalena A. García Sánchez, "El modo de vida lacustre en el valle de México, ¿mestizaje o proceso de aculturación?" in *Mestizajes tecnológicos y cambios culturales en México*, ed. Enrique Florescano and Virginia García Acosta (Mexico City: CIESAS-Miguel Ángel Porrúa, 2004), 21–90.

11. Pierre Lemonnier, "Introduction," in *Technological Choices: Transformation in Material Cultures Since the Neolithic*, ed. Pierre Lemonnier (London and New York: Routledge, 1993), 1–35. See also Pierre Lemonnier, *Elements for an Anthropology of Technology* (Ann Arbor: Museum of Anthropology-University of Michigan, 1992), ch. 3.

12. Peirce Lewis, "Axioms for Reading the Landscape: Some Guides to the American Scene," in *The Interpretation of Ordinary Landscapes: Geographical Essays*, ed. D. W. Meinig (New York: Oxford University Press, 1979), 11–32; Don Mitchell, "New Axioms for Reading the Landscape: Paying Attention to Political Economy and Social Justice," *Political Economies of Landscape Change* 89, no. 1 (2008): 29–50.

13. AGN, Desagüe, vol. 14, exp. 5, f. 5v.

14. See Karl Appuhn, *A Forest on the Sea: Environmental Expertise in Renaissance Venice* (Baltimore, MD: Johns Hopkins University Press, 2009); Chandra Mukerji, *Impossible Engineering: Technology and Territoriality on the Canal du Midi* (Princeton, NJ: Princeton University Press, 2009).

15. Michael Polanyi, *The Tacit Dimension* (London: Routledge and Kegan Paul, 1966). See also H. M. Collins, "What Is Tacit Knowledge?" in *The Practice Turn in Contemporary Theory*, ed. Theodore R. Schatzki (New York: Routledge, 2001), 107–19; Davis Baird, *Thing Knowledge: A Philosophy of Scientific Instruments* (Berkeley: University of California Press, 2004).

16. Charles Gibson, "The Pre-Conquest Tepanec Zone and the Labor Drafts of the Sixteenth Century," *Revista de Historia de América*, nos. 57–58 (January–December 1964): 136–45; Teresa Rojas, *La Organización del Trabajo para las Obras Públicas: El Coatequitl y las Cuadrillas de Trabajo* (Mexico City: INAH, 1977).

17. Such projects were obviously present in early modern Asia as well, of course. While relevant for a different comparative analysis, they are not in the Desagüe context.

## CHAPTER ONE

1. Per unit of surface area, wetlands produce more biomass than other ecosystems.

2. Jaime Durazo and R. N. Farvolden, "The Groundwater Regime of the Valley of Mexico from Historic Evidence and Field Observations," *Journal of Hydrology* 112, nos. 1–2 (December 1989): 171–90.

3. Ross Hassig, *Trade, Tribute, and Transportation: The Sixteenth-century Political Economy of the Valley of Mexico* (Norman: University of Oklahoma Press, 1985).

4. Sherburne F. Cook and Woodrow Borah, *Essays in Population History, Vol. III: Mexico and California* (Berkeley: University of California Press, 1971–79), 168–69.

5. Geohydrological setting in Luis Espinosa, "Descripción oro-hidrográfica y geológica del Valle de México," in Junta Directiva del Desagüe del Valle de México, *Memoria histórica, técnica y administrativa de las obras del Desagüe del Valle de México, 1449–1900*, 3 vols. (Mexico City: Tipografía de la Oficina Impresora de Estampillas, Palacio Nacional, 1902), I: 1; José Luis Bribiesca-Castrejón, "Hidrología histórica del Valle de México," *Ingeniería Hidráulica en México* 14, no. 3 (1960): 43–46. Development of settlement in Paul Tolstoy et al., "Early Sedentary Communities of the Basin of Mexico," *Journal of Field Archaeology* 4, no. 1 (1977): 91–106; Christine Niederberger, "Early Sedentary Economy in the Basin of Mexico," *Science* 203, no. 4376 (January 12, 1979): 131–42; Guillermo Acosta Ochoa, "Las ocupaciones precerámicas de la cuenca de México. Del poblamiento a las primeras sociedades agrícolas," *ArqueoWeb* 8, no. 2 (January 2007).

6. Ecosystemic impact of humans in Enrique Beltrán, *El hombre y su ambiente. Ensayo sobre el Valle de México* (Mexico City: Tezontle, 1958); Emily McClung de Tapia, "Pre-Hispanic Agricultural Systems in the Basin of Mexico," in *Imperfect Balance: Landscape Transformations in the Precolumbian Americas*, ed. David L. Lentz (New York: Columbia University Press, 2000), 121–46; among others.

7. Compare peasant attitudes in Richard Hoffman, "Economic Development and Aquatic Ecosystems in Medieval Europe," *American Historical Review* 101 (1996): 631–69.

8. W. T. Sanders, J. R. Parsons, and R. Santley, *The Basin of Mexico: Ecological Processes in the Evolution of Civilization* (New York: Academic Press, 1979), 323.

9. See Deborah L. Nichols, "Risk and Agricultural Intensification during the Formative Period in the Northern Basin of Mexico," *American Anthropologist* 89, no. 3 (September 1987): 596–616.

10. Bernardino de Sahagún, *Historia general de las cosas de Nueva España* (Mexico City: A. Valdés, 1829–30), vol. 3, bk. 10, chs. 12, 23, 25, 26; bk. 11, chs. 2, 3, 7, 12; Harold Wayne McBride, "Formative Ceramics and Prehistoric Settlement Patterns in the Cuauhtitlan Region, Mexico" (PhD diss., University of California, Los Angeles, 1974), pt. 3; Teresa Rojas Rabiela, *La cosecha del agua en la Cuenca de México* (Mexico City: CIESAS, 1998), 27–98; Elizabeth M. Brumfiel, ed., *Production and Power at Postclassic Xaltocan* (Mexico City: INAH/University of Pittsburgh, 2006); and elsewhere.

11. See Candiani, "Desagüe Reconsidered."

12. Deborah L. Nichols, "Pre-Hispanic Settlement and Land Use in the Northwestern Basin of Mexico, the Cuautitlan Region" (PhD diss., Pennsylvania State University, 1980), 64, 107–8.

13. Francisco Platas et al., *Evolución de la Ingeniería Sanitaria y Ambiental en México* (Mexico City: DDF, 1994), 44–47.

14. Angel Palerm, *Obras hidráulicas prehispánicas en el sistema lacustre del Valle de México* (Mexico City: INAH, 1973), 237; Deborah Nichols and Charles

Frederick, "Irrigation Canals and Chinampas: Recent Research in the Northern Basin of Mexico," *Recent Research in Economic Anthropology*, suppl. 7 (Greenwich, CT: JAI Press, 1993), 123–50.

15. Rafael A. Strauss K., "El área septentrional del Valle de México: problemas agrohidráulicos prehispánicos y coloniales," in *Nuevas noticias sobre las obras hidráulicas prehispánicas y coloniales en el valle de México*, ed. Teresa Rojas Rabiela, Rafael A. Strauss K., and José Lameiras (Mexico City: SEP, INAH, 1974), 137–74.

16. Faustino Galicia Chimalpopoca et al. trans, *Anales de Cuautitlan* (Mexico City: Anales del Museo Nacional, 1885), 54.

17. Teresa Rojas Rabiela, "Aspectos tecnológicos de las obras hidráulicas coloniales en el valle de México" (MA thesis, ENAH, 1974), 75–84.

18. Chimalpopoca et al., *Anales de Cuautitlan*, 55, 76.

19. Vera S. Candiani, "Draining the Basin of Mexico: Science, Technology and Society, 1608–1808" (PhD diss., University of California, Berkeley, 2004), 141.

20. AGN, Desagüe, vol. 6, exp. 13, f. 18v; AGI, México, leg. 328, exp. 39, f.v, 164.

21. Nichols and Frederick, "Irrigation Canals." See also Sanders, Parsons, and Santley, *Basin of Mexico*, 176–77.

22. Strauss K., "El área septentrional."

23. See tributary structure in Pedro Carrasco, *The Tenochca Empire of Ancient Mexico: The Triple Alliance of Tenochtitlan, Tetzcoco, and Tlacopan* (Norman: University of Oklahoma Press, 1999).

24. Marcos Mazari, "Algo más sobre la Isla de los Perros, el Colegio Nacional y el agrietamiento de arcillas lacustres," in *Memoria del Colegio Nacional* (Mexico City: El Colegio Nacional, 1994), 313–33. See also Alexander M. Puzrin et al., *Geomechanics of Failures* (Dordrecht and London: Springer, 2010), ch. 1.

25. Edward Calnek, "Settlement Pattern and Chinampa Agriculture at Tenochtitlan," *American Antiquity* 37, no. 1 (January 1972): 104–15.

26. Palerm, *Obras hidráulicas prehispánicas*; M. Carballal and M. Flores, "Elementos hidráulicos en el Lago de México-Texcoco en el posclásico," *Arqueología Mexicana* 12, no. 68 (July–August 2004): 28–33.

27. Margarita Vargas Betancourt, "Santiago Tlatelolco y el sistema hidráulico de la ciudad de México (1523–1610)," in *Los indios y las ciudades de la Nueva España*, ed. Felipe Castro Gutiérrez (Mexico City: UNAM, 2010), 123–40.

28. Chimalpopoca et al., *Anales de Cuautitlan*, 78.

29. Carballal and Flores, "Elementos hidráulicos."

30. See Manuel Toussaint et al., *Planos de la ciudad de México: siglos XVI y XVII* (Mexico City: UNAM, IIE, 1938), 55–84, figs. 5, 7–8. For ancient Tenochtitlan, see Eduardo Matos Moctezuma, *Tenochtitlan* (Mexico City: FCE, 2006); and Eduardo Matos Moctezuma, *The Great Temple of the Aztecs* (London: Thames and Hudson Ltd., 2006).

31. See especially W. T. Sanders, J. R. Parsons, and R. S. Santley, *The Basin of Mexico: Ecological Processes in the Evolution of a Civilization*, vol. 1 (New York: Academic Press, 1979); Andrew Sluyter, "Intensive Wetland Agriculture in Mesoamerica: Space, Time, and Form," *Annals of the Association of American Geographers* 84, no. 4 (1994): 557–84; Karl W. Butzer, "Economic Aspects

of Water Management in the Pre-Hispanic New World," *Antiquity* 70, no. 267 (1996): 200–205; Exequiel Ezcurra, *De las chinampas a la megalópolis: El medio ambiente en la cuenca de México* (Mexico City: FCE, 1995), ch. 2; Thomas M. Whitmore and B. L. Turner, *Cultivated Landscapes of Middle America on the Eve of Conquest* (Oxford: Oxford University Press, 2001), 45–49.

32. Karl W. Butzer, "Ethno-agriculture and Cultural Ecology in Mexico: Historical Vistas and Modern Implications," *Conference of Latin Americanist Geographers* 17–18 (1991–92): 139–52.

33. Melville, *Plague of Sheep.*

34. Karl W. Butzer, "Biological Transfer, Agricultural Change, and Environmental Implications of 1492," *International Germplasm Transfer: Past and Present* (1995): 3–29; B. L. Turner and Karl W. Butzer, "The Columbian Encounter and Environmental Change," in *Global Land Use Change: A Perspective from the Columbian Encounter*, ed. B. L. Turner et al. (Madrid: CSIC, 1995), 1–26.

35. Gibson, *Aztecs*, 368.

36. Stephen Tobriner, "The Mexico Earthquake of September 19, 1985—Past Decisions, Present Danger: An Historical Perspective on Ecology and Earthquakes in Mexico City," *Earthquake Spectra* 4, no. 3 (1988): 469–79.

37. Carlos Sempat Assadourian, "The Colonial Economy: The Transfer of the European System of Production to New Spain and Peru," *Journal of Latin American Studies Supplement* (January 1, 1992): 55–68.

38. Schell Hoberman, "City Planning," 23–24.

39. Palerm, *Obras hidráulicas pre-Hispánicas*, 90–94.

40. Enrico Martínez, "Relación de enrrico Martínez architeto y maestro mayor de la obra de el desagüe de la Laguna de México," in *Memoria histórica, técnica y administrativa*, II: 5–12.

41. Andrés Cavo, *Historia de México* (Mexico City: Editorial Patria, 1949), 192.

42. See Candiani, "Desagüe Reconsidered."

43. Richard Kagan, *Urban Images of the Hispanic World, 1493–1793* (New Haven, CT, and London: Yale University Press, 2000), 25; see also "Diálogo tercero: Alrededores de México," *México en 1554; tres diálogos latinos*, trans. Joaquín García Icazbalceta (Mexico City: Antigua Librería de Andrade y Morales, 1875), 163–293.

44. Palerm, *Obras hidráulicas pre-Hispánicas*, 132. The Nezahualcoyotl dam was longer and farther to the east. San Lázaro causeway was repaired in 1580 and 1604. See Ignacio González Tascón, *Ingeniería Española en Ultramar*, 2 vols. (Madrid: CEHOPU/ CEDEX/ MOPT/ Colegio de Ingenieros de Caminos, Canales y Puertos, 1992), I: 275–76.

45. José Fernando Ramírez, *Memoria acerca de las obras e inundaciones en la ciudad de México*, intro. Teresa Rojas Rabiela (Mexico City: SEP/INAH, 1976), 39n32a; 47n44a.

46. Emma Pérez-Rocha, *Ciudad en peligro: Probanza sobre el desagüe general de la ciudad de México, 1556* (Mexico City: INAH, 1996), 23.

47. AHDF Actas de Cabildo impresas, vol. 345A, November 11, 1555; December 17, 1556.

48. Luis González Obregón, "Reseña histórica del Desagüe del Valle de México," in *Memoria histórica, técnica y administrativa*, I: 62–63.

49. For the fate of these proposals, see Michael Mathes, "To Save a City: The Desagüe of Mexico-Huehuetoca, 1607," *The Americas* 26, no. 4 (April 1970): 419–38.

50. Raquel Pineda Mendoza, *Origen, vida y muerte del acueducto de Santa Fé* (Mexico City: UNAM-IIE, 2000), 84–85; Pérez-Rocha, *Ciudad en peligro*, 17–21.

51. Durazo and Farvolden, "Groundwater Regime."

52. AHDF Actas de Cabildo impresas, vol. 345A, November 26, 1555.

53. Pérez-Rocha, *Ciudad en peligro*, 20–21.

54. González Obregón, in *Memoria histórica, técnica y administrativa*, I: 67–68; Jorge Gurría Lacroix, *El desagüe del valle de México durante la época novohispana* (Mexico City: UNAM, 1978), 52–54.

55. González Obregón, in *Memoria histórica, técnica y administrativa*, I: 64–66.

56. "Capítulos que dirige al rey Pedro de Ledesma, sobre las cosas que conviene proveer en Nueva España para engrandecimiento del país y aumento de la Real Hacienda," in *Epistolario de la Nueva España, 1505–1818*, ed. Francisco del Paso y Troncoso, 16 vols. (Mexico City: J. Porrúa e hijos, 1939–42), XV: 73–80.

57. F. de Cepeda and A. Carrillo, "Relación universal legítima y verdadera del sitio en que está fundada la muy noble insigne y muy leal ciudad de México" (1637), in *Obras públicas en México: Documentos para su Historia*, 3 vols. (Mexico City: Sec. de Obras Públicas, 1976), I: 42–44.

58. Rodolfo Acuña-Soto et al., "When Half of the Population Died: The Epidemic of Femorrhagic Fevers of 1576 in Mexico," *FEMS Microbiology Letters* 240 (2004): 1–5; Hanns J. Prem, "Disease Outbreaks in Central Mexico During the Sixteenth-Century," in *Secret Judgments of God: Old World Disease in Colonial Spanish America*, ed. Noble David Cook and W. George Lovell (Norman: University of Oklahoma Press, 1992), 20–48; Miguel E. Bustamante, "Aspectos históricos y epidemiológicos del hambre en México," in *Ensayos sobre la historia de las epidemias en México*, ed. Enrique Florescano and Elsa Malvido (Mexico City: IMSS, 1982), 37–66; Elsa Malvido, "Cronología de epidemias y crisis agrícolas en la época colonial," *Historia Mexicana* 89 (1973): 96–101.

59. González Obregón, in *Memoria histórica, técnica y administrativa*, I: 72–73.

60. Gibson, *Aztecs*, 63–68, 87.

61. AGN, Tierras, vol. 2026, exp. 2, *Repartimiento que se hizo de las partes de agua que les pertenecen a los Pueblos, barrios, Haziendas, y ranchos de esta cavezera de Quautitlan*, f. 27.

62. AGN, Tierras, vol. 2026, exp. 2, ff. 27–28v.

63. AGN, Desagüe, vol. 11, exp. 5, ff. 1–7v; AGN, Tierras, vol. 2028, exp. 5, ff. 10–11v.

64. Strauss K., "El área septentrional"; Rojas Rabiela, "Aspectos tecnológicos."

65. AGN, Tierras, vol. 2028, exp. 5, f. 118.

66. AGN, Desagüe, vol. 12, exp. 5, f. 50.

67. AHA, Aguas Nacionales, 18–212–7/30.

68. AHA, Aguas Nacionales, 18–212–5/30; AGN, Desagüe, vol. 12, exp. 5, ff. 48–48v.

69. Modern plans also call each of the five canals a "river."

70. AGN, Tierras, vol. 2026, exp. 2, ff. 31–31v, and map in AGN, Tierras,

vol. 2028, exp. 5, f. 118. Also AGN, Desagüe, vol. 14, exp. 5, 67fs, f. 1v; and AGN, Desagüe, vol. 12, exp. 5, f. 50, *"Lista de los interesados en la agua de la Pila Real."* See also Israel Sandré Osorio, "Del Derecho Colonial al Derecho Municipal: la Distribución de las Aguas del Río Cuautitlan," *Boletín del Archivo Histórico del Agua* 2, no. 5 (September–December 1995): 37–49.

71. AGN, Desagüe, vol. 11, exp. 5, f. 7; AGN, Desagüe, vol. 12, exp. 13, f. 66v.

72. AGN, Indios, vol. 6.2, exp. 852, f. 208. Strauss K., "El área septentrional"; Rojas Rabiela, "Aspectos tecnológicos," ch. 3.

73. AGN, Desagüe, vol. 11, exp. 5, f. 7; AGN, Desagüe, vol. 12, exp. 13, f. 66v.

74. Susan Chimonas, "Occupational History of Pre-Hispanic Xaltocan," in *Production and Power*, ed. Brumfiel, 169–94, and others in same volume.

75. Gibson, *Aztecs*, 366–67, 446; Platas et al., *Evolución de la Ingeniería*, 46.

76. Candiani, "Desagüe Reconsidered."

77. García Sánchez, "El modo de vida lacustre."

78. Antonio Lot and Alejandro Novelo, *Iconografía y estudio de plantas acuáticas de la Ciudad de México y sus alrededores* (Mexico City: UNAM-Instituto de Biología, 2004), 112.

79. AGN, Desagüe, vol. 7, exp. 4, ff. 3v, 159v.

80. AGN, Desagüe, vol. 18, exp. 3, 9 fols. In the piedmont of the Cuautitlan area, floodwater irrigation captured seasonal runoff: Nichols, "Pre-Hispanic Settlement," ch. 4. For details on *entarquinamiento*, see Martín Sánchez Rodríguez, "Mexico's Breadbasket: Agriculture and the Environment in the Bajío," in *A Land Between Waters: Environmental Histories of Modern Mexico*, ed. Christopher R. Boyer (Tucson: University of Arizona Press, 2012), 50–72.

81. Delfina E. López Sarrelangue, "Las tierras comunales indígenas de la Nueva España en el siglo XVI," *Estudios de Historia Novohispana* 1 (January 1966).

82. As mentioned earlier, one vara is .83 meters.

83. See Woodrow Borah, *Justice by Insurance: The General Indian Court of Colonial Mexico and the Legal Aides of the Half-Real* (Berkeley: University of California Press, 1983), 136–38. On indigenous landholding over time, see Enrique Florescano, "La formación de los trabajadores en la época colonial, 1521–1750," in *La clase obrera en la historia de México: De la colonia al imperio*, ed. Carmen Valcarce (Mexico City: Siglo XXI, 1980), 9–124; Francisco de Solano, "El régimen de tierras y la significación de la composición de 1591," *Revista de la Facultad de Derecho de México* 26, nos. 101–2 (1976): 649–70; Mariano Galván Rivera, *Ordenanzas de tierras y aguas, ó sea Formulario geométrico-judicial para la designación, establecimiento, mensura, amojonamiento y deslinde de las poblaciones*, 5th ed. (Paris: Librería de Rosa y Buret, 1868), 188–206; George McBride, *The Land Systems of Mexico* (New York: American Geographical Society of New York, 1923); Raúl Lemus García, *Derecho agrario mexicano, sinopsis histórica* (Mexico City: LIMUSA, 1975), 115–16; Wistano Luis Orozco, *Los ejidos de los pueblos* (Mexico City: Ediciones El Caballito, 1975); Robert J. Knowlton and Lucrecia Orensanz, "El ejido mexicano en el siglo XIX," *Historia Mexicana* 48, no. 1 (July–September 1998): 71–96.

84. Michael C. Meyer, "The Legal Relationship of Land to Water in Northern Mexico and the Hispanic Southwest," *New Mexico Historical Review* 60, no. 1 (January 1985): 61–79.

CHAPTER TWO

1. See especially Alain Musset, *De l'eau vive à l'eau morte: Enjeux techniques et culturels dans la vallée de Mexico (XVe-XIXe s)* (Paris: Éditions Recherche sur les Civilisations, 1991).

2. Cepeda and Carrillo, *Relación Universal*, 50–57.

3. José María Marroquí, *La Ciudad de México*, 2d facsimile edition (Mexico City: Jesús Medina, 1969), 114–15.

4. AHDF Actas de Cabildo impresas, vol. 356, September 17, 1607, ff. 180–87.

5. Ibid., f. 192.

6. Ramírez, *Memoria acerca de las obras e inundaciones*, 60.

7. González Obregón, in *Memoria histórica, técnica y administrativa*, I: 76.

8. Enrico Martínez, *Breve relación de la importancia de la obra del Desagüe, Y de lo que está Hecho y Resta por hazer en ella*, in *Memoria histórica, técnica y administrativa*, II: 13–18.

9. Martínez, *Relación de enrrico Martínez*; Enrico Martínez, *Verdadera relación de la obra que se hacía para desaguar esta laguna de México* (1628), in Secretaría de Obras Públicas, *Obras públicas en México. Documentos para su historia: Relaciones del Desagüe del Valle de México, Años de 1555–1823*, 3 vols. (Mexico City: n.p., 1976), III: 27–38, 34.

10. Schell Hoberman, "City Planning." See also Mathes, "To Save a City"; Louisa Schell Hoberman, "Enrico Martínez: Printer and Engineer," in *Struggle and Survival in Colonial America*, ed. David Sweet and Gary Nash (Berkeley: University of California Press, 1981), 331–46; Francisco de la Maza, *Enrico Martínez, cosmógrafo e impresor de Nueva España* (Mexico City: UNAM, 1991), among others.

11. AGI, México, leg. 328, exp. 2, f. 2; Fray Andrés de San Miguel, *Informe dado en 1636–37 al virrey marqués de Cadereyta, acerca del Desagüe de Huehuetoca*, in *Obras públicas en México: Documentos*, III: 46, 54; Edward Halse, *A Dictionary of Spanish and Spanish-American Mining and Metallurgical Terms* (London: Charles Griffin & Company; Philadephia: J.B. Lippincott Company, 1908), 44, 178, 198, 324.

12. Enrico Martínez, *Verdadera Relación*, 27, 34.

13. AGI, México, leg. 328, exp. 3, f. 15.

14. RAH, Manuscritos de América, 9/1917, exp. 12, ff. 184–84v.

15. AGEM, Manuscritos, Pueblos del Estado de México, vol. 2, exp. 3; AHDF, Inundaciones, vol. 2272, exp. 6, f. 82.

16. RAH, Manuscritos de América, 9/1917, exp. 12, ff. 182–84.

17. Francisco del Barrio Lorenzot, *Ordenanzas de gremios de la Nueva España: Compendio de los tres tomos de la Compilación Nueva de Ordenanzas* (Mexico City: Secretaria de Gobernación, 1920), 244.

18. RAH, Manuscritos de América, 9/1917, exp. 12, f. 184.

19. AGN, Desagüe, vol. 15, exp. 4, f. 58.

20. AGN, Indiferente Virreinal, caja 3196, exp. 3.

21. Ibid., exp. 7.

22. For pre-Hispanic *llamamiento* and its relevance to sixteenth-century Spanish labor districting, see Charles Gibson, "Llamamiento General, Repartimiento, and the Empire of Acolhuacan," *The Hispanic American Historical Review* 36,

no. 1 (February 1956): 1–27; and Charles Gibson, "The Pre-Conquest Tepanec Zone and the Labor Drafts of the Sixteenth Century," *Revista de Historia de América* 57–58 (January–December 1964): 136–45.

23. Silvio Zavala, *Ordenanzas del trabajo, siglos XVI y XVII* (Mexico City: UNAM, 1947), doc. 56.

24. AGN, Desagüe, vol. 4, exp. 1, f. 105.

25. Ibid., ff. 284–96.

26. Sander Spanoghe, "El repartimiento de mano de obra indígena dentro de la construcción de la catedral de México: Cuentas de Indios," in *Orbis in orbem: liber amicorum John Everaert*, ed. Jan Parmentier and Sander Spanoghe (Ghent: Academia Press, 2001), 431–73.

27. Gibson, *Aztecs*, 241.

28. Because data for labor is incomplete, scholars use financial records as proxies: Gibson, *Aztecs*, 241; Schell Hoberman, "City Planning," 249; Alain Musset, *El agua en el valle de México: Siglos XVI–XVIII* (Mexico City: Pórtico-CEMC, 1992), 208. But this works only after showing capital investments did not replace human labor, as done in Candiani, *Draining the Basin*, ch. 6 and app. 1.

29. AGN, Desagüe, vol. 4, exp. 1, f. 10.

30. Ibid., f. 2.

31. Gibson, *Aztecs*, 240–42; Silvio Zavala and María Castelo, eds., *Fuentes para la historia del trabajo en Nueva España*, 8 vols. (Mexico City: FCE, 1939–46), IV: ix–xv.

32. AHDF, Actas Antiguas de Cabildo, Book 17, October 2, 1609, 397.

33. AHDF, Actas Antiguas de Cabildo, Book 18, July 18, 1611, 139.

34. Francisco de Garay, *El valle de México: Apuntes históricos sobre su hidrografía* (Mexico City: Oficina Tipográfica de la Sec. de Fomento, 1888), 44–45.

35. The nature of cocoliztli and matlazahuatl is still debated. See John S. Marr and James B. Kiracofe, "Was the Huey Cocoliztli a Hæmorrhagic Fever?" *Medical History* 44 (2000): 341–62.

36. Zavala and Castelo, *Fuentes*, VII: doc. 76.

37. See "composite monarchies" concept in John Elliott, "A Europe of Composite Monarchies," *Past & Present* 137 (November 1992): 48–71.

38. Juan Helguera et al., *El canal de Castilla* (Valladolid: Junta de Castilla y León, 1988), 38–40.

39. Zavala and Castelo, *Fuentes*, VI: doc. 539.

40. Ibid., doc. 560.

41. Ibid., doc. 497.

42. Robert S. Haskett, "Our Suffering with the Taxco Tribute": Involuntary Mine Labor and Indigenous Society in Central New Spain," *The Hispanic American Historical Review* 71, no. 3 (August 1991): 447–75.

43. Zavala and Castelo, *Fuentes*, VI: docs. 539, 542, 560, 497.

44. Zavala and Castelo, *Fuentes*, VI: doc. 345.

45. Ibid., docs. 498–99.

46. AGN, Desagüe, vol 8, exp. 3, f. 122v.

47. Martínez, *Verdadera Relación*, 29.

48. AHDF, Actas Antiguas de Cabildo, Book 18, November 19, 1611.

49. AGI, México, leg. 2771, exp. 2, f. 1.

50. AGI, Indiferente, leg. 2075, exp. 146. Rodríguez, who in Mexican documents appears as Juan Núñez, returned to Spain in 1615. Presumably, Boot was by then communicating on his own.

51. AGI, Contratación, leg. 5334, exp. 2, ramo 17, June 25, 1613; AGI, Contratación, leg. 5413, exp. 38, April 21, 1633.

52. AGI, México, leg. 2771, exp. 2, ff. 2–2v.

53. Ibid., ff. 5, 11.

54. Ibid., f. 28v.

55. Ibid., f. 9.

56. Ibid., ff. 10–10v.

57. Ibid., f. 14v.

58. Ibid., f. 17.

59. A haulage unit, the pipa contained about 469 liters.

60. AGI, México, leg. 2771, exp. 2, f. 24.

61. Ibid., f. 28.

62. Ibid., f. 32.

63. For Milanese canals, see William Barclay Parsons, *Engineers and Engineering in the Renaissance* (Baltimore, MD: Williams and Wilkins Company, 1939), ch. 22.

64. Ramírez, *Memoria acerca de las obras e inundaciones*, 102–9.

65. María I. Monroy Castillo, ed., *Guía de las actas de Cabildo de la ciudad de México. Siglos XVII y XVIII. Años 1601–1610* (Mexico City: DDF–Universidad Iberoamericana, 1987–88), Actas for January 31, February 5 and 14, and March 13, 1620.

66. Cepeda and Carrillo, *Relación Universal*, f. 16.

67. Ramírez, *Memoria acerca de las obras e inundaciones*, 117.

68. Zavala and Castelo, *Fuentes*, VI: doc. 542.

69. Ibid., doc. 547.

## CHAPTER THREE

1. BNE, Fondo Antiguo, Ms. 2362, ff. 267–68v.

2. BNE, Fondo Antiguo, Ms. 18660.2, f. 5.

3. For San Miguel's biography, see Eduardo Báez Macías, *Obras de Fray Andrés de San Miguel* (Mexico City: UNAM-IIE, 2007).

4. Fray Andrés de San Miguel, *Relación del sitio, trabajos y estado de la ciudad de México y su remedio,* BLAC, Andrés de San Miguel Manuscripts, G73, ff. 166–69.

5. Rojas Rabiela, "Aspectos tecnológicos," 171–76; Ramírez, *Memoria acerca de las obras e inundaciones*, 150–54.

6. San Miguel, *Relación del sitio, trabajos y estado*, f. 166v.

7. Fray Andrés de San Miguel, *Informe dado en 1636–37*, 46.

8. San Miguel, *Relación del sitio, trabajos y estado*, f. 166.

9. Ibid., f. 169. Despite his claims to mathematical dexterity, San Miguel made arithmetical errors in sections two and four and as a result also in the addition of segments.

10. AGN, Desagüe, vol. 3, exp. 6, f. 23v; San Miguel, *Relación del sitio, trabajos y estado*, f. 168.

11. Garay, *El valle de México*, 42; José Sala Catalá, "La localización de la capital de Nueva España como problema científico y técnico," *Quipu* 3, no. 3 (1986): 279–97.

12. Garay, *El valle de México*, 43.

13. Gibson, *Aztecs*, 238.

14. AGN, Obras Publicas, 1648, vol. 7, exp. 1, f. 10. Juan Serrano was also one of the master architects intervening in the Desagüe proper, where he likely also must have thought mules were desirable substitutes for humans: "que se ha de conducir con mulas, por ahorrar a su Magestad indios y dinero y corresponder el trabajo de cada mula al de seis indios." See also Pilar Martínez López-Cano, *La génesis del crédito colonial. Ciudad de México, siglo XVI* (Mexico City: UNAM, 2001), 297n74.

15. González Obregón, *Memoria histórica, técnica y administrativa*, 149–51.

16. AGN, Obras Publicas, vol. 7, exp. 1, f. 10.

17. At the time, the price of a mule was about thirty pesos. Ibid., 284n27.

18. Ross Hassig, "One Hundred Years of Servitude: Tlamemes in Early New Spain," in *Ethnohistory* (Supplement to the *Handbook of Middle American Indians*, vol. 4), ed. Ronald Spores and Patricia A. Andrews (Austin: University of Texas Press, 1986), 134–46.

19. San Miguel, *Informe dado en 1636–37*, 53–54, 69–71.

20. Ibid., 50.

21. Martínez, *Verdadera Relación*, 29.

22. Ramírez, *Memoria acerca de las obras e inundaciones*, 160.

23. Schell Hoberman, "City Planning," 245.

24. AGI, México, leg. 2771, exp. 3, ff. 10v–26.

25. San Miguel, *Informe dado en 1636–37*, 54.

26. Ibid., 54.

27. AGI, México, leg. 328, exp. 1.

28. Schell Hoberman, "City Planning."

29. Ramírez, *Memoria acerca de las obras e inundaciones*, 105.

30. Carrillo and Cepeda, *Relación universal*, 73.

31. See descriptions and images of devices in Luis Arciniega et al., "The Representation of Architecture in Construction during the Hispanic Early-Modern Period," in *The Second International Congress on Construction History*, ed. Malcolm Dunkeld et al. (Cambridge, UK: CHS, The Construction History Society, 2006), 1: 221–37; Catherine Wilkinson-Zerner, *Juan de Herrera: Architect to Philip II of Spain* (New Haven, CT: Yale University Press, 1993).

32. See George Kubler, *Mexican Architecture of the Sixteenth Century* (New Haven, CT: Yale University Press, 1948); John McAndrew, *The Open-air Churches of Sixteenth-century Mexico* (Cambridge, MA: Harvard University Press, 1965); Samuel Y. Edgerton, *Theaters of Conversion: Religious Architecture and Indian Artisans in Colonial Mexico* (Albuquerque: University of New Mexico Press, 2001); Robert Ricard, *The Spiritual Conquest of Mexico* (Berkeley and Los Angeles: University of California Press, 1966), ch. 8; Javier Gómez Martínez, *Fortalezas Mendicantes* (Mexico City: University Iberoamericana, 1997); among others.

33. See Manuel Ramos Medina, *El Carmelo novohispano* (Mexico City: Centro de Estudios de Historia de México-Carso, 2008); Nile Ordorika Bengoechea, *El convento del Carmen de San Angel* (Mexico City: UNAM, 1998); Dionisio Victoria Moreno, OCD, *Los carmelitas descalzos y la conquista espiritual de México, 1585–1612*, 2d ed. (Mexico City: Porrúa, 1983).

34. Fray Andrés de San Miguel, *Cuáles han de ser nuestros templos*, BLAC Andrés de San Miguel Manuscripts, G73, ff. 3v–5.

35. See Elías Trabulse, *Ciencia y religión en el siglo XVII* (Mexico City: El Colegio de México, 1974).

36. Bengoechea, *Convento*, 46. The last mention of an outside builder is in 1602 in the works to refurbish the church of San Sebastián in Tlatelolco, for which secular architects were hired. See Moreno, *Los carmelitas descalzos*, 85; Fray Agustín de la Madre de Dios, *Tesoro escondido en el Santo Carmelo Mexicano* (1650; reprint, Mexico City: PROBURSA-Universidad Iberoamericana, 1984), ch. 23.

37. Báez Macías, *Obras de Fray Andrés*, 105–9.

38. Diego de Sagredo, "De algunos principios de geometría necesarios y muy usados en el arte de trazar," *Medidas del Romano* (1526; 1541).

39. Andrés de San Miguel, *Figura es, según Euclides . . .* , BLAC, Andrés de San Miguel Manuscripts, G73, ff. 8–16v.

40. Báez Macías, *Obras de Fray Andrés de San Miguel*, 68.

41. *De cómo con facilidad se saca cuantos granos de semilla de col hacen un montón tan grande como toda la esfera del mundo, tierra y agua*, BLAC, Andrés de San Miguel Manuscripts, G73, ff. 28–29.

42. Ibid.

43. Louisa Schell Hoberman, "Bureaucracy and Disaster: Mexico City and the Flood of 1629," *Journal of Latin American Studies* 6, no. 2 (November 1974): 211–30.

44. See Arne Kaijser, "System Building from Below Institutional Change in Dutch Water Control Systems," *Technology and Culture* 43, no. 3 (July 2002): 521–48.

45. Schell Hoberman, "Bureaucracy and Disaster."

46. Horst Pietschmann, *Las reformas borbónicas y el sistema de intendencias en Nueva España: un estudio político administrativo* (Mexico City: FCE, 1996), 39–41.

47. Ramírez, *Memoria acerca de las obras e inundaciones*, 168–69.

48. AGI, Pasajeros de Indias, Book 11, exp. 2649.

49. Robert C. Padden, "Ordenanza del Patronazgo of 1574: An Interpretive Essay," *The Americas* 12 (1956): 333–54; Virve Piho, "La secularización de las parroquias y la economía eclesiástica en la Nueva España," *Journal de la Société des Américanistes* 64 (1977): 81–88; John F. Schwaller, "The Ordenanza del Patronazgo in New Spain, 1574–1600," *The Americas* 42, no. 3 (January 1986): 253–74.

50. AHDF, Actas Antiguas de Cabildo, Book 27, January 21, 1630.

51. See Paolo Squatriti, ed., *Working with Water in Medieval Europe: Technology and Resource Use* (Leiden; Boston; Cologne: Brill, 2000); Ellen F. Arnold, "Engineering Miracles: Water Control, Conversion and the Creation of a Religious Landscape in the Medieval Ardennes," *Environment and History* 13

(2007): 447–76; Fernando Miguel Hernández, "El sistema hidráulico en un monasterio cisterciense: Santa María de Carracedo (León)," in *El agua en zonas áridas: Arqueología e historia. Hidráulica tradicional de la provincia de Almería, ed. Lorenzo Cara Barrionuevo* (available at Instituto de Estudios Almerienses, 1989), 897–928; among others.

52. Richard Holt, "Medieval England's Water-related Technologies," in *Working with Water in Medieval Europe: Technology and Resource Use*, ed. Paolo Squatriti (Leiden; Boston; Cologne: Brill, 2000), 51–100.

53. See Mariano Cuevas, *Historia de la Iglesia en México*, 5 vols. (Mexico City: Editorial Patria, 1946), vol. 3; Kubler, *Mexican Architecture*; Sidney Markman, *Colonial Architecture of Antigua Guatemala* (Philadelphia, PA: Philosophical Society, 1966); among others, for monastic architecture in New Spain.

54. AHDF, Actas Antiguas de Cabildo Impresas, Book 27, January 24, 1630.

55. Cepeda and Carrillo, *Relación universal*, 223–30; Fray Luis Flores, *Memorial que el padre predicador Fray Luis Flores remite a su Majestad dándole cuenta del estado que tiene la obra del Desagüe de Huehuetoca* (1653), in *Obras públicas en México. Documentos*, III: 96–97; Ramírez, *Memoria acerca de las obras e inundaciones*, 164–65.

56. Helguera et al., *El canal de Castilla*, 37.

57. Flores, *Memorial*, 96–97.

58. Flores, *Memorial*, 100; Ramírez, *Memoria acerca de las obras e inundaciones*, 227.

59. AGN, Desagüe, vol. 4, exp. 1, ff. 2v, 289.

60. Cepeda and Carrillo, *Relación Universal*, 383.

61. AGN, Desagüe, vol. 7, exp. 4, f. 1; Gibson, *Aztecs*, 103–6.

62. Flores, *Memorial*, 109.

63. AGN, Desagüe, vol. 2, exp. 3.

64. AGN, Ordenanzas, vol. 1, exp. 128, f. 120v; AGN, Desagüe, vol. 4, exp. 1, f. 274.

65. Gibson, *Aztecs*, 246–49.

66. Ibid., 240.

67. For the 1607 works a property tax of 1 percent was assessed on the city, including ecclesiastics, yielding 304,013 pesos (Cepeda and Carrillo, *Relación Universal*, 81). Additionally, a tax on wine imports was assessed for most of the colonial period, and the taxes on wine consumption and butcher shops used for municipal water tubing in the City of Mexico were shifted to the Desagüe fund.

68. Gurría Lacroix, *El Desagüe*, 69.

69. Cepeda and Carrillo, *Relación Universal*, 72.

70. AGI, México, leg. 328, exp. 8, f. 10.

71. Cepeda and Carrillo, *Relación Universal*, 384.

72. Gurría Lacroix, *El Desagüe*, 69.

73. AGI, México, leg. 38, exp. 91: Baños to King, October 21, 1661.

74. AGN, Desagüe, vol. 4, exp. 1, f. 11.

75. Ibid., f. 58.

76. Ibid., f. 95.

77. AGN, Desagüe, vol. 12, exp. 13, ff. 71–71v.

78. AGN, Desagüe, vol. 20, exp. 7, f. 52.

79. AGN, Desagüe, vol. 4, exp. 1, ff. 110–12.

80. Flores, *Memorial*, 128, 110.

81. "Mémoire sur l'éboulement qui arrive quelquefois à des portions de montagnes," in *Description des projets et de la construction des ponts* (Paris: Imprimerie de François-Ambroise Didot, 1788), 632–43.

82. Manuel Orozco y Berra, *Memoria para la carta hidrográfica del valle de México* (Mexico City: A. Boix, 1864), 119.

83. AGN, Desagüe, vol. 20, exp. 7, ff. 51–52.

84. AGN, Desagüe, vol. 10, exp. 1, f. 5v.

85. Flores, *Memorial*, 85.

86. Sala Catalá, "La localización de la capital."

87. González Obregón, in *Memoria histórica, técnica y administrativa*, I: 194–95.

88. AGI, México, leg. 328, exp. 15, ff. 1–4v.

89. Cabrera, *Verdad aclarada*, 134.

90. AGN, Desagüe, vol. 7, exp. 4, ff. 114–15.

91. AGI, México, leg. 328, exp. 15, f. 5v.

92. AGI, México, leg. 328, exp. 1, f. 1.

93. AGN, Fomento Desagüe, vol. 1, f. 45. For New World shadufs, see Jacinta Palerm Viqueira et al., "Técnicas hidráulicas en México, paralelismos con el Viejo Mundo," *II Encuentro sobre Historia y Medio Ambiente*, October 24–26, 2001, Huesca, Spain.

94. AGI, México, leg. 328, exp. 42, f. 28v.

95. Cabrera, *Verdad aclarada*, 143–45.

96. Ibid., 144.

97. Cyril Stanley Smith, "Art, Technology, and Science: Notes on Their Historical Interaction," in *Perspectives in the History of Science and Technology*, ed. Duane Rolled (Oklahoma: University of Oklahoma Press, 1971), 129–44.

98. Pamela H. Smith, *The Body of the Artisan: Art and Experience in the Scientific Revolution* (Chicago: University of Chicago Press, 2004); Pamela O. Long, *Openness, Secrecy, Authorship: Technical Arts and the Culture of Knowledge from Antiquity to the Renaissance* (Baltimore, MD: Johns Hopkins University Press, 2001); among others.

99. Cabrera, *Verdad aclarada*, 151.

100. San Miguel, *Informe dado en 1636–37*, 67.

101. Ibid., 69.

102. Pierre Y. Julien, *River Mechanics* (Cambridge: Cambridge University Press, 2002), 2–3.

103. Andrés García de Céspedes's 1606 *Libro de instrumentos nuevos de geometría* has been described as containing "the most comprehensive section on [hydraulics] published in Spain during the Renaissance," but it too did not stray into hydromechanical inquiry. See Victor Navarro Brotons, "Mechanics in Spain at the End of the 16th Century and the Madrid Academy of Mathematics," in *Mechanics and Natural Philosophy before the Scientific Revolution*, ed. Walter Roy Laird and Sophie Roux (Dordrecht, The Netherlands: Springer, 2008), 239–58.

104. Benedetto Castelli, *Of the Mensuration of Running Waters: Translated from the Third Edition, with the Addition of a Second Book*, trans. and intro.

Thomas Salisbury from *Della misura dell' acque correnti* (1628; reprint, London: William Leybourn, 1661), I: 5.

105. San Miguel, *Informe dado en 1636–37*, 56.

106. Ibid.

107. Ibid., 65–66.

108. AGN, Desagüe, vol. 7, exp. 1, ff. 2–6.

109. Humboldt, *Essai politique*, II: 148–49.

110. See units in Manuel Carrera Stampa, "The Evolution of Weights and Measures in New Spain," *The Hispanic American Historical Review* 29, no. 1 (February 1949): 2–24.

111. AGN, Desagüe, vol. 4, exp. 1, f. 289.

112. Ibid., ff. 95, 98.

113. AHDF, Ordenanzas, 1736, vol. 2984, exp. 14, no. 11.

114. AGN, Reales Cédulas Originales, vol. 2, exp. 199, f. 424.

115. Mario Biagioli, *Galileo, Courtier: The Practice of Science in the Culture of Absolutism* (Chicago: University of Chicago Press, 1993), 238.

116. AGN, Desagüe, vol. 4, exp. 1, ff. 1v–2; González Obregón, in *Memoria histórica, técnica y administrativa*, I: 201. In 1636, Huehuetoca's secular priests were assigned two pesos per day of attendance in the Desagüe. AGI, México, leg. 2771, exp. 3, f. 52v.

117. AGN, Desagüe, vol. 7, exp. 4, ff. 158–60.

118. AGI, México, leg. 2772, exp. 1, f. 11.

## CHAPTER FOUR

1. AGI, México, leg. 2772, exp. 1, ff. 23–29; Manuel Carrera Stampa, *Gremios mexicanos*, 60.

2. AHDF, Arquitectos, vol. 380, exp. 1, ff. 23–29.

3. Martha Fernández, *Arquitectura y gobierno virreinal: los maestros mayores de la ciudad de México, siglo XVII* (Mexico City: UNAM, 1985), 49.

4. Francisco Barrio Lorenzot, *Ordenanzas de gremios de la Nueva España* (Mexico City: Secretaría de Gobernación, 1920), 279.

5. AGN, General de Parte, vol. 8, exp. 33, f. 17v; vol. 8, exp. 36, f. 20; vol. 9, exp. 37, f. 25; vol. 54, f. 33.

6. González Obregón, in *Memoria histórica, técnica y administrativa*, I: 157.

7. AHDF, Arquitectos, vol. 380, exp. 1, ff. 34–35v.

8. AHDF, Ordenanzas, vol. 2984, exp. 14.

9. AGN, Desagüe, vol. 7, exp. 4, f. 6.

10. Villalobos: AGN, Indios, vol. 26, exp. 127, f. 123v; Martínez: AGN, Desagüe, vol. 8, exp. 3, f. 132; Quevedo: AGN, Desagüe, vol. 8, exp. 2, ff. 4, 11; Espínola: AGN, Desagüe, vol. 8, exp. 3, f. 90; Valladolid: AGN, Desagüe, vol. 9, exp. 1, ff. 100–102, 116.

11. AGN, Desagüe, vol. 9, exp. 1, f. 31.

12. Patricia Seed, *To Love, Honor, and Obey in Colonial Mexico: Conflicts over Marriage Choice, 1574–1821* (Stanford: Stanford University Press, 1992), 62–63.

13. AGN, Desagüe, vol. 8, exp. 2, f. 9; AGN, Desagüe, vol. 9, exp. 1, ff. 115–16; AGN, Desagüe, vol. 14, exp. 1, ff. 15–15v; AGN, Desagüe, vol. 17, exp. 11, f. 27.

14. AGN, Desagüe, vol. 9, exp. 1, ff. 79, 124–124v;

15. *Memoria de los principales ramos de la policía urbana de los fondos de la Ciudad de México* (Mexico City: Imprenta de J.M. Andrade y F. Escalante, 1864), 278; AGN, Desagüe, vol. 20, exp. 7, f. 13; Cabrera, *Verdad aclarada*, 170.

16. AGI, México, leg. 328, exp. 39, f. 72; AGN, Desagüe, vol. 7, exp. 1, f. 1; AGN, Desagüe, vol. 7, exp. 4, f. 118.

17. AGN, Desagüe, vol. 9, exp. 1, f. 77v.

18. AGN, Desagüe, vol. 7, exp. 4, f. 6.

19. AGI, México, leg. 328, exp. 39, f. 73v.

20. Cabrera, *Verdad aclarada*, 158–60, 175.

21. AGN, Desagüe, vol. 7, exp. 1, ff. 1–6.

22. AGN, Desagüe, vol. 7, exp. 2, ff. 18v–19.

23. AGN, Desagüe, vol. 7, exp. 3, f. 113.

24. Ibid.

25. Mukerji, *Impossible Engineering*; Appuhn, *Forest on the Sea*.

26. Joaquín Bérchez, *Arquitectura Mexicana de los siglos XVII y XVIII* (Mexico City: Azabache, 1992), 40–42.

27. AHDF, Arquitectos, vol. 380, exp. 1, f. 117v.

28. Bérchez, *Arquitectura Mexicana*, 30.

29. Irving Leonard, *Baroque Times in Old Mexico: Seventeenth-Century Persons, Places, and Practices* (Ann Arbor: University of Michigan Press, 1966), ch. 6; "Carta relación de don Pedro Porte Casanate sobre descubrimiento del Golfo de California (1643, 1649)," in *Colección de documentos inéditos, relativos al descubrimiento, conquista y organización de las antiguas posesiones españolas de América y Oceanía*, ed. Luis Torres de Mendoza (Madrid: Imprenta de Frías y Compañía, 1868), 5–18; I. Osorio Romero, *Historia de las bibliotecas novohispanas* (Mexico City: SEP, 1986), 52–54; Elías Trabulse, *El círculo roto* (Mexico City: FCE/SEP, 1984), 33–35.

30. Francisco de Solano, *Normas y leyes de la ciudad hispanoamericana*, 2 vols. (Madrid: CSIC-Centro de Estudios Históricos, 1996), I: 194–218; Carlos Chanfón Olmos, *Historia de la arquitectura y el urbanismo mexicanos*, 2 vols. (Mexico City: UNAM/FCE, 1997) II: 200.

31. Jorge E. Hardoy, *El modelo clásico de la ciudad colonial hispanoamericana* (Buenos Aires: Instituto Torcuato di Tella, 1968), 40.

32. Carlos Blázquez Herrero and Severino Pallaruelo Campo, *Maestros del agua*, 2 vols. (Zaragoza: Gobierno de Aragón, 1999), II: 538.

33. *Los diez libros de Architectura de León Baptista Alberto traduzidos de latín en romance* (Madrid: Casa de Alonso Gómez, 1582), Libro Tercero, f. 92.

34. See Pamela O. Long, "Hydraulic Engineering and the Study of Antiquity: Rome, 1557–70," *Renaissance Quarterly* 61, no. 4 (winter 2008): 1098–138.

35. Bérchez, *Arquitectura Mexicana*, 45.

36. Chanfón Olmos, *Historia de la Arquitectura*, II: 271.

37. González Obregón, in *Memoria histórica, técnica y administrativa*, I: 123.

38. AHDF, Ordenanzas, vol. 2984, exp. 14, ff. 1–7.

39. AGEM, Manuscritos, Pueblos del Estado de México, vol. 1, exp. 46; AGI, México, leg. 328, exp. 5, f. 4v.

40. AGN, Desagüe, vol. 18, exp. 7, ff. 242–242v.

41. AGEM, Manuscritos, Pueblos del Estado de México, vol. 1, exp. 45, 13 fols.

42. AGN, Desagüe, vol. 18, exp. 9, f. 275.

43. Ibid., ff. 275–76v.

44. AGN, Desagüe Fomento, vol. 1, f. 45: Inventory of 1683 taken by guarda Fernando Chirinos Villalobos of the equipment acquired by Solís.

45. AGI, México, leg. 328, exp. 3, f. 14v.

46. AGN, Desagüe, vol. 18, exp. 9, ff. 275–76v.

47. Palerm, *Obras hidráulicas prehispánicas*, 97.

48. Arnold Pacey, *Technology in World Civilization* (Cambridge, MA: MIT Press, 1990), 61.

49. William E. Doolittle, *Canal Irrigation in Prehistoric Mexico: The Sequence of Technological Change* (Austin: University of Texas Press, 1990), 120.

50. The Cruz del Rey stood part-way on this mortar-and-stone dam, shown under the name of albarradón del Rey in Figure 5.2.

51. AGN, Desagüe, vol. 4, exp. 1, f. 12v.

52. AGN, Desagüe, vol. 20, exp. 1, f. 15.

53. González Obregón, in *Memoria histórica, técnica y administrativa*, I: 216–17.

54. AGN, Desagüe, vol. 6, exp. 13, f. 18v; AGI, México, leg. 328, exp. 39, ff. 84v, 164; AGN, Desagüe, vol. 19, f. 15.

55. Enrico Martínez, *Verdadera relación*, 33–34.

56. AGI, México, 328, exp. 15, f. 5.

57. See James Lockhart, *Nahuas and Spaniards: Postconquest Central Mexican History and Philology* (Stanford: Stanford University Press, 1991).

58. Brigitte B. de Lameiras, *Terminología agrohidráulica prehispánica nahua* (Mexico City: INAH-SEP, 1974), 29.

59. Ibid., 29–30; Alonso de Molina and Julius Platzmann, *Vocabulario de la lengua mexicana*, vols. 1–2 (1571; reprint, Leipzig: B. G. Teubner, 1880), 56, 66, 157; Rémi Siméon, *Diccionario de la lengua náhuatl o mexicana* (1885; reprint, Mexico City: Siglo XXI, 1996), 133–34.

60. AGN, Desagüe, vol. 4, exp. 1, f. 12.

61. In AGI, México, 328, exp. 39, for instance.

62. AGN, Desagüe, vol. 20, exp. 1.

63. Jerzy Rzedọwski, "An Ecological and Phytogeographical Analysis of the Grasslands of Mexico," *Taxon* 24, no. 1 (February 1975): 67–80; Jerzy Rzedowski and Laura Huerta M., *Vegetación de México* (Mexico City: Ed. Limusa, 1978), 222–33.

64. List of tules in Nahuatl and their uses in Orozco y Berra, *Memoria para la carta hidrográfica*, 163–64.

65. AGN, Desagüe, vol. 4, exp. 1, f. 12v.

66. AGI, México, 2771, exp. 2, f. 56v.

67. AGN, Ríos y Acequias, vol. 4, exp. 5.

68. This *xalocotle* (*jalocote, oyamel*) appears as "guava" in Francisco Hernández, *Cuatro libros de la naturaleza y virtudes de las plantas y animales que están recibidos en uso de medicina en la Nueva España* (1615; reprint, Morelia, Mexico: Escuela de Artes, 1888), IV: 51, 55–56. However, it was more likely the pine *Abies religiosa* described in Paul C. Standley, *Trees and Shrubs of Mexico* (Washington, DC: Government Printing Office, 1920), 59.

69. See Martha Fernández, *Cristóbal de Medina Vargas y la arquitectura salomónica en la Nueva España durante el siglo XVII* (Mexico City: UNAM-IIE, 2002), 428, 434, 457, 564.

70. Full list also in Rojas Rabiela, "Aspectos tecnológicos," 111.

71. AGN, Ríos y Acequias, vol. 4, exp. 5, f. 4.

72. AGN, Desagüe, vol. 6, exp. 13, ff. 2–18v.

73. AGN, Indios, vol. 45, exp. 123; AGN, Indios, vol. 54, exp. 5, f. 28v.

74. AGN, Indios, vol. 25, exp. 311, f. 229.

75. AGN, Desagüe, vol. 9, exp. 1, ff. 134v–38v.

76. Data in Candiani, *Draining the Basin*, 394.

77. AGN, Desagüe, vol. 9, exp. 1, ff. 93v–94v, 115v.

78. AGN, Desagüe, vol. 14, exp. 6, f. 7v.

79. AGI, México, 328, exp. 5, f. 5; AGN, Desagüe, vol. 14, exp. 6, f. 7v.

80. Giovanni Francesco Gemelli Careri, *Giro del Mondo*, vol. 6 (Naples: Stamperia di Guiseppe Roselli, 1700), 124.

81. AGN, Desagüe, vol. 4, exp. 1, ff. 96–98.

82. José Ignacio Urquiola Permisán, "Los textiles bajo el mestizaje tecnológico," in *Mestizajes tecnológicos*, 201–59; Humboldt, *Essai Politique*, III: 191.

83. According to José Antonio de Alzate y Ramírez, *cáñamo* was *pilpitzitzintli*, used by Indians for its medicinal and narcotic leaves and seeds. See *Gacetas de Literatura*, tomo 4, 1772 (Puebla, 1831 reprint), 95–101.

84. AGN, Desagüe, vol. 2, exp. 3, ff. 18–21. See also Standley, *Trees and Shrubs*, 118–20.

85. Blázquez Herrero and Pallaruelo Campo, *Maestros del agua*, II: 323.

86. Personal communication, SDP engineer Julián Zarco-Herrera, Tequixquiac, Mexico, July 2008.

87. AGN, Desagüe, vol. 4, exp. 1, f. 12v.

88. AGN, Desagüe, vol. 8, exp. 3, f. 111.

89. AGN, Desagüe, vol. 14, exp. 5, ff. 15–20v.

90. For hacienda calendars and daily routines, see Arij Ouweneel, "Schedules in Hacienda Agriculture: The Cases of Santa Ana Aragón (1765–1768) and San Nicolás de los Pilares (1793–1795), Valley of Mexico," *Boletín de Estudios Latinoamericanos y del Caribe* 40 (June 1986): 63–97.

91. AGN, Desagüe, vol. 4, exp. 1, f. 12v.

92. In the seventeenth century it was comparable to wages for hacienda labor.

93. In 1687, 1694, and 1699 tributary recounts were undertaken for the towns in the jurisdiction of Cuautitlan, including Teoloyuca. What motivated the recount was an epidemic of *cocolitzle*. In 1687–88, Teoloyuca had 346 tributarios. AGN, Indios, vol. 89, exp. 1, f. 70.

94. AGN, Desagüe, vol. 7, exp. 3, ff. 161v–62.

## CHAPTER FIVE

1. AGN, Desagüe, vol. 9, exp. 1, ff. 60–61v.

2. BNE, Fondo Antiguo, ms. 3534, "Informe de Francisco Xavier de Gamboa sobre el Desagüe de Huehuetoca," f. 216v.

3. AGN, Desagüe, vol. 15, exp. 4, ff. 49–60.
4. Ibid., 204.
5. AGN, Desagüe, vol. 14, exp. 5, f. 15v.
6. AGN, Tierras, vol. 1868, exp. 11.
7. AGN, Tierras, vol. 1124, exp. 2.
8. RAH, Manuscritos de América, 9/1917, exp. 12, ff. 196–97v. Other guardas earned 475 pesos.
9. AGN, Desagüe, vol. 17, exp. 11, ff. 28, 38–39.
10. *Extracto de los autos de diligencias y reconocimiento de los ríos, lagunas, vertientes y desagües de la capital México y su Valle* (Mexico City: Viuda de Joseph Bernardo de Hogal, 1748), 3, 5.
11. AGN, Desagüe, vol. 15, exp. 4, ff. 49–60.
12. AGI, México, leg. 328, exp. 15, f. 4v.
13. *Instrucciones y memorias de los virreyes novohispanos*, I: 591.
14. Hubert Chanson, "Embankment Overflow Protection Systems and Earth Dam Spillways," in *Dams: Impacts, Stability and Design*, ed. Walter P. Hayes and Michael C. Barnes (Hauppauge, NY: Nova Science Publishers, 2009), 101–32.
15. Espinosa, *Extracto*, 3.
16. AGN, Desagüe, vol. 24, exp. 12, f. 1.
17. AGN, Desagüe, vol. 26, exp. 15, f. 4v.
18. Ibid., ff. 6v–7.
19. Espinosa, *Extracto*, 25.
20. AGN, Desagüe, vol. 12, exp. 13, f. 71.
21. González Obregón, in *Memoria histórica, técnica y administrativa*, I: 216–17.
22. AGN, Desagüe, vol. 14, exp. 11, f. 19.
23. AGN, Tierras, vol. 2028, exp. 5, f. 6.
24. AGN, Desagüe, vol. 13, exp. 2.
25. Ibid.; AGN, Desagüe, vol. 14, exp. 5, f. 10v.
26. AGN, Desagüe, vol. 14, exp. 5.
27. Slow court culture was sometimes useful to Desagüe neighbors. When in 1728 Coyotepec Indians complained that guarda Joseph de la Barreda barred them from tilling the bed of the lake because of its detrimental effects on the Desagüe, the district magistrate granted the Indians legal possession of the lakebed in question and told Barrera to "go complain to the audiencia." By the time the court cleared up the matter, the Indians may well have harvested there for many seasons (AGN, Desagüe, vol. 11, exp. 1). Then again, this was before Trespalacios.
28. BNM, Fondo Reservado, ms. 1405, "Noticias del Desagüe de Zumpango."
29. AGN, Desagüe, vol. 4, exp. 1, f. 272.
30. AGN, Desagüe, vol. 9, exp. 1, f. 77.
31. Ibid., ff. 76v–77.
32. AGN, Desagüe, vol. 17, exp. 11, f. 38.
33. AGN, Desagüe, vol. 12, exp. 4, ff. 19–19v.
34. AGN, Desagüe, vol. 14, exp. 5, ff. 1v–2.
35. Ibid., ff. 1–2. Also AGN, Desagüe, vol. 14, exp. 6, f. 10v; AGN, Desagüe, vol. 12, exp. 12; AGN, Desagüe, vol. 18, exp. 1. Obligations were attached to the

property, not to individuals, and passed on with the sales of the properties: AGN, Desagüe vol. 14, exp. 5, f. 6.

36. Haciendas San Pedro Coamatla, Xaltipa, Corregidora, San Matheo, Cartagena, Portales; a flour mill; and ranchos Miranda, Ozumbilla, Angulo, San Joseph, Cárdenas, and Salitre. AGN, Desagüe, vol. 18, exp. 1, f. 1.

37. AHDF, Inundaciones, vol. 2272, exp. 6, f. 82.

38. RAH, Manuscritos de América, 9/1917, exp. 12, f. 189v.

39. Ibid., ff. 187v, 190v. The wine tax appears as farmed out between 1743–51, however: AGN, Desagüe, vol. 15, exp. 4, f. 39v.

40. RAH, Manuscritos de América, 9/1917, exp. 12, f. 185.

41. Ibid., ff. 190–91.

42. Robert C. West, *The Mining Community in Northern New Spain: The Parral Mining District* (1946; reprint, New York: AMS Press, 1980), 127–30.

43. West, *Mining Community*, 127.

44. AGN, Desagüe, vol. 15, exp. 4, ff. 1–4.

45. BNE, Fondo Antiguo, ms. 3534, f. 212.

46. Musset, *El agua*, 207–8; RAH, Manuscritos de América, 9/1917, exp. 12, ff. 185, 188.

47. RAH, Manuscritos de América, 9/1917, exp. 12, f. 213.

48. See Juan Miguel Muñoz Corbalán, *Los ingenieros militares de Flandes a España (1691–1718)*, 2 vols. (Madrid: Ministerio de Defensa, 1990); María Isabel Vicente Maroto and Mariano Esteban Piñeiro, *Aspectos de la ciencia aplicada en la España del siglo de oro* (Toledo: Junta de Castilla y León, Consejería de Cultura y Bienestar Social, 1991), 140, 173–76; Víctor Navarro Brotóns, "Los jesuitas y la renovación científica en la España del siglo XVII," *Studia historica. Historia moderna* 14 (1996): 15–44; among others.

49. AGI, México, leg. 46, exp. 40, ff. 1–2.

50. Horacio Capel Sáez et al., *De Palas a Minerva: La formación científica y la estructura institucional de los ingenieros militares en el siglo XVIII* (Madrid: Serbal-CSIC, 1988), 17.

51. Sebastián Fernández de Medrano, *Rudimentos Geométricos y Militares* (Brussels: Casa de la viuda Vleugart, 1677), n.p.

52. Jesús Pérez Magallón, *Construyendo la modernidad: La cultura española en el tiempo de los Novatores (1675–1725)* (Madrid: CSIC/ Instituto de la Lengua Española, 2002), 35.

53. Juan Navarro-Loidi and José Llombart, "The Introduction of Logarithms into Spain," *Historia Mathematica* 35 (2008): 83–101. In New Spain, sor Juana Inés de la Cruz and Carlos de Sigüenza y Góngora were part of this shift, but focused on astronomy. See José María López Piñero, *Ciencia y técnica en la sociedad española de los siglos XVI y XVII* (Barcelona: Labor, 1979), 450–51.

54. *Compendio Mathemático, Tomo IV* (Valencia: Imprenta de Joseph García, 1757), 312. See also Mariano Esteban Piñero and Vicente Salavert Fabiani, "Las matemáticas," in *Historia de la Ciencia y de la Técnica en la Corona de Castilla*, 4 vols., ed. José María López Piñero (Salamanca: Junta de Castilla y León, 2002), III: 231–58; *Alicia Cámara Muñoz*, "La arquitectura militar del padre Tosca y la formación teórica de los ingenieros entre Austrias y Borbones," in *Los ingenieros militares de la monarquía hispánica en los siglos XVII*

*y XVIII*, ed. *Alicia Cámara Muñoz* (Madrid: Fernando de Villaverde Ediciones, 2005), 133–58.

55. Artist and builder Teodoro de Ardemans, for instance, owned Tosca's and similar texts but modeled his career after Vitruvius, Serlio, and Alberti and understood water through Aristotle, Euclid, Ptolemy, and Galen. See Beatriz Blasco Esquivias, "Una biblioteca modélica. La formación libresca de Teodoro de Ardemans (I y II)," *Ars Longa* 5 (1994): 73–97 and 7–8; (1997): 155–75.

56. See corps history in Capel Sáez et al., *De Palas a Minerva*; José Omar Moncada Maya, *Ingenieros militares en Nueva España. Inventario de su labor científica y espacial, siglos XVI a XVIII* (Mexico City: UNAM, 1993); María Gloria Cano Révora, *Cádiz y el Real Cuerpo de Ingenieros Militares (1697–1847). Utilidad y Firmeza* (Cádiz: Universidad de Cádiz, 1994); and Juan Miguel Muñoz Corbalán, *Los ingenieros militares de Flandes a España (1691–1718)*, 2 vols. (Madrid: Ministry of Defense, 1990); among others.

57. Aurora Rabanal Yus, "El tratado de arquitectura enseñado en la Real y Militar Academia de Matemáticas de Barcelona," *Anuario del Departamento de Historia y Teoría del Arte* (Universidad Autónoma de Madrid) II (1990), 179–85. Bernard Forest de Bélidor's *La science des ingénieurs dans la conduite des travaux de fortification et d'architecture civile* and *Architecture hydraulique ou l'art de conduire, d'élever et de ménager les eaux pour les différents besoins de la vie* came out in 1729 and 1739, respectively, and provided what Tosca did not—algebraic solutions for fortifications and hydraulic design.

58. See Rafael Alcaide González and Horacio Capel Sáez, eds., *Pedro de Lúcuce, Tratado de Cosmografía del curso matemático para la instrucción de los militares (1739–1779)*, Colección Geocrítica de textos electrónicos, 1 (January 2000) http://www.ub.es/geocrit/tcport.htm.

59. Alcaide and Capel, "El Curso de Cosmografía de Lúcuce en las academias de matemáticas militares: El problema de los textos científicos y el desarrollo de la ciencia española del siglo XVIII," in *Pedro de Lúcuce, Tratado de Cosmografía.*

60. Capel Sáez et al., *Palas a Minerva*, 332.

61. AHM, Aparici, rollo 15, tomo LV, sig. 1–2–6, f. 105.

62. AGMS, Expedientes personales, legs. A-497; C-3337; C-3429, L-85; S-1737: Desagüe engineers Ricardo Aylmer, Alberto de Córdova, Miguel del Corral, Nicolás de Lafora, and Manuel Santiesteban all went through this licensing process, the latter three to marry in Mexico.

63. Moncada Maya, *Ingenieros*, 23; Calderón Quijano, *Historia de las Fortificaciones*, 107.

64. AGI, México, 60, leg. 5, exp. 24.

65. Ibid.; Calderón Quijano, *Historia de las Fortificaciones*, 105–95; Francisco Vidargas, *San Juan de Ulúa y Carlos de Sigüenza y Góngora* (Xalapa, Mexico: Instituto Veracruzano de Cultura, 1997).

66. Albert Manucy, "Founding of Pensacola: Reasons and Reality," *The Florida Historical Quarterly* 37, nos. 3–4 (January–April 1959), 223–41.

67. Calderón Quijano, *Historia de las Fortificaciones*, 107, n1.

68. Jorge Galindo Díaz, "El legado técnico de los tratados de fortificación en América hispánica," *Revista Apuntes* 17, nos. 1–2 (2005): 8–29.

69. Rojas, *Teórica y Práctica*, ff. 52–52v.

70. AGI, México, leg. 2771, exp. 2, ff. 55v, 61v.

71. AHDF, Actas Antiguas de Cabildo Impresas, Book 22, January 31, 1620; AGN, RCO, vol. 2, exp. 199, ff. 421–25.

72. Silvio Zavala, *Fuentes para la historia del trabajo en la Nueva España*, vol. 6, doc. 497.

73. AGN, Desagüe, vol. 9, exp. 1, f. 58.

74. *Rudimentos Geométricos y Militares* (Brussels: Casa de la viuda Vleugart, 1677), f. 184.

75. Sebastián Fernández de Medrano, *El architecto perfecto en el arte militar* (Brussels: Casa de Lamberto Marchant, 1700), 190.

76. *Architectvra militaris nova et aucta: oder Newe vermehrte Fortification, von regular Vestungen, von irregular Vestungen vnd aussen Wercken: von praxi Offensivâ und Defensivâ: Auff die neweste niederländische Praxin gerichtet vnd beschrieben* (Leyden, 1631), f. 66.

77. Sebastián Fernández de Medrano, *El ingeniero: primera parte, de la moderna architectura militar* (Brussels: n.p., 1687), f. 108. See also Alicia Cámara Muñoz, "Esos desconocidos ingenieros," in *Los ingenieros militares de la monarquía hispánica*, 13–29.

78. Zavala and Castelo, *Fuentes*, V: doc. 102.

79. See, for example, AGN, Indios, vol. 33, exp. 133, f. 86; AGN, Indios, vol. 54, exp. 304, f. 279.

80. AGN Tierras, vol. 1110, exp. 5, ff. 11–22v.

81. Pedro Betancourt Yáñez et al., "Pérdidas de suelo y potencial hidrológico en parcelas con coberturas vegetativas de especies forrajeras," *TERRA Latinoamericana* 18, no. 3 (July–September 2000): 263–75.

82. AGN, Desagüe, vol. 9, exp. 1, f. 64v.

83. AHM, Aparici, rollo 15, tomo 55, sig. 1–2–6, ff. 232–56, Ordinance 8.

84. Antoine Picon, *French Architects and Engineers in the Age of Enlightenment*, trans. Martin Thom (Cambridge and New York: Cambridge University Press, 1992), 155.

85. Although this is a modern formulation, the *concept* of work originated with Galileo, developing thereafter in the interaction between engineering and science. See D. S. L. Cardwell, "Some Factors in the Early Development of the Concepts of Power, Work and Energy," *The British Journal for the History of Science* 3, no. 3 (June 1967): 209–24.

86. AGN, Desagüe, vol. 9, exp. 1, ff. 60–61v.

87. Ibid., f. 64.

88. Ibid., f. 152.

89. AGN, Desagüe, vol. 10, exp. 1, ff. 4v–5v.

90. AGN, Desagüe, vol. 12, exp. 10, ff. 1–6v.

91. AGN, Desagüe, vol. 12, exp. 4, f. 17.

92. AGN, Desagüe, vol. 12, exp. 10, ff. 1–6: cuentas de gastos en la obra del Desagüe; AGN, Desagüe, vol. 10, exp. 9, f. 15v.

93. AGN, Desagüe, vol. 10, exp. 3, f. 6.

94. AGN, Desagüe, vol. 15, exp. 7, f. 306.

95. AGN, Desagüe, vol. 17, exp. 5, f. 10.

96. Alicia Cámara Muñoz, "La arquitectura militar y los ingenieros de la

monarquía española: Aspectos de una profesión (1530–1650)," *Revista de la Universidad Complutense* (1981): 3, 255–69.

97. AGN, Desagüe, vol. 7, exp. 3, f. 113.

98. Ibid., f. 116.

99. Ibid., f. 116v.

100. See Ernst H. Gombrich, *La imagen y el ojo: Nuevos estudios sobre la psicología de la representación pictórica* (Madrid: Alianza Editorial, 1987), 163.

101. On shadows in plans, see Patricia Zulueta Pérez et al., "El uso de la sombra en los planos y dibujos de los científicos de la Ilustración," XVI International Congress of Graphic Engineering, Zaragoza, June 2–4, 2004.

102. AHM, Aparici, rollo 15, tomo 55, sig. 1–2–6, ff. 232–56, Copia de las ordenanzas dadas al Cuerpo de Ingenieros en 1718.

103. AGN, Desagüe, vol. 18, exp. 6, f. 14.

104. Eugene S. Ferguson, *Engineering and the Mind's Eye* (Cambridge, MA: MIT Press, 1999), 3–4, 42.

105. AGN, Historia, vol. 346, exp. 4, f. 120.

106. AHM, Aparici, rollo 15, tomo 55, sig. 1–2–6, ff. 77–80.

107. AGN, Desagüe, vol. 9, exp. 1, f. 56v.

108. Ibid., f. 63v.

109. María Giménez Prades et al., "El color y su significado en los documentos cartográficos del Cuerpo de Ingenieros Militares del siglo XVIII," *Ge-conservación* (2009): 141–60.

110. Christopher Duffy, *Fire and Stone: The Science of Fortress Warfare, 1660–1860* (London and Vancouver: David and Charles Limited, 1975), 34.

111. AHM, Aparici, rollo 15, tomo 55, sig. 1–2–6, f. 368.

112. AGN, Desagüe, vol. 7, f. 113v.

113. Cabrera, *Verdad aclarada*, 165–66.

114. AGI, México, leg. 328, exp. 2, f. 3v.

115. This Jesuit is not listed in Zambrano's *Diccionario bio-bibliográfico de la Compañía de Jesús*, in Félix Osores's *Noticias bio-bibliográficas de alumnos distinguidos del colegio de San Pedro, San Pablo y San Ildefonso*, or Décorme's *La obra de los jesuitas en México durante la época colonial.*

116. AGN, Desagüe, vol. 19, exp. 19, ff. 12–13.

117. *Compendio Mathemático, Tomo IV* (Valencia, Spain: Imprenta de Joseph García, 1757), 381.

## CHAPTER SIX

1. AGN, Desagüe, vol. 17, exp. 3.

2. AGN, Desagüe, vol. 16, exp. 11, f. 29.

3. Ibid., ff. 2v–3v.

4. Ibid., f. 24–24v.

5. Ibid., f. 29v.

6. Ibid., f. 24.

7. For details on this process, see González Tascón, *Ingeniería Española en Ultramar*, I: 28–35.

8. AGN, Desagüe, vol. 13, exp. 7, f. 37.

9. AGN, Desagüe, vol. 13, exp. 9, ff. 18–21.
10. Ibid., f. 20v.
11. AGN, Desagüe, vol. 13, exp 7, f. 37.
12. AGN, Reales Cédulas Originales, vol. 105, exp. 85.
13. For example, AGN, Desagüe, vol. 17, exp. 4, f. 16.
14. AGN, Desagüe, vol. 26, exp. 1, f. 160.
15. AGI, México, leg. 2772, exp. 29, ff. 7v–9.
16. Mark A. Burkholder and D. S. Chandler, *From Impotence to Authority: The Spanish Crown and the American Audiencias, 1687–1808* (Columbia: University of Missouri Press, 1977, 1982), 51, 63.
17. AGI, México, leg. 2772, exp. 52, f. 10.
18. AGEM, Manuscritos, Pueblos del Estado de México, vol. 2, exp. 8; AGN, Desagüe, vol. 17, exp. 14, f. 11.
19. AGN, Desagüe, vol. 17, exp. 14, f. 5v.
20. Ibid., f. 11v.
21. Burkholder and Chandler, *From Impotence to Authority*, 51, 63.
22. AGN, Desagüe, vol. 18, exp. 6, ff. 18–19.
23. AGI, México, leg. 2772, exp. 29, ff. 12v–13.
24. AGI, México, leg. 2772, exp. 51, f. 78v.
25. Humboldt, *Essai politique*, II: 227.
26. Ernesto de la Torre Villar, *Instrucciones y memorias de los virreyes novohispanos*, 2 vols. (Mexico City: Porrúa, 1991), II: 1112.
27. AGN, Desagüe, vol. 17, exp. 11, f. 2.
28. Ibid., f. 4v.
29. Elías Trabulse, *Francisco Xavier Gamboa: Un político criollo en la Ilustración Mexicana, 1717–1794* (Mexico City: El Colegio de México, 1985), 45–48.
30. The frontispiece of Bancroft Library's copy, for instance, states that "the date of the original is said to have been around 1706 in a copy made in 1719 from which this copy was made." Bancroft, M-M 95.
31. Francisco Xavier de Gamboa, *Comentarios a las ordenanzas de minas* (Madrid: Joaquín Ibarra, 1761), 168–82.
32. Juan José Saldaña, "Nacionalismo y Ciencia ilustrada en América," in *Ciencia, Técnica y Estado en la España Ilustrada*, ed. Joaquín Fernández Pérez and Ignacio González Tascón (Zaragoza, Spain: *Sociedad Española de Historia de las Ciencias y de las Técnicas*, 1990), 115–29.
33. Gamboa, *Comentarios*, 464–65.
34. AGI, México, leg. 2772, exp. 52, f. 32.
35. For example, José Joaquín de Ariscorreta owned the haciendas San Pedro Huamatla and San Nicolás Lanzarote in Cuautitlan. See Guillermina del Valle Pavón, "Intereses del Consulado de comerciantes en la reconstrucción de las calzadas de la ciudad de México, siglo XVIII," *Entorno Urbano: Revista de historia* 2, no. 4 (July–December 1996): 7–24.
36. AGN, Desagüe, vol. 18, exp. 8, ff. 4–5.
37. For instance, AGN, Desagüe, vol. 17, exp. 14, f. 5v.
38. This was not to deprive the workers of their Desagüe rations, but to ensure against the practice by some master architects, guardas, and others who took advantage of their positions in the Desagüe to establish veritable *repartimientos de*

*mercancías* on Desagüe premises, and to appropriate the Desagüe repartimiento laborers for their private businesses.

39. For the classic and revised views on *repartimiento de mercancías*, see Brian R. Hamnett, *Politics and Trade in Mexico, 1750–1821* (Cambridge: Cambridge University Press, 1971); and Jeremy Baskes, *Indians, Merchants, and Markets: A Reinterpretation of the Repartimiento and Spanish-Indian Economic Relations in Colonial Oaxaca, 1750–1821* (Stanford: Stanford University Press, 2000).

40. AGI, México, leg. 2772, exp. 1, ff. 9v–10v.

41. AGN, Desagüe, vol. 18, exp. 1, 6 fols.

42. AGN, Desagüe, vol. 19, exp. 7, f. 4v, and exp. 9.

43. AGN, Desagüe, vol. 17, exp. 11, f. 39.

44. AGN, Desagüe, vol. 18, exp. 8, ff. 1–2v.

45. AGI, México, leg. 2772, exp. 1, f. 9.

46. AGN, Indios, vol. 26, exp. 127, cuaderno 2, f. 123v; AGN, Padrones, 4, f. 240 (map).

47. AGN, Desagüe, vol. 18, exp. 9, f. 259.

48. AGN, Desagüe, vol. 19, exp. 14, f. 6.

49. AGN, Desagüe, vol. 18, exp. 10, ff. 1–19.

50. For example, AGN, Tierras, vol. 2028, exp. 5; AGN, Desagüe, vol. 12, exp. 5, f. 50; AGN, Desagüe, vol. 19, exp. 14.

51. AGI, México, leg. 2772, exp. 26, ff. 89v–90v; AGN, Desagüe, vol. 20, exp. 7, f. 53.

52. For Alzate's career, see *Periodismo científico en el siglo XVIII: José Antonio de Alzate y Ramírez*, ed. Patricia Acévez Pastrana (Mexico City: UNAM, 2001); Jorge Cañizares-Esguerra, *How to Write the History of the New World* (Stanford: Stanford University Press, 2001); Alberto Saladino García, *Dos científicos de la ilustración hispanoamericana* (Mexico City: UNAM, 1990); among others.

53. AGN, Desagüe, vol. 17, exp. 10, f. 17v.

54. Musset, *El agua*, 45–46.

55. Leopoldo Sequeiros, "El geocosmos de Athanasius Kircher: Una imagen organicista del mundo en las ciencias de la naturaleza del siglo XVII," *Llull* 24, no. 51 (2001): 755–807.

56. Thomas F. Glick, "On the Influence of Kircher in Spain," *Isis* 62, no. 3 (autumn 1971): 379–81.

57. Ignacio Osorio Romero, *La luz imaginaria: Epistolario de Atanasio Kircher con los novohispanos* (Mexico City: UNAM, 1993), 7–8.

58. Paula Findlen, "A Jesuit's Books in the New World. Athanasius Kircher and His American Readers," in *Athanasius Kircher: The Last Man Who Knew Everything*, ed. Paula Findlen (New York and London: Routledge, 2004), 316–50; Osorio Romero, *La luz imaginaria*, 17–23.

59. BNM, Fondo Reservado, Lafragua, vol. 936, doc. 7, *Máquina para Desaguar las Minas, inventada, fabricada, examinada, y comprobada por D. Juan Antonio de Mendoza y González* (Mexico City: Imprenta de Joseph Bernardo de Hogal, 1727). This machine may have been inspired by the Newcomen "fire pump" deployed in Real del Monte, as analyzed in Carlos Sempat Assadourian, "La bomba de fuego de Newcomen y otros artificios de desagüe: Un intento de

transferencia de tecnología inglesa a la minería novohispana, 1726–1731," *Historia Mexicana* 50, no. 3 (January–March 2001): 385–457.

60. Horacio Capel, *Organicismo, fuego interior y terremotos en la ciencia española del XVIII* (Barcelona: Facultad de Geografía e Historia, Universidad de Barcelona, 1980).

61. AGN, Desagüe, vol. 17, exp. 10, f. 15v.

62. Ibid., f. 21.

63. Roberto Moreno, "El agua en la Nueva España ilustrada," in *Obras hidráulicas en América colonial* (Madrid: CEHOPU, 1993), 73.

64. AGN, Desagüe, vol. 17, exp. 4, ff. 22–29.

65. See Luis Jáuregui, *Historia económica de México: Los transportes, siglos XVI al XX* (Mexico City: UNAM, 2000), 21–70.

66. AHDF, Desagüe, vol. 740, exp. 11, "Proyecto del Padre Alzate sobre hacer un canal de México a Totolsingo."

67. Ibid., ff. 7–10.

68. Ibid., ff. 12v–14.

69. See Antonio Lafuente and Nuria Valverde, "Linnaean Botany and Spanish Imperial Biopolitics," in *Colonial Botany: Science, Commerce, and Politics in the Early Modern World*, ed. Londa Schiebinger and Claudia Swan (Philadelphia: University of Pennsylvania Press, 2004), 134–47; Patricia Acévez, "La difusión de la ciencia en la Nueva España en el siglo XVIII: La polémica en torno a la nomenclatura de Linneo y Lavoisier," *Quipu* 4, no. 3 (1987): 357–85. For social science, see Cañizares-Esguerra, *How to Write the History*, ch. 5.

70. In his "Arquitectura hidráulica. Economía" (*Observaciones Sobre la Física*, Issue 8, June 15, 1787), Alzate returned to the sumps, now drawing support for his pyrophyliacic solution from Betancourt's *Theatro Mexicano*, where this author allegedly described how the catastrophic flood of 1629 had ended thanks to an earthquake.

71. Juan José Saldaña, "Ciencia y felicidad pública en la ilustración americana," in *Historia social de las ciencias en América Latina*, ed. J. J. Saldaña (Mexico City: UNAM-Miguel Angel Porrúa, 1996), 151–207.

72. J. H. Elliott, *Empires of the Atlantic World: Britain and Spain in America 1492–1830* (New Haven, CT: Yale University Press, 2007), 303–4; Antonio Lafuente, "Enlightenment in an Imperial Context: Local Science in the Late-Eighteenth-Century Hispanic World," *Osiris*, 2d series, vol. 15 (2000): 155–73.

73. On Alzate as diffuser of utilitarianism, see José Enrique Covarrubias, *En busca del hombre útil: un estudio comparativo del utilitarismo neomercantilista en México y Europa, 1748–1833* (Mexico City: UNAM, 2005), 393–402.

74. AGN, Desagüe, vol. 17, exp. 10, f. 17v.

75. Ibid., f. 17v.

76. AGN, Desagüe, vol. 19, exp. 17, ff. 1–19; AGN, Desagüe, vol. 32, exp. 10.

77. Maurice Daumas, *Les instruments scientifiques aux XVIIe et XVIIIe siècles* (Paris: Presses Universitaires de France, 1953), 343.

78. AGN, Desagüe, vol. 19, exp. 17, ff. 15–16v.

79. Ibid.

80. Ibid.

81. Ibid.

82. "Circular a los Ingenieros Directores de las distintas provincias de 29 de diciembre de 1737 para que los que se admitiesen para ingenieros sea mediante examen y que los que se hubiesen admitido sin hacerlo se les examine," AHM, Aparici, rollo 15, tomo LV, sig. 1–2–6, ff. 77–80.

83. Jorge Galindo Díaz, "La enseñanza de la perspectiva como parte de la fortificación en el siglo XVIII: El caso de la Real Academia de Matemáticas de Barcelona," *Varia Historia* 24, no. 40 (July–December 2008): 465–82.

84. AGN, Desagüe, vol. 18, exp. 6, ff. 1–2.

85. Ibid., ff. 14–15.

86. AHDF, Desagüe, vol. 740, exp. 35.

87. BNE, Fondo Antiguo, ms. R/38236, no. 1, no. 3.

88. *Estatutos de la Real Academia de San Carlos de Nueva España* (Mexico City: Imprenta Nueva Mexicana de Felipe de Zúñiga y Ontiveros, 1785), no. 13, 20, 29.

89. Elizabeth Fuentes Rojas, *La Academia de San Carlos y los Constructores del Neoclásico* (Mexico City: UNAM, 2002), 34; Omar Moncada Maya, "El Ingeniero militar Miguel Costanzó en la Real Academia de Bellas Artes de San Carlos de la Nueva España," *Scripta Nova* VII, no. 136 (March 2003); available online at http://www.ub.es/geocrit/sn/sn-136.htm.

90. Moncada Maya, "Miguel Costanzó en la Real Academia."

91. AHDF, Desagüe, vol. 740, exp. 41.

92. See Teresa Sánchez Lázaro, *Carlos Lemaur y el Canal de Guadarrama* (Madrid: Colegio de Ingenieros de Caminos, Canales y Puertos, 1995); Juan Helguera, "Aproximación a la historia del canal de Castilla," in *El canal de Castilla* (Valladolid, Spain: Junta de Castilla y León, Consejería de Cultura y Bienestar Social, 1988), 11–162; *Memoria histórica del canal imperial de Aragón: noticias sobre las utilidades que produce* (Madrid: Imprenta de D.J. Palacios, 1833); and Conde de Sástago, *Descripción de los canales imperial de Aragón y Real de Tauste* (n.p., 1790). In the drawings that accompany Sástago's work it is evident that the multiple aims were imbricated in the design, with numerous references to irrigation structures such as side canals and intakes, for example.

93. Guillermina del Valle Pavón, "Antagonismo entre el Consulado de México y el virrey Revillagigedo por la apertura comercial de Nueva España, 1789–1794," *Estudios de historia novohispana* 24 (2001): 111–37.

94. AGN, Desagüe, vol. 13, exp. 9, ff. 18–21.

95. *Instrucciones y memorias de los virreyes*, II: 1114.

96. Espinosa, *Extracto*, 17; Max L. Moorhead, *New Mexico's Royal Road: Trade and Travel on the Chihuahua Trail* (Norman: University of Oklahoma Press, 1958), 106.

97. Valle Pavón, "Intereses del Consulado."

98. AGI, México, leg. 2772, exp. 1, f. 12.

99. Revillagigedo to Branciforte, in *Instrucciones y memorias de los virreyes*, II: 1114. See also Ernest Sánchez i Santiró, "Comerciantes, mineros y hacendados: la integración de los mercaderes del Consulado de la ciudad de México en la propiedad minera y azucarera de Cuernavaca y Cuautla de Amilpas (1750–1821)," in *Mercaderes, comercio y consulados de Nueva España en el siglo XVIII*, ed. Guillermina del Valle Pavón (Mexico City: Instituto Mora, 2003), 159–90.

100. Christiana Borchart de Moreno, "Los miembros del Consulado de la

Ciudad de México en la época de Carlos III," *Jahrbuch fur Geschichte von Staat, Wirtschaft, und Gesellschaft Lateinamerikas* 14 (1977): 134–60.

101. Valle Pavón, "Antagonismo."

102. Ibid.

103. AHDF, Desagüe, vol. 740, exp. 13, f. 12.

104. AGN, Desagüe, vol. 24, exp. 2, f. 16.

105. Ibid., f. 3.

106. Humboldt, *Essai Politique*, II: 252–54.

107. AGN, Desagüe, vol. 24, exp. 2, ff. 16–17v.

108. Valle Pavón, "Antagonismo."

109. They indeed lost this exclusivity in 1795, thenceforth sharing with the Guadalajara and Veracruz consulados.

110. BNE, Fondo Antiguo, ms. 3534, ff. 212–15; AGN, Desagüe, vol. 24, exp. 3, f. 2v.

111. Revillagigedo to Consulado, July 9, 1790, AHDF, Desagüe, vol. 740, exp. 13, ff. 1–1v.

112. *Instrucciones y memorias de los virreyes*, I: 1064–65.

113. *Instrucción reservada que el Conde de Revilla Gigedo dió a su sucesor en el mando, Marqués de Branciforte sobre el gobierno de este continente en el tiempo en que fue su virrey* (Mexico City: Imprenta de Agustín Guiol, 1831), 46–47.

114. BNM, Fondo Reservado, ms. 1405, "Noticias del desagüe de Zumpango," f. 16v. "If [Branciforte] provided defenses for the Kingdom/If the latter's and the King's wealth he increases/with measures taken with refined talent/No less did he prevent risks and ills/for this City with works that he imagined,/which shall be celebrated in its Annals:/such is that of the Desagüe where he approves/ what its superindendant proposes."

115. Ibid., f. 15.

116. Valle Pavón, "Intereses del Consulado."

117. Valle Pavón, "Antagonismo."

## CHAPTER SEVEN

1. As an echo from the seventeenth century, mathematics for most still included areas such as astronomy, hydrostatics, and hydrodynamics.

2. Tutino, *Making a New World.*

3. AHDF, Desagüe, vol. 740, exp. 13, ff. 12–12v.

4. AGN, Tierras, vol. 287, exp. 6, f. 30.

5. Orozco y Berra, *Memoria para la carta hidrográfica*, 171.

6. For the difference between the old and new dynastic regimes, see Colin M. MacLachlan, *Spain's Empire in the New World: The Role of Ideas in Institutional and Social Change* (Berkeley: University of California Press, 1991).

7. Stanley J. Stein, "Bureaucracy and Business in the Spanish Empire, 1759–1804: Failure of a Bourbon Reform in Mexico and Peru," *The Hispanic American Historical Review* 61, no. 1 (February 1981): 2–28.

8. Spain, *Real Ordenanza para el establecimiento e instrucción de intendentes de ejército y provincia en el reino de la Nueva España*, 1786, Art. 28, 35–36; Isabel Gutiérrez del Arroyo, "El nuevo régimen institucional bajo la real orde-

nanza de intendentes de la Nueva España (1786)," *Historia Mexicana* 39, no. 1 (July–September 1989): 89–122.

9. Trabulse, *Francisco Xavier Gamboa*, 134.

10. BNE, Fondo Antiguo, ms. 3534, ff. 211–32; AGN, Desagüe, vol. 24, exp. 2.

11. Susan Deans-Smith, "'A Natural and Voluntary Dependence': The Royal Academy of San Carlos and the Cultural Politics of Art Education in Mexico City, 1786–1797," *Bulletin of Latin American Research* 29, no. 3 (2010): 278–95.

12. AGN, Desagüe, vol. 29, exp. 1, ff. 3v–4.

13. AGN, Desagüe, vol. 29, exp. 10, ff. 2v–5v.

14. Trabulse, *El Círculo Roto*, 101–2.

15. Juan Manuel Espinosa Sánchez, "Newton en la Ciencia Novohispana del Siglo XVIII" (PhD diss., UAM-Iztapalapa, Mexico, 2006), 60–62.

16. BNM, Fondo Reservado, Lafragua, vol. 936, doc. 6.

17. Espinosa Sánchez, "Newton en la Ciencia," 84–85.

18. AGN, Desagüe, vol. 29, exp. 10, ff. 29–37.

19. Ibid., ff. 59v–61.

20. Ibid., f. 62.

21. Ibid., f. 92.

22. Ibid., ff. 91, 96.

23. Ibid., f. 60. In the 1770s, Antonio de Ulloa remarked on the lack of woods in the vicinity of the Desagüe, which he attributed to aridity of the soil. See *Descripción geográfico-física de una parte de la Nueva España in Francisco de Solano, Antonio de Ulloa y la Nueva España* (Mexico City: UNAM, 1987), 48.

24. AGN, Desagüe, vol. 29, exp. 10, f. 95v.

25. Ibid., f. 61.

26. AGN, Desagüe, vol. 19, exp. 19, ff. 2–3.

27. AGN, Desagüe, vol. 31, exp. 1, ff. 3–5.

28. AHDF, Desagüe, vol. 740, exp. 18, n.p.

29. AGN, Desagüe, vol. 31, exp. 1, ff. 3–5.

30. AGN, Desagüe, vol. 31, exp. 1 and 6; AGN, Desagüe, vol. 30, exp. 5.

31. AGN, Desagüe, vol. 33, exp. 8, ff. 3–12.

32. AHDF, Desagüe, vol. 740, exp. 18, n.p.

33. AGN, Desagüe, vol. 31, exp. 1, ff. 19v–20.

34. Ibid., f. 31v.

35. Ibid., f. 44v.

36. AHDF, Desagüe, vol. 740, exp. 32, "Informe emitido por Don Antonio Velázquez sobre el reconocimiento que hizo al nuevo canal para Desagüe de la laguna de San Cristóbal."

37. AGN, Desagüe, vol. 31, exp. 1, ff. 40v–41.

38. AGN, Desagüe, vol. 29, exp. 10, f. 120.

39. *Estatutos de la Real Academia de San Carlos de Nueva España* (Mexico City: Imprenta Nueva Mexicana de Felipe de Zúñiga y Ontiveros, 1785), art. 10.

40. See Thomas Brown, *La Academia de San Carlos de la Nueva España*, 2 vols. (Mexico City: Sepsetentas, 1976); Eduardo Flores Clair, *Minería, educación y sociedad: El Colegio de Minería, 1774–1821* (Mexico City: INAH, 2000), 93.

41. *Gazeta de México*, vol. XI, no. 18, Sept., 17, 1802; and vol. XI, no. 21, Oct. 29, 1802.

42. AGN, Desagüe, vol. 28, exp. 6, ff. 4–5.

43. Ibid., 9 fols. AGN's "Argena" database attributes this report to Miguel de Costanzó; the handwriting and style are Castera's.

44. Pascual Ignacio Apecechea, *El Museo Mexicano* (Mexico City, 1815), I: 312–30; Garay, *El valle de México*, 50–52.

45. AGN, Desagüe, vol 40, f. 3v.

46. AGN, Desagüe, vol. 35, exp. 1, ff. 38–40; AGN, Desagüe, vol. 40, exp. 1, ff. 13–14v.

47. AGN, Desagüe, vol. 40, exp. 1, f. 38. The salt residue was tequesquite, the result of the evaporation of lake waters, which continually concentrated the natural salts dissolved in the water.

48. Ibid., ff. 29, 38–46, 65–69.

49. Roberto Moreno, *Joaquín Velázquez de León y sus trabajos sobre el valle de México* (Mexico City: UNAM, 1977), intro.

50. Santiago Ramírez, *Datos para la historia del Colegio de Minería* (Mexico City: Imprenta del Gobierno Federal, 1890), 20.

51. Walter Howe, *The Mining Guild of New Spain* (Cambridge, MA: Harvard University Press, 1949), 40–44.

52. "Ordenanzas para el cuerpo de minería. Escritas por el virrey de Nueva España y revisadas y expedidas por rey de haberlas recibido en carta del virrey fechada 26 agosto 1779," in Eusebio Ventura Beleña, *Recopilación sumaria de todos los autos acordados en la Real Audiencia y Sala del Crimen de esta Nueva España y providencias de su Superior Gobierno*, 2 vols. (1787; reprint, Mexico City: UNAM, 1981), II: 214–92.

53. Eduardo Flores Clair, "El Colegio de Minería: una institución ilustrada en el siglo XVIII novohispano," *Estudios de Historia Novohispana* 20, no. 20 (1999): 33–65; Juan José Saldaña, "The Failed Search for 'Useful Knowledge': Enlightened Scientific and Technological Policies in New Spain," in *Cross Cultural Diffusion of Science: Latin America (Proceedings of the XVII ICHS, 1985)*, ed. Juan José Saldaña, 1987, 33–58; Trabulse, *Francisco Xavier Gamboa*, 109–23; Ramírez, *Datos para la historia*, 53–117; Brading, *Miners and Merchants*, 126–27.

54. The Born method of amalgamation with mercury differed little from that explained by Alvaro Alonso Barba in his 1640 *Arte de los Metales*, written to address the decline in productivity of Potosí, from where it diffused to other mine districts. See Arthur P. Whitaker, "The Elhuyar Mining Missions and the Enlightenment," *The Hispanic American Historical Review* 31, no. 4 (November 1951): 557–85.

55. Francisco Omar Escamilla González, "Luis Fernando Lindner (Schemnitz, ca. 1763–México, 1805): Catedrático de química y metalurgia del Real Seminario de México," *Jahrbuch für Geschichte Lateinamerikas* 41 (2004): 167–97.

56. Ramírez, *Datos para la historia*, 57–59.

57. Ibid., 55, 61–73.

58. José Alfredo Uribe Salas, "Labor de Andrés Manuel del Río en México: Profesor en el Real Seminario de Minería e innovador tecnológico en minas y ferrerías," *Asclepio* LVIII, no. 2 (July–December 2006): 231–60.

59. J. J. Saldaña, "Ciencia y felicidad pública en la ilustración americana," in *Historia social de las ciencias en América Latina*, ed. J. J. Saldaña (Mexico City: Porrúa, 1996), 151–207.

60. Roberto Moreno de los Arcos, *Ensayos de historia de la ciencia y la tecnología en México* (Mexico City: UNAM, 1986), 111–22.

61. None of the twenty-three science professors working between 1788 and 1808 appear in Desagüe documentation. See names in Flores Clair, *Minería, educación y sociedad*, 162.

62. Ibid., 69.

63. Trabulse, *Francisco Xavier Gamboa*, 124.

64. Flores Clair, *Minería, educación y sociedad*, 40.

65. Daniela Bleichmar, *Visible Empire: Botanical Expeditions and Visual Culture in the Hispanic Enlightenment* (Chicago: University of Chicago Press, 2012); Appuhn, *Forest on the Sea*.

66. L. E. Harris, "Land Drainage and Reclamation," in *A History of Technology*, 5 vols., ed. Charles Singer et al. (Oxford: Clarendon Press, 1957), III: 300–23.

67. AGN, Desagüe, vol. 20, exp. 7, ff. 21v–22.

68. Ibid., ff. 22v–23.

69. Moreno, *Joaquín Velázquez de León*, 131–32.

70. For all this, Ignacio Castera was no less an instrument for Viceroy Revillagigedo's ordering and rationalization of the City of Mexico, likening him to Charles III's Francisco Sabatini, the prominent neoclassical architect who erected major public buildings in Madrid. See Guillermo Tovar de Teresa, *La Ciudad de los Palacios*, 2 vols. (Mexico City: Vuelta, 1992), I: xix.

71. Estatutos de la Real Academia, arts. 20 and 13.

72. Brown, *La Academia de San Carlos*, II: 83–85. To pass the recertification, master architects needed to show original plans for a major public building.

73. BNE, Fondo Antiguo, ms. 3534, f. 212.

74. AGN, Desagüe, vol. 34, exp. 1, ff. 18v–23v.

75. AHDF, Desagüe, vol. 741, exp. 45.

76. AGN, Desagüe, vol. 30, exp. 4.

77. AGN, Desagüe, vol. 33, exp. 6, ff. 5v–6.

78. AGN, Desagüe, vol. 33, exp. 6, ff. 3v–5v.

79. Barbara H. Stein and Stanley J. Stein, *Edge of Crisis: War and Trade in the Spanish Atlantic, 1789–1808* (Baltimore, MD: Johns Hopkins University Press, 2009), 3.

80. AGN, Desagüe, vol. 33, exp. 6, ff. 5–5v. See also Enrique Florescano, *Precios del maíz y crisis agrícolas en México (1708–1810)* (Mexico City: El Colegio de México, 1969), 159–63.

81. AGN, Desagüe, vol. 31, exp. 6, 5 fols.

82. AGN, Desagüe, vol. 33, exp. 9, "Ocurso de los Governadores de los Partidos, y Repúblicas que componen la Jurisdicción de Zempoala sobre que se releven los Yndios dela Jurisdicción del Travajo en el Desagüe de Huehuetoca"; AGN, Desagüe, vol. 12, exp. 4, ff. 19–19v; AGN, Desagüe, vol. 30, exp. 5; AGN, Desagüe, vol. 33, exp. 8, f. 3v.

83. AGN, Desagüe, vol. 33, exp. 6, ff. 5–5v, 12–17; AGN, Desagüe, vol. 33, exp 7.

84. AGN, Desagüe, vol. 36, exp. 7, 9 fols.

85. AGN, Desagüe, vol. 33, exp. 8, f. 8.

86. Ibid., f. 12.

87. AGN, Desagüe vol. 32, exp. 1, ff. 74–77v.
88. AGN, Desagüe, vol. 38, exp. 10, 10 fols.; AGN, Desagüe, vol. 38, exp. 11, 6 fols.
89. AGN, Desagüe, vol. 33, exp. 9.
90. AGN, Desagüe, vol. 8, exp. 3, f. 122v.
91. AGN, Desagüe, vol. 33, exp. 9.
92. As Bruce Castleman suggests happened in road construction, in *Building the King's Highway* (Tuscon: University of Arizona Press, 2005), 38–42.
93. AGN, Desagüe, vol. 31, exp. 5, 3 fols.
94. AGN, Indios, vol. 52, exp. 119, f. 124.
95. AGN, Indios, vol. 53, exp. 90; vol. 54, exp. 35, ff. 27–28v; AGN, Indios, vol. 61, exp. 183, ff. 167v–68.
96. AGN, Desagüe, vol. 37, exp. 2, f. 4v.
97. Ibid., f. 5.
98. Ibid., ff. 5v–8.
99. Tutino, *Making a New World*, 7, 30.

## CHAPTER EIGHT

1. AHDF, Desagüe, vol. 740, exp 28.
2. For an assessment of the epoch, see Eric Van Young, *The Other Rebellion: Popular Violence, Ideology, and the Mexican Struggle for Independence, 1810–1821* (Stanford: Stanford University Press, 2001), 385–91.
3. AGN, Indiferente Virreinal, caja 2849, exp. 8.
4. AGN, Indiferente Virreinal, caja 4321, exp. 25.
5. Humboldt, *Essai politique*, II: 232.
6. RAH, Jesuitas, 9/3682, exp. 8, "Proposición que la Ciudad de México hizo en su consistorio de 28 de setiembre [de 1630]," ff. 27–27v.
7. AGN, Desagüe, vol. 3, exp. 6, f. 3.
8. Platas et al., *Evolución de la Ingeniería*, 117.
9. Mazari, "Algo más."
10. AHDF, Actas Antiguas de Cabildo Impresas, Book 23, January 31, 1620.
11. See Marcos Mazari and Francisco Platas, "Cuatro grandes en el salvamento de la ciudad de México ante sus inundaciones," *Memoria del Colegio Nacional* (Mexico City: El Colegio Nacional, 1999), 165–219.
12. AHDF, Inundaciones, vol. 2272, exp. 30, f. 352, Maestro mayor José del Mazo.
13. BNM, Fondo Reservado, ms. 1394, ff. 1–13; González Obregón, in *Memoria histórica, técnica y administrativa*, I: 244–45.
14. BNM, Fondo Reservado, ms. 1394, ff. 7v–8.
15. See Mazari, "Algo más," figs. 2 and 3.
16. J. M. Lesser, "El hundimiento del terreno en la ciudad de México y sus implicaciones en el sistema de drenaje," *Ingeniería Hidráulica* 13, no. 3 (1998): 13–18; Marisa Mazari Hiriart and Marcos Mazari, "Efectos Ambientales Relacionados con la Extracción de Agua en la Megaciudad de México," *Agua Latinoamérica* 8, no. 2 (January 2008).
17. AHDF, Inundaciones, vol. 2272, exp. 6, ff. 62–65. The haciendas were

Patera, La Leona, San Pablo, Camarones, and Pantaco; the unnamed ranchos were contiguous to them.

18. Ibid., ff. 66–68.

19. AGN, Desagüe, vol. 18, exp. 4, f. 46. The technique is the same that was used in terrace building in the region, where rows of maguey are planted in the harsh soils of slopes to trap eroded materials and create better soil deposits over time.

20. AHDF, Inundaciones, vol. 2272, exp. 30, f. 362.

21. Ibid., f. 353.

22. AGN, Desagüe, vol. 18, exp. 4, ff. 44–45.

23. For a complete analysis along these lines, see Candiani, "Desagüe Reconsidered."

24. Enrique Florescano and Susan Swan, *Breve Historia de la Sequía en México* (Xalapa, Mexico: Universidad Veracruzana, 1995), append. 3; Arij Ouweneel, "Se quedó pachacate. Sobre las sequías en el altiplano central de México durante las últimas décadas del siglo XVIII," *Ciclos interrumpidos*, 67–96.

25. For eighteenth-century demographic growth in the basin, see Arij Ouweneel, "Growth, Stagnation, and Migration: An Explorative Analysis of the Tributario Series of Anahuac (1720–1800)," *The Hispanic American Historical Review* 71, no. 3 (August 1991): 531–77.

26. This process described by Luna Leopold, *A Primer on Water* (Washington, DC: Government Printing Office, 1960), pt. I; Charles J. Vörösmarty and Dork Sahagian, "Anthropogenic Disturbance of the Terrestrial Water Cycle," *BioScience* 50, no. 9 (September 2000): 753–65; Mark Cioc, *The Rhine: An Eco-biography, 1815–2000* (Seattle: University of Washington Press, 2002), 150–58; among others.

27. Geoff Petts, "Forested River Corridors: A Lost Resource," in *Water, Engineering, and Landscape: Water Control and Landscape Transformation in the Modern Period*, ed. Dennis Cosgrove and Geoff Petts (London: Belhaven Press, 1990), 12–34; McBride, "Formative ceramics," 29.

28. Concept explained in Gretchen Daily, *Nature's Services: Societal Dependence on Natural Ecosystems* (Washington, DC: Island Press, 1997), 1–17.

29. AGN, Indios, vol. 6:1, exp. 230, f. 58v.

30. AGN, Tierras, vol. 2026, exp. 3, f. 3v.

31. AGN, Desagüe, vol. 13, exp. 6, f. 7.

32. AGN, Desagüe, vol. 26, exp. 1, f. 169.

33. AGN, Tierras, vol. 1110, exp. 5, ff. 1–35v; AGN, Tierras, vol. 1124, exp. 2, ff. 1–8; AGN, Desagüe, vol. 19, exp. 14, f. 6v. See Lipsett-Rivera, *To Defend Our Water with the Blood in Our Veins* (Albuquerque: University of New Mexico Press, 1999), for analogous patterns of struggle in Puebla.

34. See Candiani, "Desagüe Reconsidered."

35. See, for instance, Sonya Lipsett-Rivera, "Puebla's Eighteenth-Century Agrarian Decline: A New Perspective," *The Hispanic American Historical Review* 70, no. 3 (August 1990): 463–81; Martín Sánchez Rodríguez, "Las disputas por el agua en el Bajío Mexicano a fines del siglo XVIII," *Relaciones* 22, no. 87 (summer 2001): 159–74; and especially, Eric Van Young, *La crisis del orden colonial: Estructura agraria y rebeliones populares de la Nueva España, 1750–1821* (Mexico City: Alianza Editorial, 1992).

36. AHDF, Inundaciones, vol. 2272, exp. 6, f. 67.

37. AGN, Desagüe, vol. 18, exp. 4, f. 15.
38. Ibid., f. 46.
39. Ibid., ff. 46v, 48.
40. See Chapter 6, note 92, above.
41. AGN, Desagüe, vol. 40, exp. 1, f. 65.
42. AGN, Tierras, vol. 1110, exp. 5, f. 14.
43. AGN, Desagüe, vol. 18, exp. 5, f. 6.
44. AGN, Desagüe, vol. 40, exp. 1, f. 14.
45. See George Novack, "The Law of Uneven and Combined Development and Latin America," *Latin American Perspectives* 3, no. 2 (1976): 100–106.
46. For a classic analysis of "routinism" among Latin America's ruling classes, see Stanley J. Stein, *Vassouras: A Brazilian Coffee County, 1850–1900: The Roles of Planter and Slave in a Plantation Society* (Princeton, NJ: Princeton University Press, 1986), chs. II, VI, and IX.
47. Richard Kagan, "Universities in Castile 1500–1700," *Past and Present* 49 (November 1970): 44–71.
48. See the oeuvre of scholars such as David Goodman, José Luis Peset Reig, Ursula Lamb, José María López Piñero, and Ignacio González Tascón, as well as more recent works by Jorge Cañizares-Esguerra, María M. Portuondo, Daniela Bleichmar, and Paula de Vos, among others.
49. Magdalena Chocano Mena, *La fortaleza docta: Elite letrada y dominación social en México colonial (siglos XVI-XVII)* (Barcelona: Ediciones Bellaterra, 2000).
50. See especially Karen Harvey, ed., *History and Material Culture: A Student's Guide to Approaching Alternative Sources* (London: Routledge, 2009); and Leora Auslander, "Beyond Words," *American Historical Review* 110, no. 4 (October 2005): 1015–45.
51. Mukerji, *Impossible Engineering.*
52. Harris, "Land Drainage and Reclamation."
53. Fernand Braudel, *La Méditerranée et le Monde Méditerranéen à l'époque de Philippe II* (Paris: Colin, 1949), 619.
54. J. R. Ravensdale, *Liable to Floods: Village Landscape on the Edge of the Fens, AD 450–1850* (London: Cambridge University Press, 1974), 39–63; David Hall and John Coles, *Fenland Survey: An Essay in Landscape and Persistence* (London: English Heritage, 1994), Archaeological Report 1; Carolyn Merchant, "Hydraulic Technologies and the Agricultural Transformation of the English Fens," *Environmental Review* 7, no. 2 (summer 1983): 165–78; and elsewhere.
55. Robin Butlin, "Drainage and Land Use in the Fenlands and Fen-edge of Northeast Cambridgeshire in the Seventeenth and Eighteenth Centuries," in *Water, Engineering, and Landscape*, 54–76.
56. Keith Lindley, *Fenland Riots and the English Revolution* (London: Heinemann Educational Books, 1982), esp. chs. 4–6; Mark E. Kennedy, "Charles I and Local Government: The Draining of the East and West Fens," *Albion* 15, no. 1 (spring 1983): 19–31.
57. Joan Thirsk, *English Peasant Farming* (London: Routledge & Kegan Paul, 1957), 116–19.
58. For an overview, see Louis Edouard Marie Hippolyte Dienne, *Histoire du desséchement des lacs et marais en France avant 1789* (Paris: H. Champion, 1891).

59. Jean-Michel Derex, "Pour une histoire des zones humides en France (XVIIe-XIXe siècle). Des paysages oubliés, une histoire à écrire," *Histoire et Sociétés Rurales* 15 (January–June 2001): 11–36.

60. Nadine Vivier, quoted in Derex, "Pour une histoire des zones humides."

61. For exceptions, see Jean-Paul Billaud, "L'État nécessaire? Aménagement et corporatisme dans le marais poitevin," *Etudes rurales* 101/102: *L'État en perspective* (January–June 1986): 73–111, for instance.

62. Mark E. Kennedy, "Fen Drainage, the Central Government, and Local Interest: Carleton and the Gentlemen of South Holland," *The Historical Journal* 26 (1983): 15–37; Eugen Weber, *Peasants into Frenchmen: The Modernization of Rural France, 1870–1914* (Stanford: Stanford University Press, 1976), 45–46.

63. In John Tutino, "The Revolution in Mexican Independence: Insurgency and the Renegotiation of Property, Production, and Patriarchy in the Bajío, 1800–1855," *The Hispanic American Historical Review* 78, no. 3 (August 1998): 367–418. The locally oriented economy and the position of rancheros, many of them women, was strengthened vis-à-vis the hacendados during the putatively economically destructive period of insurgence. For analogous local implications of macro decline, in the Spanish empire's "espacio económico peruano," see Carlos Sempat Assadourian, *El sistema de la economía colonial: Mercado interno, regiones y espacio económico* (Lima: Instituto de Estudios Peruanos, 1982).

## EPILOGUE

1. José María Luis Mora, *Memoria que para informar sobre el origen y estado actual de las obras emprendidas para el desagüe de las lagunas del Valle de México* (Mexico City: Imprenta de la Aguila, 1823), 43.

2. AGN, Fomento Desagüe, 1830, vol. 3, exp. 10. Informe del coronel José Rincón sobre estado de Huehuetoca.

3. Lucas Alamán, *Documentos diversos (inéditos y muy raros)* (Mexico City: Editorial Jus, 1945), 429.

4. Outlook shared in the *Memoria del Desagüe del Valle de México*, ed. Junta Directiva del Desagüe del Valle de México, 3 vols. (Mexico City: Tipografía de la Oficina impresora de estampillas, 1902).

5. John Tutino, *From Insurrection to Revolution in Mexico: Social Bases of Agrarian Violence, 1750–1940* (Princeton, NJ: Princeton University Press, 1986), 213–41. For contrasting regional views of the crisis of the elite, see John H. Coatsworth, *Los orígenes del atraso* (Mexico City: Alianza, 1990); and Margaret Chowning, *Wealth and Power in Provincial Mexico* (Stanford: Stanford University Press, 1999).

6. Israel Sandré Osorio, *Documentos sobre posesión de aguas de los pueblos indígenas del Estado de México, siglos XVI al XVIII* (Mexico City: CIESAS, 2005), 30–33; Luis Aboites Aguilar, *El agua de la nación* (Mexico City: CIESAS, 1998), 51–52.

7. For the creation and performance of the rurales, see Paul J. Vanderwood, *Disorder and Progress: Bandits, Police, and Mexican Development* (Lanham, MD: Rowman & Littlefield, 1992).

8. See Claudia Agostoni, *Monuments of Progress: Modernization and Pub-*

*lic Health in Mexico City, 1876–1910* (Calgary, Canada: University of Calgary Press, 2003).

9. See Alejandro Tortolero Villaseñor, "Transforming the Central Mexican Waterscape: Lake Drainage and its Consequences during the Porfiriato," in *Territories, Commodities and Knowledges: Latin American Environmental History in the Nineteenth and Twentieth Centuries*, ed. Christian Brannstorm (London: Institute for the Study of the Americas, 2004), 121–47.

10. See Priscilla Connolly, *El contratista de don Porfirio: obras públicas, deuda y desarrollo desigual* (Mexico City: FCE, 1987); and Manuel Perló Cohen, *El paradigma porfiriano: Historia del desagüe del Valle de México* (Mexico City: UNAM, 1999).

11. AHDF, Pozos artesianos, vol. 3700, exp. 1; L. Río de la Loza and E. Craveri, *Opúsculo sobre los pozos artesianos y las aguas naturales de más uso en la Ciudad de México* (Mexico City: Murguía y Compañía, 1854).

12. Platas et al., *Evolución de la ingeniería*, 259–60.

13. Nabor Carrillo, "Influence of Artesian Wells in the Sinking of Mexico City," *Comisión Impulsora y Coordinadora de la Investigación Científica*, Anuario 47, vol. "Nabor Carrillo" (1969, trans. of 1947), 7–14.

14. Juan Manuel Lesser Illades and Miguel Angel Cortés Pérez, "El hundimiento del terreno de la Ciudad de México y sus implicaciones en el sistema de drenaje," *Ingeniería Hidráulica en México* 13, no. 3 (September–December 1998): 13–18.

15. See full description in Platas et al., *Evolución de la ingeniería*, 335–48.

16. As told by Julián Zarco-Herrera, SDP engineer, Tequixquiac, July 1, 2008.

17. See their functioning in Comisión Nacional del Agua, "Mapa interactivo del funcionamiento del funcionamiento del drenaje en la Ciudad de México"; available online at http://www.conagua.gob.mx/CONAGUA07/Temas/InteractivoDrenaje.swf.

18. See Raúl J. Marsal and Marcos Mazari, *El Subsuelo de la Ciudad de México*, 2d ed. (Mexico City: UNAM, 1969).

19. Jaime J. Carrera-Hernández and S. J. Gaskin, "Water Management in the Basin of Mexico: Current State and Alternative Scenarios," *Hydrogeology Journal* 17, no. 6 (2009): 1483–94.

20. Dalia del Carmen Ortiz Zamora, "Origen y evolución de un nuevo lago en la planicie de Chalco: Implicaciones de peligro por subsidencia e inundación de áreas urbanas en Valle de Chalco (Estado de México) y Tláhuac (Distrito Federal)," *Investigaciones Geográficas, Boletín del Instituto de Geografía, UNAM*, 64 (2007), 26–42.

21. Marisa Mazari and Douglas M. Mackay, "Potential for Groundwater Contamination in Mexico City," *Environmental Science and Technology* 27, no. 5 (1993): 794–802.

22. John Tutino, "Globalizaciones, autonomías y revoluciones: Poder y participación popular en la historia de México," in *Crisis, Reforma y Revolución. México: historias de fin de siglo*, ed. Leticia Reina and Elisa Servín (2001; reprint, Mexico City: CONACULTA, INAH, Taurus, 2002), 25–86.

# Index

Note: Page numbers in italic type indicate figures and tables.

Kolibri 360 GmbH
Ferdinand-Jühlke-Straße 7
99095 Erfurt, Deutschland
produktsicherheit@kolibri360.de